DATE DUE

NO 25 '99			
OC 15 96			
AP 2 '03			
MY 27 03			

DEMCO 38-296

How do engineers respond to ethical dilemmas that occur in practice? How do they view their individual and collective responsibilities? How do they make decisions before all the facts are in?

Using the space shuttle program as the framework, this book examines the role of ethical decision making in the practice of engineering.

In particular, the book considers the design and development of the main engines of the space shuttle as a paradigm for how individual engineers perceive, articulate, and resolve ethical dilemmas in a large, complex organization. A series of in-depth case studies shows engineers at work on various stages of the project as they balance budgets, deadlines, and risks.

By documenting the historical development of a single system, the book provides a unique opportunity to explore the complex interactions between political, organizational, and technical pressures and engineering and management decisions. Both students and engineers will gain valuable insight into ethical dilemmas common to engineering practice and to ethical decision making.

ENGINEERING
ETHICS

ENGINEERING ETHICS

Balancing Cost, Schedule, and Risk – Lessons Learned from the Space Shuttle

Rosa Lynn B. Pinkus
University of Pittsburgh

Larry J. Shuman
University of Pittsburgh

Norman P. Hummon
University of Pittsburgh

Harvey Wolfe
University of Pittsburgh

CAMBRIDGE
UNIVERSITY PRESS

OF THE UNIVERSITY OF CAMBRIDGE
Cambridge CB2 1RP, United Kingdom

CAMBRIDGE UNIVERSITY PRESS
The Edinburgh Building, Cambridge CB2 2RU, United Kingdom
40 West 20th Street, New York, NY 10011–4211, USA
10 Stamford Road, Oakleigh, Melbourne 3166, Australia

© Cambridge University Press 1997

First published 1997

Printed in the United States of America

Typeset in Ehrhardt Monotype

Library of Congress Cataloging-in-Publication Data
Engineering ethics : balancing cost, schedule, and risk – lessons
learned from the space shuttle / Rosa Lynn B. Pinkus . . . [et al.].
p.   cm.
Includes bibliographical references and index.
ISBN 0-521-43171-9 (hardback). -- ISBN 0-521-43750-4 (pbk.)
1. Engineering ethics 2. Decision-making (Ethics) 3. Space
shuttles – Propulsion systems – Design and construction – Case studies.
I. Pinkus, Rosa Lynn B.
TA157.E673   1997   96–12332
174'.962 – dc20   CIP

A catalog record for this book is available from the British Library.

ISBN   0-521-43171-9 hardback
ISBN   0-521-43750-4 paperback

In loving memory of Beatrice and Maurice Brothman
and to the bright future of
Rebecca Anne Pinkus
Zak and Jessica Shuman
Peter, Amanda, and Daniel Hummon
Andy, Roni, Bonnie, Stephanie, Shoshanah, and Ian and Molly Wolfe

Contents

Part III: Postscript

Figures and Tables

Figures

Tables

Preface

We have written what we intend to be an eminently practical book. Its concern is to engage engineers and managers, as well as engineering students and professional ethicists, in a reflective look at the everyday decisions that are made as technology is designed, tested, built, and set into operation. This critical reflection – about why one acts and what one does – is defined here as ethics. We recognize that this reflective investigation into decision making in engineering practice cannot be limited to the individual practitioner. The organizational context and the political parameters shaping the organization's options are also of concern. Basing our approach to engineering ethics on the emerging analytical consensus reached in medical ethics over the past thirty years, we studied how individual engineers perceived, articulated, and resolved moral dilemmas that arose in practice. The development of the main engines of the space shuttle (SSME) provided the focus for our work.

Engineering is often a heuristic skill practiced before "all the facts are in." Estimating risks and balancing budgets and deadlines characterizes most engineering decisions. Who decides what the level of risk for a given technology should be? The engineer? The organization? The legal system? How and when should the public be involved in these decisions? What role do professional codes of ethics play in guiding practitioners when dilemmas arise? Are there principles common to engineering practice that describe the "ethical engineer" and the "ethical organization"?

Answering these questions is the primary objective of this book. The role of ethical decision making in the practice of engineering is illustrated through an interdisciplinary analysis of one design and development component of the space shuttle program, SSME. Particular emphasis is directed at how engineers perceive, articulate, and resolve ethical dilemmas that arise when complex, advanced technologies like the shuttle's main engines are developed. The role that the organizational context plays in this decision-making process is also explored and the topic of "political" ethics is introduced. By examining engineering decision making, we have identified three basic principles that together provide a framework for articulating and resolving ethical issues in practice: competence, responsibility, and Cicero's Creed II (an updated version of engineering's oldest ethic, to "insure the safety of the public"). Each principle pertains to both the individual engineer and the organization as a whole.

Five engineering case studies are presented with the aim of clarifying and documenting the tacit ethical dimension in engineering practice. The cases are based on data contained in official NASA reports, newspaper articles, congressional records, interoffice memos, phase A, B, C contracts, and contemporary engineering texts. After identifying and reviewing these materials, we ordered the events describing the building of the SSME chronologically and wrote the case studies. When read in sequence, the cases provide a descriptive narrative rich in engineering detail. Our framework has resulted from a careful analysis of the cases. Thus, the basic principles were derived from tacit rules and assumptions underlying engineering practice. Traditional approaches to organizational theory and engineering were innovatively combined and also directed our analysis. We adapted Herbert A. Simon's Nobel Prize theory on organizational behavior and used it as a model for understanding how NASA's organization and formal structure relates to individual engineers and the ethical dilemmas they encounter.

Of primary interest with the SSME are those design decisions involving activities such as scale-up, the extrapolation of performance specifications, and mathematical representations of phenomena. Since these design activities involved the analysis of uncertainty and risk, they created situations in which the exercise of engineering judgments and, concomitantly, values were specifically relied on. Our studies have identified and documented, in both engineering and ethics terminology, where these junctures occurred. By presenting the completed cases to practicing engineers and engineering students for both validation and identification of additional issues, we constructed a normative description of how ethical decisions are perceived and resolved. This normative description was then tailored into the framework and used throughout the text.

The book is divided into three sections. Part I: Theoretical and Practical Background includes the normative framework and two theoretical chapters. Specifically, a brief overview of both the scholarly and practical importance of the book is included in Chapter 1. Chapter 2 includes the framework. Chapters 3 and 4 offer an in-depth theoretical justification, in both engineering and medical ethics and in organization theory, for our framework and for our analytical approach. Ever cognizant of the practical nature of engineering decision making, we include the theoretical chapters as a complement to, but not a necessary prerequisite for, interpreting the historical case narrative.

Part II: Political and Technical Background, plus the cases, consisting of Chapters 5 through 13, provides both the necessary technical and ethical background and the case studies. Three descriptive chapters provide a context for understanding the political environment in which the decision to build the space shuttle occurred (Chapter 5), review the technical aspects of how the SSME works (Chapter 6), and evaluate the testing procedure termed "all-up" that was unique to NASA and used throughout the building of the shuttle (Chapter 7). The cases consider how the framework relates to the individual engineer as illustrated by examining A. O. Tischler's role in NASA (Chapter 8); an example of trade-offs between cost, risk, and schedule as they occurred when a test stand was constructed (Chapter 9); detailed congressional investigations of NASA's assumptions regarding testing and its veracity in reporting the progress of the SSME's development to Congress (Chapter 10); the role of judgment in the repair of a hydrogen leak and the subsequent six-month delay in a scheduled launch of the *Challenger* (Chapter 11); a summary of specific issues illustrating the effect of organizational decisions on the ethical judgment of engineers (Chapter 12); and an in-depth discussion of risk assessment – based on the findings of a special national Research Council Study convened in 1988 to reexamine NASA's risk assessment and management post-*Challenger* – that specifically establishes the linkages between organizational framework and individual ethical-technical decisions (Chapter 13).

Part III is composed of a postscript, offering a detailed analysis of the decision to launch the *Challenger,* based on the long-term perspective presented in the book (Chapter 14). Our final case, it further illustrates our ethical framework.

We have benefited from pertinent work in medical ethics. In particular, questions concerning how one applies a theoretical ethics framework to professional endeavors were of significant interest. What is the importance of theory in this endeavor? Can "middle-level" principles in engineering comparable with medical ethics "autonomy, beneficence, and justice" be identified and used to begin to sort through the complex turf of engineering practice? Are these principles "action guides" or "checklists" against which moral behavior can be gauged? What is the role of case analysis? Are there paradigm cases in engineering that should become common knowledge as reference points when one deliberates about the moral nature of the profession?

We focused attention on issues unique to the engineering profession. How, for example, are "collective" and "individual responsibility" defined in an innovative, highly visible and political, federally funded project such as the space shuttle? Who is the "client": the public, NASA, individual corporations, the astronauts, or all these parties? How is risk calculated and what role does the organization have in this process? As we teased out the answers to these questions, we began to appreciate that while not traditionally addressed within medical ethics, given the changing environment in which physicians currently practice, the issues represented an avenue for investigation for medicine as well. In a "managed care" environment, the loyalties of physicians and other health-care workers to individual patients are being challenged by the organizational practices of hospitals and the cost-cutting tactics of third-party payers. In drawing comparisons between medicine and engineering, we hope that we have contributed to a greater understanding of the field of applied ethics.

Essentially, we agreed that a knowledge of ethical theories and principles proved important to our own communication process. An interdisciplinary group, we needed a common language to discuss the complex technical ethical dilemmas we identified. The principles of competence, responsibility, and Cicero's Creed II, however, are grounded in engineering practice, not in abstract theory. While their definitions have been embellished by using select theoretical sources, tacit rules of professional behavior informed their development. We also believe that analyzing, recognizing, and resolving an ethical dilemma is an important exercise for each reader. Thus the descriptive cases are written in technical and common "everyday" language. Our goal has been to convey to the reader what it would have been like to "be there." The framework aids in analysis but is not meant to preempt the individual's "moral imagination" or represent a perspective based on a "moral expert." Rather, we concede that "moral expertise" is within the grasp of each practicing engineer and offer our framework as a way to communicate ethical concerns among professionals.

While we have benefited enormously from the insights and work of others, we take full responsibility for the interpretations and conclusions recorded herein.

Acknowledgments

This book reflects a collaborative effort in the true sense of the word. There are a number of special people who provided valuable assistance during its research, editing, and production phases. We would not have been able to undertake or complete the work without their help, guidance, and patience.

Major research and support for this book was provided by a grant from the EVIST Program/National Science Foundation. Our project officer, Rachelle Hollander, was extremely supportive and patient both throughout the conduct of our grant and afterward. She provided us with two important forums to present our ideas and critiqued our work in progress. A first task in our grant was to choose a topic around which the study would focus. Drs. Max Williams, Vivian Weil, and Ken Schaffner, consultants, served as our expert panel and unanimously suggested that we focus on the space shuttle main engine. Max provided much insight into design engineering, particularly in the aerospace arena, and served to keep us "honest"; Vivian helped establish a groundwork for understanding various approaches to engineering ethics; Ken was able to assess the importance of the work from the perspective of medical ethics and encouraged us to pursue our comparison with engineering.

Gerald Kayten, truly an ethical aeronautical engineer, helped us focus our initial ideas and provided insight into the environment in which the NASA engineers worked. He also assisted with locating engineers involved with the development of the space shuttle main engine. Don Musa, research associate at the University of Pittsburgh, helped us with much of our initial literature search into engineering ethics. Phyllis Kayten, then at NTSB, provided valuable assistance in locating documents and experts to help clarify particular points of interest. She was always available to discuss or critique particular aspects of the project.

Frank Stevenson provided an important overview of the early development of the main engine and the problems faced by the design engineers and helped us locate several early NASA reports that served as source material for this study. Drs. Joel Peterson and Michael Kolar from the Mechanical Engineering Department, University of Pittsburgh, provided additional insight into engineering design. Joel was particularly helpful in providing an understanding to liquid chemical rocket engine design and functioning in terms that our interdisciplinary team could understand and appreciate; Mike provided insight into the problem of cracking turbine blades and how a similar problem was resolved in the nuclear power industry. Two aerospace historians, Drs. Alex Roland, Duke University, and Edward Constant, Carnegie Mellon University, provided important perspectives for the study. We are particularly indebted to Alex, who made a special trip from Gettysburg, where he was completing the last week of his sabbatical, to meet with our group in Pittsburgh. James Wilson, a congressional staff consultant, provided a number of key insights into House and Senate proceedings and hearings related to NASA, and directed us to a number of important documents. A. O. Tischler kindly read the entire manuscript and critiqued specific issues concerning the SSMEs. Moral philosopher Larry Churchill read and commented on original drafts. His comments regarding our third concept were most appreciated. We have liberally benefited from his definition of our approach to the study of ethics.

We have also benefited greatly from the scholarship of others working in this field. We wish to thank the authors, publishers, and organizations who have granted us permission to use portions, as noted in the text, of their copyrighted material. These include:

Accreditation Board for Engineering and Technology, *Code of Ethics of Engineers,* October 5, 1977.

Bell, Trudy E. "Managing Murphy's Law: Engineering a Minimum-risk System," *IEEE Spectrum,* June, 1989. © 1989 IEEE.

Cassel, Christine K. "Deciding to Forego Life-Sustaining Treatment Implications for Policy." This article originally appeared in 6 **Cardozo L. Rev.** 287 (1984).

Clouser, K.D. "Bioethics and Philosophy." *Hastings Center Report* (Special Supplement). Nov.-Dec. 1993, 510–11. Reproduced by permission. © by the Hastings Center.

Feynman, Richard P. *What Do You Care What Other People Think? Further Adventures of a Curious Character,* as told to Ralph Leighton. New York: W. W. Norton and Co., 1988. Copyright © by Gweneth Feynman and Ralph Leighton.

Institute of Electrical and Electrical and Electronic Engineers, *Code of Ethics,* 1990.

King, N. M., Churchill, L. R., and Cross A. *The Physician as Captain of the Ship: A Critical Reappraisal.* Boston, MA: Reidel Publishers, 1986, p. xi. Reprinted by permission of Kluwer Academic Publishers

Martin, Mike W., and Schinzinger, Roland. *Ethics in Engineering.* 2nd ed. New York: McGraw-Hill Book Co., 1989. Reproduced with permission of the McGraw-Hill Companies.

McCurdy, Howard E., *Inside NASA: High Technology and Organizational Change in the American Space Program.* Baltimore: Johns Hopkins University Press, 1993. Reprinted by Permission of the Johns Hopkins University Press.

Simon, Herbert A. *Administrative Behavior.* 3rd ed. New York: Free Press, 1976. Reprinted with permission of The Free Press, an imprint of Simon & Schuster Inc.

Smith, Robert W. *The Space Telescope: A Study of NASA, Science, Technology and Politics.* New York: Cambridge University Press, 1989.

Thomson, Dennis F. *Political Ethics and Public Office.* Cambridge, MA: Harvard Univesity Press, 1987, p. 4. Reprinted by permission of the publishers from *Political Ethics and Public Office* by Dennis F. Thompson, Cambridge, MA: Harvard Univesity Press. Copyright © 1987 by the President and Fellows of Harvard College.

Personnel at the NASA History Office, particularly Lee Sagasser, were indispensable in enabling us to assemble much of the primary source material we have drawn upon. We could not have done the extensive literature searches without the assistance of the University of Pittsburgh's NASA Industrial Applications Center (NIAC). NIAC personnel assisted with our computerized searches, ordered hard-to-find documents for us, and provided guidance into the different NASA databases. Joe Wujak from University of California at Berkeley reviewed the chapter on the *Challenger* and provided a number of valuable comments and insights into the accident and its subsequent investigation.

We particularly want to thank the four student research assistants, all of whom made major contributions to this project. Chris Fitzmartin and Mary Ann Holbein offered insights that helped us construct our applied ethics framework. Chris's knowledge of philosophy provided us with a rational definition of principles of competence, responsibility, and nonmalevolence (which has since become Cicero's Creed II). Mary Ann's practical engineering experience insured that the framework was understandable to engineers. She was largely responsible for writing first drafts of the chapters on "All-Up Testing" and "Back to the Drawing Board." Mary Ann,

Janet Legal, and Anne Schreiber did much of the research work and initial writing of the case studies. This report could not have been completed without them. They made it possible to construct five case studies rather than the one we had originally intended to prepare as part of what was to be a "pilot" study. They also formed what became a relatively large but very collegial team. Often, interdisciplinary projects fail because the varied perspectives of the investigators inhibit rather than foster fruitful exchange. Our constant meetings, discussions, and sincere efforts to bridge these interdisciplinary barriers provided a rich environment for us to learn from each other. We hope that this integrated perspective is reflected in this book.

A special acknowledgment goes to Frederick I. Ordway III, who has graciously allowed us to use a number of the illustrations from *Blueprint for Space*, edited by Frederick I. Ordway III and Randy Leibermann (Washington, DC: Smithsonian Institution Press, 1992). Mr. Ordway kindly provided us with a number of the original prints to ensure the best possible reproduction.

Technical assistance for this project has also been an interdisciplinary effort and has been generously supported by the Department of Medicine, Division of Internal Medicine; the Associate Dean's Office in the School of Engineering; and the Center for Medical Ethics, Consortium Ethics Program at the University of Pittsburgh. Ms. Patricia Costa in the Associate Dean's Office, School of Engineering, typed early drafts. Ms. Jody Chidester in the Department of Medicine and Marcia Lasky in the Engineering School patiently revised copy. Alan Joyce in the Center for Medical Ethics, Consortium Ethics Program, was responsible for the final typing and revisions. Rebecca A. Pinkus provided technical editing of the entire manuscript and compiled a first listing of the index. To each of them, we offer our sincere thanks and appreciation. Florence Padgett, our editor at Cambridge, could not have been more supportive. She never pressed us and always offered words of encouragement. The book clearly would not have been published without her patient faith.

On a personal note, Richard Simmons, Barbara Shuman, Kathy Hummon, and Elaine Wolfe deserve mention, as do Diane Coveny, Elizabeth Ela Beeson, and Arthur R. Brothman. Their support and encouragement throughout this project are greatly appreciated.

Introduction

This book is about practical ethics. Its choice of the building of the main engines of the space shuttle as a thematic case, however, raises issues that are anything but practical. The building of a reusable rocket to shuttle people into space has been a fascination for hundreds of years (Figure I.1). What would Mars look like? How could we better view Saturn's rings (Figure I.2)? Travel to the moon and beyond has fascinated writers of fantasy and fiction (Figures I.3, I.4, I.5) since the seventeenth century, but Jules Verne's *De la terre à la lune*,[1] published in 1865, has often been cited as the first technically and scientifically sophisticated description of a rocket launch to the moon (Figures I.6, I.7). For our purposes, we cite it as a prototype also for preempting moral, political, and organizational dilemmas.

In its ideal construction, the story identifies human concerns that we also cited as the saga of the building of the space shuttle main engines (SSME) unfolded. However, Verne's adventure had some basic parameters that dramatically differed from those surrounding the SSME. First, the decision to build a rocket to the moon was made amid a post–Civil War boredom with peace, not a continuation of cold war competition with Russia. During the Civil War, "an influential club" was founded in Baltimore for men interested in the science of ballistics. Cannons, howitzers, and mortars were the pride of American Yankee artillery. Yet after the war, there was no challenge. The club's 1,833 members and 30,575 corresponding members had no reason, barring another war, to meet. No reason, that is, until the club's president, Impey Barbicane, announced that after studying the question from every angle, he calculated that "a projectile endowed with an initial velocity of 36,000 feet per second, and aimed at the moon, is bound to reach it." With that announcement, the experiment was embraced. Not having to satisfy competitive bidding, the first step was to insure the scientific credibility of the project. That was done by enlisting the aid of one authority: the Cambridge Observatory, which assured the Gun Club president with precise detail that his project could succeed. A "phase A" feasibility study completed,

1

Figure I.1 Fouché's *Le terre du ciel,* sunrise canals on Mars; black and white etching; courtesy Frederick I. Ordway III Collection.

the date for the launch was to be "December 1st, at thirteen minutes, twenty seconds before eleven o'clock at night."[2]

With a scientific stamp of approval and a specific target date, Barbicane set out to resolve the "cosmographic, geological, political and moral" aspects of the experiment. Money was no object. He enlisted donations internationally and within days accrued funds to more than cover the costs. With all this support and financial backing, what were the dilemmas? One was political. Where should the launch originate? It was understood that the chosen location would benefit economically. Representatives from the two leading contenders, Florida and Texas, traveled to Baltimore to fight for

Figure I.2 Saturn's rings (tinted slide); courtesy Frederick I. Ordway III Collection.

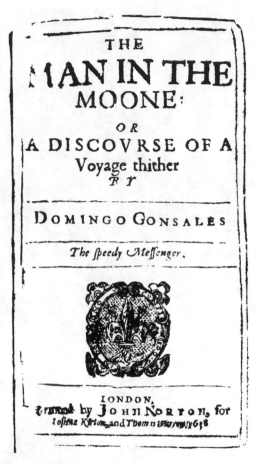

Figure I.3 Title page: Domingo Gonsales, *The Man in the Moone . . .* (1638), courtesy
Frederick I. Ordway III Collection.

the contract. The battle was vicious and, when Tampa, Florida, was chosen,
the Texas delegates were literally placed on a special train and "railroaded"
back home.[3]

Technical problems were a challenge, but Barbicane was assured by the
engineers that the technology was available – it would just have to be used
in an innovative way. A combination of cannons, howitzers, and mortar was
chosen and a cannon, one-half mile long, was to be built. The cannon (not
the moon rocket) was given a name: Columbiad. The material chosen to
build Columbiad was cast iron. It satisfied the criteria of cost and durability.
But while the technical know-how existed, this was clearly not "off-the-

Figure I.4 Title page: Mrs. A. Behn, *The Emperor of the Moon: A Farce* (1687); courtesy Frederick I. Ordway III Collection.

shelf" technology. The technical capabilities of the engineers were challenged to their limits and the casting of the cannon – an unprecedented event – was an "all-up" testing procedure.

Heretofore attributed as being unique to NASA, Verne's method of casting the cannon and his reasons for doing so are cited here as the first documented example of a "success-oriented" approach. The December 1 launch date had been precisely calculated by scientists at the Cambridge Observatory. If these specifications were not met, the moon shot could not take place for another "eighteen years and eleven days." There was no time to proceed in a step-by-step fashion and the Gun Club members began "at

Figure I.5 Title page: "The Discovery of a World in the Moone" (1638); courtesy Frederick I. Ordway III Collection.

once the preparations" needed.[4] The casting was to take place at the site of the launch. A huge shaft sixty feet across and nine hundred feet deep was dug into the ground. The digging alone took eight months to complete. Masonry work reinforced the shaft and, at a designated depth, an oak disk was firmly bolted. The disk had a hole in the middle that matched the diameter of the outer rim of the Columbiad. A cylinder, nine hundred feet high and nine feet across, was placed inside the "shaft so that it would exactly fill the space reserved for the bore of the Columbiad." Composed of a mixture of clay, earth, and sand to which straw and hay had been added, the space left between the mold and the masonry was to be filled with molten metal.

The underground mold was surrounded by twelve hundred blast furnaces, which could each melt 114,000 pounds of iron ore (135 million pounds in total). The ore had been transported in sixty-eight ships and

Figure I.6 Cannon from Jules Verne's *De la terre à la lune* (1866); courtesy Frederick I. Ordway III Collection.

Figure I.7 Lunar-bound projectile from Jules Verne's *De la terre à la lune* (1866); engraving from the 1872 edition; courtesy Frederick I. Ordway III Collection.

Figure I.8 Verne's spaceship nears the Moon, rockets firing. *Autour de la lune* (1870); courtesy Frederick I. Ordway III Collection.

then delivered to the site. At twelve noon on the designated date, each furnace simultaneously released its molten load into the shaft. The heat from the furnaces produced atmospheric disturbances that, by comparison, dwarfed the effects of tornadoes, volcanoes, or hurricanes. So much for "off-the-shelf" technology! The engineers and moon seekers had to wait over a month for the ground to cool to see if their casting operation had been successful.[5]

In both the building of the shaft and in the casting process there had been serious accidents and lives lost. Unfortunate as this was, the high risk of the enterprise was understood, predicted, and calculated. "Because of his [Barbicane's] intelligence, his useful intervention in difficult cases and humane capacity, the accident rate did not exceed that of other European countries that exercised great precaution. It was comparable, in fact, to that of France, where 'there is an average of one accident for every 200,000 franc's worth of work.'"[6]

The grand experiment was successful. Success, however, was bittersweet. Verne's story concludes with three adventurers being launched into space and never heard from again. (The addition of the three – arch rivals Barbicane and Nichol and a French adventurer – in the capsule was an afterthought.) The travelers were spotted – after inclement weather cleared

Figure I.9 Recovery of the spaceship: The projectile has returned from its lunar trip and landed in the Pacific where it was met by the *Susquehanna* (a U.S. naval corvette), one of whose boats retrieves the returning heroes; courtesy Frederick I. Ordway III Collection.

– orbiting the moon and only faithful followers assured their return. Five years later, in Verne's sequel, *Autour de la lune,* the capsule finally returned and the heroes were rescued (Figures I.8, I.9).

So in Verne's story the risky all-up approach to building the Columbiad was not prompted by a shortage of funds. The fixed launch date dictated the schedule. Combined with the enormity of the undertaking, it left no alternative to the success-oriented construction. The probability of loss of life during the construction phase and the actual launch were not underestimated. Space flight, even in the sophisticated, technical world of a master storyteller, was admittedly expensive and risky. With appropriate funding, this risk could be minimized but not erased. In vastly simplified forms, these trade-offs between cost, risk, and schedule produced the practical and moral dilemmas we observed in our in-depth examination of the building of the SSMEs. They are dilemmas that, we assume, will follow the space program on whatever path it takes as long as it is committed to launch vehicles that include the transport of a human cargo. How the dilemmas are resolved and which solutions are morally defensible constitute essentially the thrust of this book.

Notes

1. Jules Verne, *From the Earth to the Moon,* trans. Jacqueline Baldick and Robert Baldick (London: J. M. Dent; New York: E. P. Dutton, 1970).
2. Ibid., pp. 1–16.
3. Ibid., pp. 18–61.
4. Ibid., p. 26.
5. Ibid., pp. 86–98.
6. Ibid., p. 92.

PART I

Theoretical and Practical Background

1

Low-Bid Ethics: Decision Making about Cost, Safety, and Deadlines

Philosophizing about space flight while awaiting blast-off atop the space shuttle *Columbia*, astronaut Alan B. Shepard jokingly commented that it was a humbling experience knowing that his fate depended on a vehicle built by the lowest bidder![1] Teasing out the truth and relevance to engineering ethics in this offhand comment is essentially the topic of this book. It is a political reality that projects seeking federal funding are always in competition. Although cost is clearly a factor in gaining federal support, it is only one-third of the equation that makes up good engineering practice. Engineers, admired for their ingenuity, are constantly being challenged to create solutions to problems within a given cost and time frame without jeopardizing the safety of the users of the technology. Shepard, awaiting blast-off, presumably relied upon the integrity of the engineering decisions – embedded within the low-bid reality – not to compromise the "safety of the public." If he had not maintained a high level of trust in the engineering profession, in NASA's organizational complex, and in the thousands of composite individual decisions that combined to create the space vehicle in which he was about to ride, one could assume that he would not complacently sit by philosophizing about his fate.

Yet, what are the limits to "low-bid ethics"? When do organizational decisions regarding budget and schedule so compromise individual engineering decisions as to make them ethically suspect? In complex, federally funded projects like the building of the space shuttle, what is the responsibility of Congress, NASA, aerospace companies, and individual engineers to insure the overall integrity of the technology?

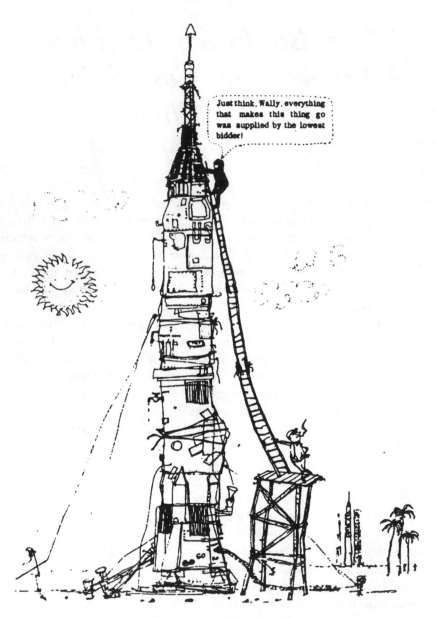

Figure 1.1 Built by the lowest bidder. (From Lloyd S. Swenson, *This New Ocean: A History of Project Mercury* [Washington, DC: NASA, 1966], p. 468.)

Everyday – Not Disaster – Ethics

These questions are unique neither to the space shuttle nor to other federally funded projects. In 1961, Alvin M. Weinberg coined the term "Big Science" for what was then a new type of endeavor.[2] To Weinberg, it was characterized by coalition building among professions, scientists, government, and industry. It involved large multidisciplinary teams that were organized hierarchically. The projects chosen as central to these endeavors were, in Weinberg's opinion, comparable in terms of their importance to society as were the building of cathedrals during the Renaissance. Scientists involved in the modern-day projects were, he observed, transformed into hardware and software managers, journalists, and lobbyists to win money. Wary of the complexity of such endeavors and of the blurring of traditional roles and their attendant responsibilities, Weinberg cautioned that "if not checked, Big Science would lead to the decline of science itself."[3]

Robert W. Smith, a historian at the National Air and Space Museum, has written a perceptive study of the development of the Hubble Space Telescope and has analyzed that project in terms of Weinberg's concept.[4] His goal was to critique the role politics played in that development. He reported that

> the shaping of the telescope's scientific capabilities was an integral part of its political pilgrimage: the telescope's design, the program to build up, and the claims made on its behalf continually had to be revised and refashioned as part of the effort to come up with a telescope that would be politically feasible. . . . on many occasions, because of cost and scheduling problems, there was great pressure to agree to reductions in the telescope's scientific capabilities. . . . NASA allowed more risk to be introduced into the instrument's construction than had originally been intended.[5]

This book focuses on one interrelated but overlooked aspect of doing "Big Science:" The impact it has on the ethical decision making of its participants. Using both historical description and ethical analysis, we examined the individual and organizational dilemmas that occurred as one component of the space shuttle, the main engines (SSME), was being built. We have studied how engineers perceived, articulated, and resolved ethical dilemmas that arose when complex, advanced technology was developed, but we have not focused solely on what philosopher Michael Pritchard has termed "disaster ethics."[6] These, according to Pritchard, are headline events that are exemplified by the explosion of the *Challenger,* the Three Mile Island Nuclear Power Plant malfunction, and the recall of the Ford Pinto.[7] Our work is directly related to the *Challenger*'s fatal launch; yet, neither that decision nor the failure of the O-rings is our primary concern. We do include a postscript on this topic, but even there we concentrate on the myriad everyday decisions made by engineers and others that preceded the *Challenger* accident. The disaster, we conclude, was not a single event. Rather, the decision by Congress to fund the space shuttle program at a "cut-rate" price and the acceptance by NASA to proceed with plans to build the shuttle set the stage for individual engineers continually to struggle to balance safety, cost, and timing. As our cases illustrate, safety, while always a part of the equation, did not consistently "trump" the other variables. Clearly, in both sheer size and complexity, the building of the SSME qualifies as "Big Science." The main engines' development, however, also provides an opportunity to study a range of mundane engineering decisions in depth.

The Historian as a Storyteller

To capture the essence of these everyday dilemmas, we relied on historical methodology for both our data collection and case description. The space shuttle is a federally funded, prototype advanced technology. As such, every aspect of its design, testing, and operation is extensively documented in the NASA history office. Primary source materials such as interoffice memos, original news clippings, official reports, technical studies, and government documents in this collection, combined with key secondary sources, provided us with data to construct case studies in intricate detail.

The historical narrative, in fact, provides a unique basis for ethical analysis. First, history is an explanatory enterprise that "induces understanding."[8] Not only a display of facts, it "enables the reader to see connections between them." The explanatory force of history, then, is a combination of the facts that the historian presents and the manner in which they are juxtaposed. Patterns are identified; trends emerge over time. Fascinated with the "multitude of particulars [and] . . . idiosyncratic details" the historian also attends to generalities but only to describe the trends that the actors themselves were influenced by. Of all storytellers, historians seek what

Thomas Kuhn describes as "an autonomy (and integrity) of historical understanding."[9] That is, the facts speak for themselves, but only if the narrative construction is crafted in a specific way, one that observes the disciplinary methods of history.

From the Decision to Build to the Launch of the *Challenger:* A Case Study

The first phase of our work then entailed writing a history of the building of the SSME. The time frame for this history spans roughly from 1969, when the decision to build the space shuttle was made, to 1986, when the *Challenger* exploded. Within this period, the following occurred: conceptual plans for a reusable, cost-effective engine were developed; Congress funded the shuttle project for approximately half the amount of money requested and NASA accepted its terms; NASA adopted a high-risk but cost-effective overall testing philosophy known as "all-up"; a test stand for the turbo pumps was constructed; a Senate task force investigated why the failure rate in engine testing was so high and why NASA repeatedly requested high sums of money to supplement its SSME program. Numerous problems arose with the engine both before and after the first successful launch of *Columbia* in 1981. A costly four-month delay of the *Challenger* launch was needed in 1983 in order to repair a hydrogen leak. In 1986, the *Challenger* exploded and, in its aftermath, the Rogers Commission publicly disclosed the highly critical stance of NASA.

Using this detailed history as a base, we constructed five separate case studies. Each describes an engineering activity undertaken as part of the space shuttle project. Specifically, design decisions that involved scale-up, the extrapolation of performance specification, and mathematical representations of phenomena were highlighted. This focus was premised on a definition of engineering practice as a heuristic skill.[10] We recognized that engineers combine scientific facts, experience, and judgment to produce technologies that have an "acceptable" level of risk. We rejected the traditional view of engineering that defined it as an applied science that has as a goal the creation of a risk-free environment. How that level is decided, by whom, what role the organization has in determining it, and what the responsibility of the individual engineer is to the public in this process are the moral issues we explored. The titles of the cases and the dates they occurred are:

- A. O. Tischler – "An Ethical" Engineer (Alternatives to Whistleblowing) (1969–72)
- Cost versus Schedule versus Risk: The Band-Aid Fix: A $3.5 million Overrun in Test Stand Construction (1974)
- Back to the Drawing Board or Normal Errors: The Senate Investigation of the SSME Development Problems (1978)
- A Judgment Call or Negligence: Maiden Voyage of the *Challenger* Delayed Six Months (1983)
- The Decision to Launch the *Challenger* (1986)

So as to provide a context and informational background to understand specific political decisions and technical approaches made and used during the SSME's construction, three descriptive chapters were written to accompany the cases. Taken together, the cases and chapters provide a history of the building of this specific component.

Why and How We Relied on Medical Ethics

While the case description, hence analysis, was influenced by methods in historical research,[11] analysis of the cases also relied on methodology developed and refined by philosophers working in the area of medical ethics over the past thirty years.[12] As noted,

> History is an explanatory enterprise; yet its explanatory functions are achieved with almost no recourse to explicit generalizations. . . . The philosopher, on the other hand, aims principally at explicit generalizations and at those universal in scope. He is no teller of stories, true or false. His goal is to discover and state what is true at all times and places rather to impart understanding of what occurred at a particular time and place.[13]

Ethics, as a branch of philosophy, shares these goals; hence philosophers working in bioethics tend to look to theories and principles for answers to ethical dilemmas. Yet, as philosophers have become engaged in actually resolving "real-life" dilemmas, their goals for constructing generalities and their reliance on theories for answers have been tempered by practical concerns. Our use of the term "practical," rather than "applied ethics," reflects this tempering.[14] Briefly, it is generally agreed that neither theories nor principles in ethics can be "applied" in a straightforward manner to resolve a real-life ethical dilemma. Rather, bioethics is regarded as an interdisciplinary endeavor. "Any solidly grounded discipline of ethics involves obtaining relevant factual information, assessing its reliability and mapping out alternative solutions to problems that have been identified. Ethical theory," conceded Beauchamp and Childress "is but one vital contributor" among other disciplines.[15] This recent recognition regarding the limits of reliance on theories and principles (referred to as "principalism") is reflected in how our framework to engineering ethics is conceived. We discuss the development of methodology in medical ethics and engineering extensively in Chapter 3.

A Case-Based Approach

The practical ethics literature in both medicine and engineering relies on the case-based approach as a method for teaching.[16] Our use of historical description and the presentation of a chronology of cases is based on the recognition that real-life ethical dilemmas are "complex, uncertain and ambiguous."[17] They do not come neatly labeled or stylized. Given this, a first goal of practical ethics is to assist practitioners in defining the dilemmas. As philosopher John Arras reports, "real life does not announce the nature of problems in advance."[18] How, then, can practical ethics help? Arras promotes "casuistry," which relies upon a comparison of cases with a well-recognized paradigm as a method for ethical analysis.[19] The cases, according to Arras, should not be hypothetical and should avoid schematic presentations. This directive comments on a common approach philosophers use as they construct a case. As discussed, their goals are often to demonstrate a theoretical stance rather than to solve a practical dilemma; hence their cases tend to be brief and are constructed so as to highlight one or two opposing principles. Urging ethicists not to follow this analytic philosophy approach, Arras emphasizes that the cases be "long, rich, messy and comprehensive; represent a variety of perspectives; [and] present complex sequences of cases that sharpen [the] student's analogical reasoning skills."[20]

While Arras relies on the method of casuistic case analysis, he has concerns about it as well. His cautions summarize the current thinking regarding the role theories, principles, and cases play in ethical analysis and reflect the approach we used here. First, Arras recognizes that these real-life cases will serve the pedagogic purpose of stimulating the moral argument. Yet, he stresses that the "mere comparing and contrasting them [casuistry's main approach] must be supplemented and guided by appeals to ethical theory; moral norms embedded in our traditions and social practices."[21] The cases presented in this book satisfy many, but not all, of Arras's suggestions. They are clearly complex and detailed. As will be seen, it would be difficult to construct hypotheticals that depict ethical dilemmas with the same drama or irony we found in the historical record. Each case is followed by commentary and questions that promote a reflective examination of the story told. We have not referred to a paradigm case for comparison's sake but offer three concepts that were generalized from our historical description to guide this examination. Each concept has implications for both the individual and the organization.

Creating a Normative Framework

We presented the cases to practicing engineers and engineering students for both validation of engineering techniques and for identification of ethical issues. Although not empirically collected, the information we obtained from this review enabled us to examine our historical narrative with an eye to describe tacit rules, maxims, or principles engineers identified as they attempted to resolve the particular dilemmas. The normative description in the narrative had several recurring themes and these were confirmed by our informal review. We categorized these themes into three core concepts: *competence* – the fact that the engineer is a knowledge expert; *responsibility* – the recognition that knowledge has "power" and must be used wisely; and *safety* – the recognition that engineers should be cognizant of, sensitive to, and strive to avoid the potential for harm. They should also opt for "doing good." We refer to this third aspect, an updated version of engineering's oldest ethic, to "insure the safety of the public," as *Cicero's Creed II*. The concepts apply to both the individual engineer's performance and to the organization's governance.

We interpreted these three core concepts – competence, responsibility, and Cicero's Creed II – as "checklists" or "chapter headings" against which moral behavior can be gauged. As we discuss fully in Chapter 3, we were cautious about labeling the concepts as "principles" in the formal analytic sense of being action guides or a logical composite of statements derived from an overarching philosophical theory.[22] We offer our three concepts as a framework within which ethical dilemmas can be articulated and discussed. This process of articulation and discussion, we hope, will contribute to both the identification and resolution of ethical dilemmas.[23]

As mentioned, one of the goals of any practical ethic – medical or engineering – is to identify ethical dilemmas. Another is to structure the issues in a way that clarifies them for practitioners. It has been our experience in teaching ethics in a practical setting that students often define ethics and the terms used to discuss it in a variety of ways. Given the interdisciplinary nature of bioethics, an agreed-upon definition of key terms by discussants can greatly facilitate the identification and resolution of issues. Here, we give the terminology of Beauchamp and Childress that we relied on as we developed our normative ethics framework.[24] These terms can be used as an aid to practitioners engaged in the identification and structuring of ethical issues, a task K. Danner Clouser describes as "conceptual geography."[25]

Ethics: A "generic term for several ways of examining the moral life," that is, critical reflection on what one does and why one does it. Some approaches to ethics are descriptive and others are normative.

Morality: Social conventions about right and wrong human conduct that are so widely shared they form a stable (although usually incomplete) communal consensus.

Descriptive ethics (nonnormative): Factual investigation of moral behavior and beliefs – the study not of "what people ought to do but how they reason and how they act."

Normative ethics (general): "The field of inquiry that attempts to answer the question, 'Which action guides' are worthy of moral acceptance and for what reasons?" Types of action guides are theories, principles, rules, and maxims. They are used to assess the morality of actions.

Normative ethics (applied): "Refers to the use of ethical theory and methods of analysis to examine moral problems in the professions." Most professions articulate a professional code of ethics to specify role norms or obligations that professions attempt to enforce. Sometimes etiquette and responsibilities are spelled out. Theory is not actually applied but rather is involved to help develop specific action guides.

Metaethics (nonnormative): The analysis of language of crucial ethical terms such as virtue, right, obligation, and responsibility. It examines the logic and the patterns of moral reasoning.

Tacit Ethic: Unsaid, unspoken rule of practice.[26]

This book is primarily concerned with descriptive and normative ethics. Our case studies and accompanying background chapters provide a rich historical account of the ongoing or previously existing codes, standards, rules, and individual judgments that were relied upon as NASA managers and engineers built the SSME's. Technical facts and details are also included. This complex historical narrative provided us with data from which we identified several recurring themes. These themes were labeled and provided the basis for our normative ethical framework.

Engineering Ethics: New Directions for Medicine

Engineering ethics boasts an impressive body of work that is relied upon throughout this text. Methods used and goals articulated for the field are similar to those found in medical ethics, yet methods per se have not been the focus of concerted study. Martin and Schinzinger, authors of the leading text in engineering ethics, comment that the field is much "younger" than medical ethics and borrow several concepts from medicine in their approach. While they use the term "applied ethics," it could be interchanged with our definition of "practical ethics." They define it as being concerned with "uncovering cogent moral reasons for beliefs and actions, as opposed to accepting uncritically whatever beliefs or actions might happen to strike one's fancy as being correct at a given moment." They contrast "applied ethics" and "general ethics" and clarify that "general ethics tends to emphasize theoretical knowledge and examines practical cases only in order to illustrate and test theories." Applied ethics, as we contend, "focuses upon concrete problems for their own sake, and invokes general theory where helpful in dealing with those problems."[27]

Charles E. Harris, Michael Pritchard, and Michael Rabins, in their recent book *Engineering Ethics: Concepts and Cases*, outline goals for the study of engineering ethics. Similar to what we have identified in medical ethics, these include stimulating the moral imagination, helping to recognize moral issues, and helping to analyze key ethical concepts. Based on their observation of professional engineers, they add engaging engineers' sense of responsibility and helping engineers address unclarity, uncertainty, and disagreement about moral issues.[28] These goals recognize both the general concerns of ethics and the specific nature of moral dilemmas in engineering.

The ethical dilemmas engineers are faced with center on how personal, professional, and organizational values affect moral decision making in engineering practice. How an HMO or a hospital as an organization affects a physician's ethical responsibilities toward his or her patients is an issue that medicine is just beginning to grapple with. Recent developments in managed competition and managed care have brought these to the forefront.[29] In contrast, engineering first drew scholarly attention to these types of questions in the early 1960s when the engineer's role in designing and building nuclear power plants and space vehicles, and their "contributions" to despoiling the environment, initially were investigated. This inquiry was part of a concern for ethical and social issues in public life in general. How does one define engineering? What does one consider the relationship of the engineer to the organization to be? How do the organization and the larger society interact?

In the late 1970s several events occurred that marked the beginning of a new field to investigate these issues systematically: a first interdisciplinary conference in engineering ethics at Rensselaer Polytechnic Institute and a scholarly bibliography in 1980; the first scholarly journal, *Business and Professional Ethics*, in 1981. "This late development of the discipline is ironic," concluded Martin and Schinzinger, given that, numerically, the engineering profession "affects all of us in most areas of our lives."[30] Thirty years later, concern has expanded to include engineers' accountability for the safety of common structures like houses, bridges, resort hotels, automobiles, and airplanes. Questions of engineering ethics today recognize that engineers make decisions that are ubiquitous to the safety and well-being of the public. Yet, they are primarily educated in a technical language that avoids explicit reference to the value-laden aspect of decision making. Do engineers perceive their decisions as having an ethical component? If they do, how do they resolve dilemmas that occur?

Methodological Importance:
New Directions in Engineering Ethics

Using the building of the SSME as a paradigm case, we examined three interrelated aspects of ethical decision making in engineering practice and one central to the general field of practical ethics. The first was to identify the types of ethical dilemmas engineers encountered as being troublesome and common in their work. Throughout the course of the development of the space shuttle program in general and in the development of the engines in particular, individual engineers and engineering managers, at both NASA and its industrial contractors, faced a number of critical decisions with serious ethical implications. How these critical decisions were perceived, articulated, and acted upon by engineers provided the data we needed to understand ethical decision making in engineering practice.

We recognized that personal values and judgments affect individual engineering decisions, and considered these to be the unique factors that characterize the engineer as a "moral agent." Professional codes, federal regulations, rules, and laws provide a framework to identify moral obligations. It is the individual engineer, however, who chooses to pursue these ideals. How he or she pursues them in the face of competing demands – primarily budget, deadlines, and safety – was of interest. Overall, our concern was to understand better how conflicts that involved individual professional responsibility, organizational goals, and societal well-being were resolved.

Second, we were concerned with how engineers defined risk and how that definition was influenced by the organization. Each engineer's "risk function" is to varying degrees dependent on the organizational structure and culture within which he or she is operating. We used a modified version of Herbert A. Simon's Nobel Prize–winning model of organizational behavior to interpret published sources, which document both the formal and informal organizational structure of NASA.[31] How did this structure affect recommendations, decisions, and actions taken? We particularly singled out engineers in our case studies who were, to varying degrees, cognizant of NASA's cultural influence. Some, like A. O. Tischler, director of NASA's Office of Advanced Research and Technology from 1964 to 1974, exercised considerable effort to combat it. How the organization responded to these efforts and how individuals coped with their moral concerns was of primary interest to us.

Third, we assumed that the political context in which the decision to fund the space shuttle project was made set the parameters for ethical dilemmas to occur. Both the decision by Congress to fund the space shuttle at $5.5 billion (instead of the $10 to $14 billion initial estimates) and the decision by NASA to accept this budget had an effect on virtually every aspect of shuttle development – particularly the main engines. A related concern then was to document the interaction between government, NASA, the aerospace industry, and individual engineering decisions so as to understand better how that interaction affected the ethical decision making of those involved.

Our cases document that in constructing space policy, "problems of efficiency, cultural pluralism, political procedures, [and] uncertainty about risk"[32] were central. The tacit principles of good engineering practice provided a "moral background" to the policy but the realities of political life shaped it. How those realities meshed, collided, or matched with the organizational priorities of NASA and its engineers was a theme we explored.

Finally, we viewed this interdisciplinary examination as being important to practical ethics in general. The field has primarily been influenced by analytic philosophy. This approach has made great strides in providing a reasoned, principled examination of complex ethical issues and has responded to sharp criticism voiced in recent years for not having practical import to "real-life" situations. "The highest morality," writes engineer and humanist Samuel Florman, "starts not with ethical maxims, but rather with a recognition of life's complexities. . . . if technologists are to be humanistically sensitized, I would have them study literature and history. The humanities are most true to themselves when they stress the pulsating diversity of life rather than the search for moral imperatives."[33] Our approach to engineering ethics *incorporates* the benefits of analytic philosophy and moves beyond that field's contributions by incorporating the richness that history can add to this endeavor.

Practical Importance:
What Is the Future of Space Policy?

The interdisciplinary ethical analysis presented in this book has practical as well as academic importance. There is still concern about the continued operation of the space shuttle for safety reasons. Indeed, a reevaluation of the future of manned space flight in general is in progress. On December 4, 1993, two months after a successful repair of the space telescope, the *New York Times* reported that "despite the successful launching of the space shuttle, Endeavor, on its mission to repair the Hubble space telescope this week, new debate is swirling over the degree of danger posed by the space shuttle's solid fueled rocket boosters. . . . It was a faulty booster that touched off the Challenger disaster in January 1986."[34] The boosters, apparently, had shown repeated and alarming variations in their thrust. NASA officials who conducted "quiet" investigations of the problem denied any danger, stating that the "thrust variations do not represent a serious problem." A NASA manager is reported as saying, "It's safe to fly . . . we've looked at the worst case, at the maximum thrust increment as a result of pressure perturbations, and the shuttle program has determined that it would cause no violations of safety factors." An aeronautical engineer disagreed. "It's death threatening" he claimed. "They haven't actually exceeded the safety margins. But they have no way of knowing of what the upper limit of that spike is going to be."[35]

This particular incident was just one of many technical problems that have repeatedly plagued the shuttle. In January 1993, the House oversight committee responsible for monitoring the shuttle requested that "the Aerospace Safety Advisory Panel (ASAP) create a temporary task force of propulsion experts to conduct a thorough assessment of the Space Shuttle Engine (SSME)." This temporary task force was to include propulsion experts drawn from academia and elsewhere. Its charge was to "assess the risks that the SSME posed to the safe operation of the space shuttle; iden-tify and evaluate safety improvements that could eliminate or reduce these risks; and recommend a set of priorities that the task force believed should be followed on implementing specific improvements."[36] The Committee sought to attain an unbiased assessment that could answer various ques-tions. How safe is the SSME? Is it safe enough? If not, what improvements need to be made? How quickly should these improvements be brought into the operational inventory? What will be the cost of making these improve-ments? What are the likely risks and consequences if these improvements are not made? The deadline set for the final report was "no later than" February 1, 1993.

NASA was directed to provide the temporary task force with all neces-sary data, access to facilities, records, analyses, and personnel; and financial and administrative support that would be required for the temporary task force to comply with this congressional request. The temporary task force completed its charge on schedule and concluded that it was "safe to fly the SSME provided that all special controls were scrupulously followed." However, they identified specific aspects of the main engines that needed to be modified to improve its safety and reliability. Reducing reliance on peo-ple and processes for safety by shifting to achieve "inherent ruggedness and operating margins of the hardware" was the strategy proposed. The single-tube heat exchanger, alternate high-pressure oxidizer turbopump, and large-throat main combustion chamber were categorized as priority I, while the alternate high-pressure fuel turbopump and the two-duct powerhead were designated as priority II. Both priority changes were to be imple-mented as soon as possible, but in an economical manner, preferably as a set of "block changes" rather than as the currently proposed serial changes. A detailed study of costs, schedule, and technical aspects of both approaches (block versus serial) was to be made. Finally, the team explained, "it can be expected that anomalies and new phenomena will continue to occur as operating and test experience is gained. A competent, sustaining engineering function should be maintained to ensure thorough investigation of all such occurrences. Efforts to develop improved fabrication and inspection techniques for the SSME should be continued and encouraged."[37]

The main engines of the space shuttle are not the only NASA technology undergoing close scrutiny. The complications surrounding the failure of mirrors in the Hubble Space Telescope have been the focus of headline news reports and two book-length publications. Both books document how scientific and technical engineering decisions were affected by the politics of NASA. Indicative of what this interaction was like, astrophysicist Eric J. Chaisson entitled his work *The Hubble Wars!*[38] A December 7, 1993, article in the *New York Times* describes the situation well. It reports that engineers and scientists who constructed the main mirror of the Hubble Space Telescope were "haunted for years that its supremely smooth surface held a major flaw. But they did nothing to alert the government of their apprehensions or to stop the launching of the most complex scientific instrument ever put into space."[39]

Evidently, independent optical experts who reviewed the government's evidence "found the actions of the contractor incredible, unconscionable, grossly negligent, criminal, irresponsible, deliberate and reckless." Perkin-Elmer Corporation, whose Optics Division made and polished the Hubble mirror, denied any wrongdoing and said, "the government was fully informed of all data, needed for judging whether the telescope's main light-gathering mirror had any imperfections." But new disclosures about the federal investigation appear to raise doubts about that assertion. Did Perkin-Elmer make false representations to NASA? Were anomalies encountered by the contractor properly disclosed? Does NASA have any legal recourse against the contractors?[40]

These concerns about two of NASA's most visible technologies are indicative of questions regarding the future direction of space policy in general. The fate of NASA as a federal organization and, with it, the dream of building a space station are seriously in danger of ever becoming a reality.[41] Changes at the top administrative level of NASA and close scrutiny of both its program and budget by the National Space Council (NSC) evidence the disquiet with an agency once thought to be indestructible. The council criticized NASA's bureaucratic structure and charged that it inhibited the "quick and cheap" launching of missions. The council successfully pressured NASA to redesign its plan to monitor climate change on Earth by launching a large number of relatively inexpensive satellites, instead of a half-dozen huge ones, which would have cost $3 billion each. "[It] also recruited a veteran of the Pentagon's Strategic Defense Initiative to head the agency's exploration unit and plan a series of cheap, quick, unmanned missions."[42]

Conclusion

Has NASA, as a federal agency, undergone an inevitable cultural maturation process?[43] Or could alternative management decisions have kept the agency intact? Will an entirely new administrative organization have to be created to guide space policy successfully?[44] Are the days and dreams of a manned space program being replaced by pragmatic technical solutions? Perhaps the historical and ethical commentary in this book will add a dimension to understanding the interaction of politics, science, and engineering and provide added insight to guide current policy dilemmas. The space shuttle, as an example of "Big Science," can be considered a paradigm case for future federally funded technologies. It is also relevant to the examination of everyday engineering decision making. Questions such as how cost constraints affect the quality of the technology, how the political ups and downs of different administrations (be they government or private) impact on the continuity of a project, even how competitive bidding influences the proposed work are salient to all future engineering endeavors. The building of the main engines of the space shuttle was chosen in part because the rich database available to chronicle its development provides a unique opportunity to recreate detailed cases to be used in understanding how "low-bid" engineering decisions are made.

Notes

1. Lloyd S. Swenson, *This New Ocean: A History of Project Mercury* (Washington, DC: NASA, 1966), p. 468. Several newspaper articles provide evidence that this was a common recognition among astronauts and others during the early years of space flight. See, for example, William Hines, "Produced by the Lowest Bidder," *Evening Star* (Washington), September 9, 1965, p. 350A.

2. Alvin M. Weinberg, "Impact of Large Scale Science," in *Reflections on Big Science* (Cambridge, MA: MIT Press, 1967), pp. 161–164. For an authoritative note on sources related to Big Science, see Robert W. Smith, *The Space Telescope: A Study of NASA, Science, Technology and Politics* (Cambridge: Cambridge University Press, 1989), p. 426.
3. Weinberg, "Impact of Large Scale Science."
4. Smith, *The Space Telescope*, pp. 373–95.
5. Ibid.
6. Michael Pritchard, "Beyond Disaster Ethics," *Centennial Review* 34(2) 1990: 295–318.
7. Each of these examples has received extensive attention in the engineering ethics literature. Mike W. Martin and Roland Schinzinger, *Ethics in Engineering*, 2d ed. (New York: McGraw-Hill, 1989), have specific cases and study questions on Three Mile Island and Chernobyl (pp. 145–59); and the *Challenger* (pp. 79–86). For the Ford Pinto, see Francis T. Cullen, William J. Maakestad, and Gray Cavender, *Corporate Crime under Attack: The Ford Pinto Case and Beyond* (Cincinnati: Anderson, 1987); Dennis A. Gioia, "Pinto Fires and Personal Ethics: A Script Analysis of Missed Opportunities," *Journal of Business Ethics* 11(5-6), 1992: 379–89. For Three Mile Island, see Sandra M. Wood and Suzanne Shultz, *Three Mile Island: A Selectively Annotated Bibliography* (New York: Greenwood Press, 1988). For the *Challenger*, see Hans Mark, "The Challenger and Chernobyl," in *Traditional Moral Values in the Age of Technology*, ed. Lawson W. Taitte, Andrew R. Cecil Lectures on Moral Values in a Free Society (Austin: University of Texas Press, 1987).
8. Thomas S. Kuhn, "The Relations between the History and Philosophy of Science," in *The Essential Tension: Selected Studies in Scientific Tradition and Change* (Chicago: University of Chicago Press, 1977), pp. 4–20.
9. Ibid., pp. 18, 19.
10. Taft H. Broome Jr., "Engineering Responsibility for Hazardous Technologies," *Journal of Professional Responsibility in Engineering* 113(2), 1987: 139–49. While Broome never uses the term heuristic, his definition of engineering practice clearly satisfies the components of such a definition.
11. Kuhn, "The Relations between the History and Philosophy of Science," pp. 4–20.
12. For an overview of theory versus practice in medical ethics, see Barry Hoffmaster, Benjamin Freedman, and Gwenn Fraser, eds., *Clinical Ethics: Theory and Practice* (Cambridge: Cambridge University Press, 1989). Also reading the introductions to the four editions of Thomas L. Beauchamp and James F. Childress's book, *Principles of Biomedical Ethics* (New York: Oxford University Press, 1978, 1982, 1989, 1994), one can gain a sense of how the field of medical ethics has evolved.
13. Kuhn, "The Relations between the History and Philosophy of Science," p. 5.
14. Beauchamp and Childress, *Principles of Biomedical Ethics*, 4th ed.
15. Ibid.
16. For the most recent contribution to the case-based approach in engineering, see Charles E. Harris Jr., Michael Pritchard, and Michael J. Rabins, *Engineering Ethics: Concepts and Cases* (New York: Wadsworth, 1995). Other texts referenced in note 7 also rely on cases.
17. John D. Arras, "Getting Down to Cases: The Revival of Casuistry in Bioethics," *Journal of Medicine and Philosophy* 161(1), 1991: 37.
18. Ibid., p. 5.
19. Albert R. Jonsen and Stephen Toulmin, *The Abuse of Casuistry: A History of Moral Reasoning* (Berkeley: University of California Press, 1988).

20. Arras, "Getting Down to Cases," p. 5.
21. Ibid.
22. K. Danner Clouser and B. Gert, "A Critique of Principalism" *Journal of Medicine and Philosophy* 15, April 1990: 219–36.
23. Larry R. Churchill, "Theories of Justice," in *Ethical Problems in Dialysis and Transplantation*, ed. C. M. Kjellstrand and J. B. Dossetor (London: Kluwer Academic Publishers, 1992), p. 22.
24. Beauchamp and Childress, *Principles of Biomedical Ethics*, 3d ed., pp. 9–11.
25. K. Danner Clouser, "Medical Ethics: Some Uses, Abuses, and Limitations," *New England Journal of Medicine* 293(8), 1975: 384–7.
26. For a definition of the tacit ethic in medicine, see Larry P. Churchill, "Tacit Components of Medical Ethics: Making Decision in the Clinic," *Journal of Medical Ethics* 3, 1977: 129–32. Also see Larry P. Churchill, "Reviving a Distinctive Medical Ethic," Hastings Center Report 19(3), 1989: 28–34.
27. Martin and Schinzinger, *Ethics in Engineering*, p. 5.
28. Harris, Pritchard, and Rabins, *Engineering Ethics*, p. 12.
29. For further discussion of this, see Chapter 3.
30. Martin and Schinzinger, *Ethics in Engineering*, p. 5.
31. Herbert A. Simon, *Administrative Behavior*, 3d ed. (New York: Free Press, 1976).
32. Beauchamp and Childress, *Principles of Biomedical Ethics*, 3d ed., p. 14.
33. Samuel C. Florman, "Moral Blueprints," in *Blaming Technology* (New York: St. Martin's Press, 1981), pp. 162–80.
34. William J. Broad, "Booster Rockets in Shuttle Fleet Cause Concern," *New York Times*, December 4, 1993, pp. 1, 6.
35. Ibid., p. 5.
36. Walter C. Williams, *Report of the SSME Assessment Team* (Washington, DC: NASA, January 1993), pp. 28, 29.
37. Ibid.
38. Eric J. Chaisson, *The Hubble Wars: Astrophysics Meets Astropolitics in the Two-Billion-Dollar Struggle over the Hubble Space Telescope* (New York: Harper Collins, 1994).
39. William J. Broad, "Some Feared Mirror Flaws Even before Hubble Orbit," *New York Times*, December 7, 1993, pp. B7–11.
40. Ibid.
41. Barbara Rosewicz, "Clinton Plan to Redesign the Space Station Might Inadvertently Wind Up Killing the Program," *Wall Street Journal*, June 7, 1993, p. A16.
42. Bob Davis, "Bush Fires NASA Administrator Truly after Dispute over Management, Policy," *Wall Street Journal*, February 13, 1992, p. B10.
43. Howard E. McCurdy, *The Space Station Decision: Incremental Politics and Technological Choice* (Baltimore: Johns Hopkins University Press, 1990).
44. Chaisson, *The Hubble Wars*, pp. 348–52.

2

Engineering Ethics Framework

An ethical engineer is one who is competent, responsible, and respectful of Cicero's Creed II. Cicero's Creed, engineering's oldest ethic, directed engineers to place the safety of the public above all else. We have added specificity to this creed and assert that an ethical engineer be knowledgeable regarding risk assessment and failure characteristics of a given technology.[1] No matter how skilled, knowledgeable, or moral a single engineer is, he or she cannot create or manage modern technology without being a member of an organization. Our framework recognizes this and defines its three principles at two levels: that of the individual engineer and of the organization. For Cicero's Creed II, the organizational ethic involves managing technology so as not to betray the public trust. The concept of "stewardship" for public resources is included here, and embodies the intent of Cicero's original ethic.

The Principle of Individual Competence

An engineer is a knowledge expert specially trained to design, test, and assess the performance characteristics of technologies within his or her realm of expertise. If an engineer works for a company, he or she is relied upon in this capacity. In an effort to attain the status of knowledge expert with respect to a given technological problem area, the engineer should acquire requisite information that is reliable, relevant, and adequate. This includes an understanding of not only the performance characteristics of the final "product" but also those of its individual components.

The engineer then is, or aspires to be, a knowledge expert and is being relied upon as such. To become knowledgeable about a given technology, the engineer must inquire into its performance characteristics. To do so

insufficiently, or to do so in a faulty manner, either knowingly or unknow-
ingly, nullifies his or her position as being adequately informed. Thus, to
act incompetently impeaches the engineer's status of being a knowledge
expert. (Of course, intentional incompetence is a more serious offense than
the unintentional.) The incompetent engineer is still "doing engineering,"
but has not discharged his or her duty.

We can expand our understanding of an engineer's *competence* by examining Figure 2.1. The large oval-shaped "blob" in Figure 2.1 represents the
entire body of knowledge required to design a technology. The smaller
shapes within the oval-shaped "blob" depict the components of one engineer's competence with respect to the overall technology: that is, empirically tested knowledge, theoretically derived knowledge, known missing
knowledge, and unknown missing knowledge.

As a knowledge expert, the engineer works with both empirical and theoretical knowledge. The first two knowledge components usually have different sources. The engineer acquires theoretical knowledge through

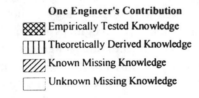

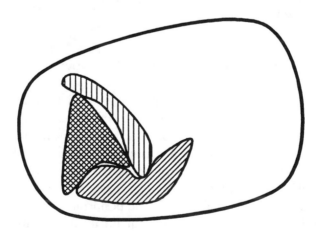

Figure 2.1 Knowledge base (empirical, theoretical, and known missing knowledge) required to design a technology with one engineer's contribution. Note unknown missing
knowledge.

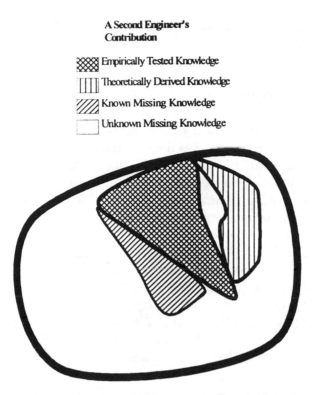

A Second Engineer's
Contribution

▨ Empirically Tested Knowledge

⫿⫿⫿ Theoretically Derived Knowledge

▨ Known Missing Knowledge

☐ Unknown Missing Knowledge

Figure 2.2 Knowledge base required to design a technology with a second engineer's contribution. Note the different knowledge base compared with the first engineer.

formal training and education, through library research, and perhaps through mathematical and other modes of formal derivation. He or she acquires empirical knowledge about a technology through experience, experimental testing, record keeping, and generally through the use of a particular technology. A *competent* engineer must also acknowledge what he or she does not know about a technology. As we will explain, an understanding of the limits of one's knowledge is important in at least two ways. Finally, the residual category of knowledge in Figure 2.1 represents the unknown missing knowledge of a single engineer.

Figure 2.2 presents a similar diagram of a second engineer's competence. In a rational organization, one based on the principle of division of labor, the components of competence for this second engineer cover different portions of the oval-shaped "blob." Together, the knowledge of these two engineers comes closer to what is required to design the technology than either could provide singly. This observation leads directly to our second principle.

The Principle of Organizational Competence

An organization is *competent* if the engineers it employs collectively have the requisite knowledge to design the technology, to the extent this knowledge domain is known. Figure 2.3 presents the "blob" diagram for a *competent organization*. Each engineering member of the organization contributes specialized knowledge to the solution of the total design problem.

Figure 2.4 presents another view of *organizational competence*. In this diagram, gaps exist in the collective knowledge required to design the technology. The crucial question concerning these gaps is whether they are known or unknown components in the collective knowledge of the engineers. If the gaps are known by at least one engineer, the organization fails to meet the requirement of the principle of organizational competence. Unknown gaps in the collective knowledge base represent a technological risk attributable to ignorance. Although organizations have the capability to be more rational than individuals, they are certainly not omniscient.

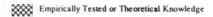 Empirically Tested or Theoretical Knowledge

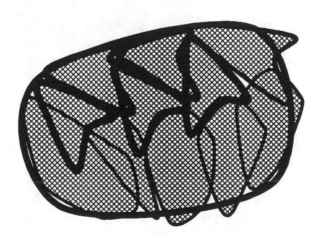

Figure 2.3 Organizational competence: Case I – each engineer contributes specialized knowledge required for the problem's solution.

Figure 2.4 Organizational competence: Case II – note the gaps that exist in the collective knowledge compared with Case I.

We have discussed these diagrams of individual and organizational competence as if they were static descriptions of the status of knowledge about the technology. This obviously is not the case for engineering as an intellectual activity. The status of knowledge changes throughout the design process. Individual engineers expand their competence with respect to the technology as they work on the design. Organizational competence changes both with the knowledge of individuals and by adding, or removing, engineers on the project. During the initial stages of an engineering design project, we would expect gaps to exist at both the individual and organizational levels. As the design progresses, the engineers, both individually and collectively, fill in the missing pieces. When the design process succeeds, the outcome resembles the first case in Figure 2.3. Gaps in the process can occur in two ways: Either known gaps are not resolved even after additional resources, time, or effort have been expended; or the technology requires knowledge beyond the competence of the organization.

The Principle of Individual Responsibility

To play the role of knowledge expert in the decision-making process implies that one has the duty to make information readily available to the other participants and to take a critical attitude toward assessing management decisions from an engineering standpoint. One must, in short, develop and communicate evidence to support one's opinion.

Our discussion of the components of individual competence immediately suggests one class of knowledge that should be communicated by a *responsible* engineer: that part of the knowledge base the engineer knows is missing – for example, a known knowledge gap should be or must be communicated. Without this information, the organization cannot rationally achieve competency.

Not to communicate what one knows is to abdicate willingly one's proper and unique role. A crucial premise currently guiding this analysis places engineers in the venerable role of functioning as trained experts (like a physician, lawyer, architect, etc.). The burden of obligation then stems from the proper discharge of this capacity from both the technical and engineering standpoints (*competence*) and in the larger context of participating in an organization's decision-making process (*responsibility*). While this following principle concerns the engineer as knowledge expert, participating in managerial decisions, it also applies to all engineers, for "knowledge is power," and with most forms of power comes responsibility.

The Principle of Organizational Responsibility

The counterpart to the principle of individual responsibility is the principle of organizational responsibility. If this principle is to work on the individual level, the organization must be responsive to the engineer who communicates a concern. This does not mean that the organization must act on every concern raised by a responsible engineer. It does mean that the organization must listen to and consider reported concerns.

We offer examples of how an engineering organization satisfies this principle of responsibility. Engineer A identifies a knowledge gap. Another member of the organization may know that engineer B can help A solve the problem. This would be communicated to A so that A and B can work together to resolve the concern. Engineer C expresses a concern about the durability of a mechanical part he or she is designing. The engineer believes that further testing will reduce the uncertainty about the design. The organization authorizes further testing, and C creates the new knowledge that resolves the concern. Finally, engineer D designs a component that he or she believes will do the job. However, on subsequent testing, the engineer sees indications that the part is not working as well as desired. The engineer voices a concern to management, requesting time to redesign the part. Management declines this request because the part is working, and the entire engineering group is on a tight schedule, and will be hard pressed to meet current design milestones. The engineer, while not completely satisfied, understands the situation and gets on with work as a member of the group.

Cicero's Creed II: The Individual

Cicero's earliest creed obligated the engineer "to insure the safety of the public." In philosophical terms, this can be described in the positive form as beneficence: doing good. It also covers the negative aspect: do not harm, or nonmalevolence. A strong version of such an idea would state that intentionally and/or gratuitously causing (serious) harm in any of its guises is universally prohibited. "Harm" as understood from the perspective of the individual engineer refers to the engineer's ability to assess the potential risks of the technology. We recognize this basic prerequisite to fulfilling Cicero's Creed and specify it as Cicero's Creed II: *The engineer should be cognizant of, sensitive to, and strive to avoid the potential for harm and opt for doing good.* With respect to a given project, in an effort to acquire reliable, relevant, and adequate information, an engineer should include as part of this process an inquiry into and an assessment of the safety, risk, and failure characteristics of the given technology of concern. Part of the information regarding a product consists of these characteristics. They are used to evaluate its potential for causing harm.

Our definition of Cicero's Creed II has already been institutionalized in many areas of technology. Engineers use "safety factors" when designing components to attempt to insure that components are a factor X times stronger than necessary. They conduct tests to determine "mean time to failure" statistics for systems and components. These practices assist engineers in framing issues of risk and safety.

As before, there is a concomitant responsibility for the engineer to communicate what he or she knows when participating in the decision-making process. In this case the "knowledge" involved has to do with critical assessments and judgments of the safety of a product based primarily on test results. The argument is the same: To remain silent in the face of worrisome data is a de facto abdication of one's role as a knowledge expert inside an organization.

Cicero's Creed II: The Organization

The risks associated with a technology have an organizational dimension. Just as the individual engineer does not possess sufficient knowledge to design the technology, the individual engineer cannot comprehend the totality of risks associated with a modern technical system. This requires an organizational-level analysis of risk. Organizations meeting this principle assess risks, and where harm may result, either make those risks known or allocate resources to reduce risks, or both. By contrast, the unethical organization fails to assess risks or ignores the potential for harm. A term commonly used to describe this principle is stewardship; for ultimately, the organization has an obligation to use resources wisely and in a just manner.

To illustrate this principle, consider the following simple example. A group of engineers diligently adheres to the principle of Cicero's Creed II. Each engineer assesses the risks for the components and systems that he or she designs or manages. Suppose these risks varied greatly, say with failure probabilities ranging from 10^{-5} to 10^{-2}. According to a "weakest link in the chain" argument, the technology would be at the risk of the largest failure probability. This simple example serves to illustrate our main concern. Thus, although many of the components of the technology meet acceptable risk criteria, those components with high failure probabilities jeopardize the whole. To satisfy this principle, the organization must attempt to improve the performance of those high-risk components. For certain technologies, this may not be possible, and the organization should articulate the high-risk nature of the venture to those immediately affected by it, assess the availability of lower-risk options, and evaluate the high-risk option in terms of organizational aspects of Cicero's Creed II (i.e., stewardship).

Summary of the Framework

We ground our framework in the observation that the engineer is a knowledge expert who works within an organizational structure. As such, he or she must satisfy a principle of competence. Competence involves both the acquisition of relevant knowledge, and the recognition of what is not known. The competent organization selects and manages engineers so as to provide or create the knowledge base for the whole technology. The responsible engineer communicates concerns about the design of the technology, and the responsible organization either acts on those concerns, or explicitly decides otherwise. An ethical engineer combines competence and responsibility with the understanding of the risks associated with his or her components of the technology. The organization, likewise, compiles such data to assess the overall risks associated with the use of the technology. These risks feed into the decision process of the organization such that the organization seeks both to reduce the harm that may result from the technology and to provide an overall benefit to society. The recognition by both the individual engineer and the organization of the need to assess and minimize risk is labeled in our framework as Cicero's Creed II.

Note

1. We include here two engineering codes of ethics. The first was approved by the Accreditation Board for Engineering and Technology in October 1977.

Code of Ethics of Engineers

The Fundamental Principles

Engineers uphold and advance the integrity, honor, and dignity of the engineering profession by:

I. using their knowledge and skill for the enhancement of human welfare;
II. being honest and impartial, and serving with fidelity the public, their employers;
III. striving to increase the competence and prestige of the engineering profession; and
IV. supporting the professional and technical societies of their disciplines.

The Fundamental Canons

1. Engineers shall hold paramount the safety, health, and welfare of the public in the performance of their professional duties.
2. Engineers shall perform services only in the areas of their competence.
3. Engineers shall issue public statements only in an objective and truthful manner.
4. Engineers shall act in professional matters for each employer or client as faithful agents or trustees, and shall avoid conflicts of interest.
5. Engineers shall build their professional reputation on the merit of their services and shall not compete unfairly with others.
6. Engineers shall act in such a manner as to uphold and enhance the honor, integrity, and dignity of the profession.
7. Engineers shall continue their professional development throughout their careers and shall provide opportunities for the professional development of those engineers under their supervision.

The second code was approved by the Institute of Electrical and Electronics Engineers (IEEE) in August 1990.

Code of Ethics

We, the members of the IEEE, in recognition of the importance of our technologies affecting the quality of life throughout the world, and in accepting a personal obligation to our profession, its members and the communities we serve, do hereby commit ourselves to the highest ethical and professional conduct and agree:

1. to accept responsibility in making engineering decisions consistent with the safety, health and welfare of the public, and to disclose promptly factors that might endanger the public or the environment;
2. to avoid real or perceived conflicts of interest whenever possible, and to disclose them to affected parties when they do exist;
3. to be honest and realistic in stating claims or estimates based on available data;
4. to reject bribery in all its forms;
5. to improve the understanding of technology, its appropriate application, and potential consequences;
6. to maintain and improve our technical competence and to undertake technological tasks for others only if qualified by training or experience, or after full disclosure of pertinent limitations;
7. to seek, accept, and offer honest criticism of technical work, to acknowledge and correct errors, and to credit properly the contributions of others;
8. to treat fairly all persons regardless of such factors as race, religion, gender, disability, age, or national origin;
9. to avoid injuring others, their property, reputation, or employment by false or malicious action;
10. to assist colleagues and co-workers in their professional development and to support them in following this code of ethics.

We do not intend our framework to replace such codes, but introduce it into the literature to facilitate discussion and recognition of ethical dilemmas in practice.

3

Engineering Ethics: The Common Ground between Individual Whim and Legality

Our engineering ethics framework is based on the assumption that the engineer is a moral agent. That is, as an autonomous professional, the engineer is a valuable resource for moral decision making both to the organization and to the society.[1] This assumption, however, is not without its critics. Samuel Florman has been most vocal in this regard. In his oft quoted article "Moral Blueprints," he asserts, "as a professional, I abide by established standards. . . . As a human being I hope that I deal adequately with each day's portion of moral dilemmas. But between legality on the one hand and individual predilection on the other, there is hardly any room for the abstraction called 'engineering ethics.'"[2] Florman assumes that engineers have a particular moral bias and concludes that to "rely on them for social well-being is not rational policy. . . . Society," he contends "would be ill served if it made technical decisions in accordance of the personal whims of engineers."[3] In this chapter, we describe what we believe to be the common ground between "individual whim" and "legality." It is within this common ground that a definition of engineering ethics can be constructed.

Medical Ethics and Engineering Ethics: A Two-Way Street

In framing our definition of engineering ethics, we relied on the methodological foundations developed in biomedical ethics to guide our analysis. Our interdisciplinary team defined key terms according to those offered by leading scholars in medical ethics. We relied on that field's emerging consensus regarding what a practical ethic is, and on ways it has been used in a professional setting.[4] Philosophers Dan Brock and Alan Buchanan provide some insight into why this common ground has developed. "Ethics matters in the real world," they conclude, "and reflection on real world ethical problems can also deepen our understanding of largely theoretical issues."[5] The increased and shared experience of ethicists working and teaching in the clinical setting, an achievement that has yet to be attained in engineering, has provided a common point of reference for both practical and theoretical reflection.

Medical ethics has made practical and sustained inroads into both clinical decision making and the conduct of research ethics. To date, approximately 80 percent of acute care hospitals have ethics committees.[6] Granted, these committees do not provide a panacea for resolving all ethical disputes, and criticism has arisen from the academic front.[7] Nevertheless, the committees provide guidance for practitioners, patients, and others stymied by any one of a number of recognized moral dilemmas occurring in health care. In recognition of the importance of medical ethics, the 1992 guidelines of the Joint Commission on Accreditation of Healthcare Organizations (JCAHO) provide a mandate for institutions to create an "ethics forum" to constructively deal with issues.[8] Ethics consultation services are becoming commonplace in hospitals. Institutional and professional policies in health care and medical research, from informed consent to forgoing life-sustaining treatment, incorporate ethical principles into their texts, which are in turn relied upon for making decisions in the clinic. Each of these areas – committee development, ethics consultation, policy writing, informed consent, and decisions near the end of life – has its own impressive body of literature.

During the past thirty years, moreover, the focus for study in medical ethics has moved from primarily examining moral issues intrinsic to the doctor–patient relationship to the exploration of broader social issues, many of which have long been central to the practice of engineering. Brock and Buchanan's authoritative study, *Deciding for Others: The Ethics of Surrogate Decision Making,* was written to provide the ethical and conceptual framework for enlightened public policy and individual choice when persons are not capable of exercising that right. One of the fundamental tasks of ethics, they explain, "is to demarcate the proper boundary between matters of individual choice and responsibility and those of collective and social policy." Moving well beyond the traditional doctor–patient relationship, they regard the issues as "institutionally embedded."[9] Health-care reform proposals suggest that this trend will continue. An emerging literature in medical ethics focuses solely on how physicians can ethically "balance traditional duties to individual patients against increasing pressure to serve broader societal and institutional goals. To cope with reform," writes Susan M. Wolf, "medical ethics must clarify physicians' moral obligations, change existing ethical codes, and develop an ethics of institutions."[10] The 1994 JCAHO guidelines include a requirement for hospitals to acquire a knowledge and expertise in institutional ethics.[11] While commentators suggest courses in business ethics to supplement the well-defined curricula in medical ethics,[12] engineering ethics can also be viewed as a resource.

We recognize that because of the rapidly changing organizational structure and goals of medicine, a comparison between medical and engineering ethics yields a two-way street. Physicians, after all, have traditionally been the central figure in medical practice. They conduct their professional duties with an autonomy unparalleled in the engineering profession. Typically a "team player," an engineer clearly has had a different range of moral responsibilities to his or her client than a physician has had to a patient. The physician's skills are directly related to an individual outcome; the engineer's skills are mediated by the products he or she creates, and typically affect large groups of people. While a physician can work in a variety of settings – private practice, a hospital, a health maintenance organization – he or she has not historically been bound to the institution for financial success. An engineer, on the other hand, is tied to a business environment and its organizational structure for advancement.

The gradual changes in the constellation of practice patterns available to and forced upon physicians over the past twenty years have changed the role of the physician as the "captain of the ship."[13] Thus, even though the professions of medicine and engineering are characterized historically by a different "professional ethic,"[14] when one takes these health-care reform changes into account, similarities emerge. King, Churchill, and Cross wrote as early as 1986:

> Twenty years ago, a single legal metaphor accurately captured the role that American society accorded the physician: they were each "captain of the ship." They were in charge of the clinic, the operating theater, and the healthcare team. They were held responsible for all that happened within their domain, and this grant of responsibility carried with it a concomitant delegation of authority: the physician was answerable to none in the practice of his art. Few would agree that, nowadays, no matter how compelling the metaphor, the mandate accorded to the medical profession by society is changing.[15]

Christine K. Cassel described the change in 1984:

> We are in the midst of a fundamental shift in the socioeconomic basis
> of the healthcare profession – a shift that has major implications for
> the ethics, as well as the ethos of my vocation. I am speaking of the
> change from a special professional ethos, with its idealized characteris-
> tics of autonomy, self-regulation and service to the public, to a corpo-
> rate or management ethos, with its emphasis on cost–benefit analysis
> and the business model of productivity.[16]

Both medicine and engineering, moreover, rely on heuristics to resolve
their practical problems.[17] Intensive care physicians' problem-solving
strategies strikingly resemble those of design engineers. Each is faced with
deciding how to balance the risks and benefits of a specific treatment or
testing procedure amid the uncertainties of outcome. Composite data may
exist to guide decision making, but how a particular procedure affects an
individual patient or how a specific material bears up once in use is the
result of a complex web of circumstances. Finally, engineers have, as do
surgeons, a range of technical skills which must be mastered. As discussed
in Chapter 10, the way surgeons categorize errors may be a useful scheme
also for engineers.

　　More to the point of this book, the professions of both engineering and
medicine have, since the mid 1960s, recognized that moral problems and
dilemmas are commonplace in their everyday practice. They have struggled
to create what philosophers Beauchamp and Childress define as "practical
ethics"[18] – that is, they sought to use ethical inquiry to examine the moral
problems of their respective professions. We contend that the methodologies developed in medical ethics, the success of practical applications in
that field, and even the problems studied have evolved over the last thirty
years in a way that bears relevance and importance to a study of engineering ethics.

The Conventional View of Ethics

When medical ethics first appeared in the late 1960s, a conventional view of
ethics prevailed. That view advised that (actual) actions, decisions, and policies were to be understood in terms of, and chosen (justified) on the basis
of philosophical theories. In practice, one would choose a controversial

moral issue, and then consult any number of canonical ethical theories for the purpose of describing the case, and for arguing over it. The "open–minded" ethicist would take an "objective" view by displaying how alternate stances are theorems of alternate ethical theories. Textbooks in medical ethics constructed fictitious cases that embodied identifiable conflicts and engaged those analyzing the cases in debates familiar to philosophers. The goal of the debate was not primarily to resolve the practical issue at hand. Rather, it was to teach critical thinking by engaging the reader in articulating issues in a rigorous fashion. A theoretical stance was then consulted for alternate and justified resolutions.

However, when this conventional approach was called upon to resolve actual dilemmas occurring in practice, those using it recognized that it had significant drawbacks. First, it was not clear how to bring a philosophical theory (which is necessarily abstract and highly generalized) to bear on specific situations. These situations are rife with complexity and ambiguity. A fair amount of ad hoc maneuvering, and possibly idiosyncratic case interpretation by individual theorists, are needed to correct theory and case. There are, moreover, a number of moral theories to choose from – for example, the natural law theory of Aristotle, nineteenth-century utilitarian and deontological theories most often associated with John Stuart Mill and Immanuel Kant, contract theories of Locke and Rousseau, intuitionism of Ross, and the twentieth-century theories of justice, most notably that attributed to John Rawls.[19] Each theory appeals to moral persuasions that have "significant plausibility as well as significant internal and external problems"[20] when applied to concrete situations.

The most influential work in medical ethics promoting the importance of moral theory and principles is the now classic *Principles of Biomedical Ethics* by Beauchamp and Childress. First published in 1978, the book provided an in-depth justification for the use of principles derived from moral theory as "action guides." Beauchamp and Childress identified "principles" as the proper level of abstraction and discussed four as being particularly relevant to medical ethics: autonomy, beneficence, nonmaleficence, and justice. Combined, the principles created a common language or framework to identify moral issues in medicine. They provided a route of entry to the resolutions suggested in overarching theories.[21] The principles, however, still required a fair amount of maneuvering if they were to be applied to a specific case. One of the strengths of the third edition of Beauchamp and Childress's book is that it provided numerous case examples followed by analysis of how this maneuvering can take place. The fourth edition clarified and defended a limited – albeit important – use of theories and principles and responded to several issues raised by critics.[22]

The Conventional View Challenged

By the mid-1980s, challenges to Beauchamp and Childress's principled approach as a basis for medical ethics decision making were gaining a consensus. Baruch Brody has articulately identified the problems faced with the four principles. Each has a common root in the five traditional moral theories – hence their appeal. Brody clarified, however, that each of the five traditional moral theories faced both a "specification problem" (how to weight individual consequences, duties, or principles) and a "theory-to-concrete-judgment problem" (how to identify additional premises not specified in the theory so as to entail a specific judgment). Until these "internal" problems are resolved, Brody advised, reliance on principles and their respective theories only highlighted inconsistencies. There was no way to resolve them. Given this recognition, he defended using "a model of conflicting appeals" to analyze specific medical ethics cases. This model combined the perspectives of each of the five theories and, in so doing, articulated reasons for a particular resolution to "real-life" complex conflicts. Summing up, Brody had confidence that each of the problems he identified would be remedied by utilizing a corrected "overarching moral theory." Until that theory was created, however, he offered his theory of conflicting appeals as a "working" remedy.[23]

Perhaps the most consistent challenge to the principle-based approach has been made by philosopher Albert Jonsen and historian of science Stephen Toulmin. They contend that practical reason – not theories or principles – is the cornerstone of bioethics. *Casuistry* is the methodological rival to the traditional theory-based view. Historically, it was regarded as a valid method for resolving moral dilemmas from approximately 1500, when it was in its prime, to 1850, when it was ignored as a reliable source of moral knowledge. Since that time it has been held suspect by academic philosophers. Jonson and Toulmin recalled their experiences in 1968 when they served on the prestigious President's Commission for Ethical Issues in Biomedical Research. The process of reasoning used by the interdisciplinary team of commissioners, whose charge was to write a policy to guide the conduct of ethical research in biomedicine, was, according to Jonsen and Toulmin, "casuistic." That is, the participants proceeded to compare and contrast cases and eventually agreed upon ethical guidelines to support acceptable research practice. "The ethic, however, was derived from a common agreement regarding the particulars of the cases. . . . The resulting principles were not justified by our overarching theory."[24] The results of this endeavor, *The Belmont Report,*[25] a document still in use, cited the principles of autonomy, beneficence, and justice as the moral basis for biomedical research. Nonetheless, Jonsen and Toulmin report that the principles were derived from a careful comparison of the practical cases.

Their book, *The Abuse of Casuistry: A History of Moral Reasoning,*[26] was published in 1988 and has had a marked impact in medical ethics. Its aim was "to resurrect casuistry" as a respected method of reasoning in applied ethics. The casuist sees individual cases as the appropriate level of moral analysis, and breaks the analysis of cases into four components.[27] First, one must identify paradigms, or cases in which a moral maxim is clearly applicable. The casuist assumes that maxims will inevitably spring forth from a given situation. Such maxims originate from religious, ideological, and personal points of view. From the paradigm case, one focuses on the maxim and moves to analogies, cases in which, due to different circumstances, the maxim is less suitable, more open to rebuttal, and more susceptible to exception. In making the comparison one must carefully examine the circumstances – the set of concrete facts extant in a given situation. Given this examination, one determines not only the relevant moral species, but also the strength or weakness of the analogy in relation to the primary paradigms. In the casuistic approach, the circumstances make the cases, not a priori philosophical classifications.

In sum, it is crucial to classify and identify cases by maxims. A casuist puts forth what is called an "arguable case," with an addendum indicating its persuasive force, for example, certain, probable, or less probable. The aim of an argument is not to lead the mind through ineluctable steps of formal argument to a compelling truth, but to reveal, by a juxtaposition of maxims and circumstances, an opinion that could be reasonably entertained. The final result is a recommendation (indicating its own relative merit) that serves to counsel an actor in his or her present situation.

The obvious drawback to this approach is the strength of the analytic tradition. As Brody and Arras both insist, we do want a conclusive and compelling decision or justification procedure for settling (once and for all) disputes. The casuist retorts that what is lost in rigor is amply paid back in practicability, and reminds that ethics is practical reason after all. The traditionalist cries "sophist" while the casuist cries "academic" and the dispute rages on.

The Emerging Consensus

This debate between the use of theories and principles versus cases and its relevance to clinical ethics is still carried on in the bioethics field. Yet there is evidence that a common ground is emerging. Physician–philosopher Edmund Pellegrino reviews the various stages in the development of medical ethics. The first was what he called the "Quiescent Period," spanning the ages from Hippocrates to the early 1960s. It was characterized by genuine ethical precepts such as confidentiality, beneficence, and an appeal to the physician's "life of virtue," and Pellegrino traces its roots to Greek and Judeo-Christian traditions. "Phronesis," or practical judgment, was a central virtue and provided a key role in a physician's ability to resolve ethical dilemmas. As mentioned several times, the 1960s ushered in a new wave of professional medical ethics. Pellegrino, recognizing trends we have cited, calls the years 1960–80 the "Period of Principlism." He has a harsher view of the success of this approach and insists that the failure of the use of principles led to a third phase, "Antiprinciplism"; and the time we are presently in is one of "chaos." His is not quite the positive assessment of medical ethics we present in this book; still, Pellegrino is hopeful that a resolution among the current approaches will enable the field of "clinical bioethics" to "provide a reality check" and achieve a compromise between philosophy and day-to-day clinical practice.[28]

Beauchamp's recollection of his own work is instructive in understanding that this compromise is at hand.[29] "It has become fashionable in recent years to criticize the 'principle approach' as too limited," he writes. Answering his critics, he clarifies that his view is based on the belief that "a careful analysis and specification of principles is consistent with a wide variety of types of ethical theory, including virtue theory, . . . communitarian ethics, casuistical and theories concerned with the ethics of care. . . . A principle-based approach," he stresses, "is not a one-sided exclusionary . . . approach to bioethics." The four-principle approach he and Childress developed is not a "canon," or "scripture," to interpret authoritatively the complex body of judgments, rules, standards of virtue that are included in "bioethics." "Morality," Beauchamp and Childress agree, "is a complex social institution that draws from law, religion, government, and institutional roles, all of which use general action guides to prescribe behavior. It is fruitless to seek a precise and decisive set of criteria that will delineate the difference between moral guides and all other action-guides. Part of understanding morality is to appreciate the broad scope of the concept and how pervasively it is infused into the expectations we have for proper social practices and institutional behavior."[30]

Albert Jonsen reminds us that bioethics was born of interdisciplinary roots. "Basic ethical principles" gave the various scholars who came to the field "a clear framework for a normative ethics that had to be practical and productive." The principles gained dominance in bioethics because they provided a language for the ethicist – a term that did not exist until 1971 – to speak with physicians, nurses, and other health care professionals.[31]

Finally, K. Danner Clouser, who actually pioneered the applied field of medical ethics, also adds some balance to the relevance of the debate. Describing his observations regarding the birth of bioethics, Clouser recalls that

> Philosophy provided the push toward systematization, consistency, and clarity as problems within medicine increasingly erupted into moral dilemmas. The maneuvers, ploys, and strategies of philosophy have been important for bringing system and organization into medical ethics. It asks probing questions; it understands how to discover and work with assumptions, implications and foundations. Conceptual analysis, which is central to the "doing of philosophy" has been central also to the doing of medical ethics.[32]

Clouser did not claim that his discipline had a monopoly on bioethics. He recognized, as others we have quoted, that "Bioethics essentially concerns a particular arena of human activity and the morality relating to it. Many disciplines have morally relevant insights into these activities, as well as organizing concepts." Philosophy's expertise was in showing how crucial concepts were related to each other and in defining key terms used in analysis.[33]

Because of bioethics' interdisciplinary foundation, Clouser and others have consistently stressed that it "is the hard reality of its cases . . . [based] on real situations, thick with details, which is the challenge of the field." He clarifies that philosophers typically use hypothetical cases, invented or highly abstracted from real circumstances. The cases are constructed to defend selectively the particular point the philosopher wants to substantiate. Pointing again to the theme of the interrelationship between theory and practice, Clouser recognized, as did Arras, that philosophers often use cases to illustrate a theory rather than to test it. "But when solving the moral problem is the main point, the relentlessness of details becomes readily apparent. There is no refuge; rather, there is one quagmire after another; retreating to the theory is not a viable option." The disciplined, reflective questioning of the philosopher is invaluable here, but so are perspectives of law, history, sociology, psychology, and medicine. Churchill refers to disciplined, reflective questioning as "theorizing" and contends that while theories may not bear directly on practical reasoning, theorizing is essential.[34] It is the cases, then, based on real-life dilemmas, which provide a key to the common ground in applied ethics. If maxims or rules can be derived from the cases, principles and theories can be referred to as one reflects on whether the maxim is ethically justified.

Our Framework and Cases

The goal of this book is within the realm of practical, not theoretical, ethics. We recognize the value of a rigorous analytic questioning of assumptions and a definition of concepts. The complexity of the issues practical ethics faces and its interdisciplinary underpinnings is evident in this work. Taking these considerations into account, we developed a series of complex cases. During our interdisciplinary discussions we relied on basic definitions of terms cited in Beauchamp and Childress as a baseline for our discussions. Adopting such definitions greatly facilitated our own understanding of ethical analysis. We were, after all, an interdisciplinary team plagued by our own professional perceptions.[35]

We queried our historical data with reference to a "principle-based approach." This helped us to understand diverse points and to communicate among ourselves the complexity we encountered in our in-depth historical narrative. Our principles, however, are not formalized routes of entry into theory.[36] They are more akin to Clouser and Gert's "checklists" for morally acceptable behavior.[37] Concepts of competency, responsibility, and Cicero's Creed II were each derived from and defined by data included in the historical record. While we referred to theory for ways to clarify our definitions, we were not concerned with settling the debate regarding the foundations of bioethics. Rather, we had a common goal of making explicit the tacit normative code of ethics implicit in our cases. Reference to or development of an overarching moral theory to justify the professional ethic was not a question we addressed.

The framework we created is based on three professional values tacitly evidenced in our in-depth historical review. When we defined these concepts in our framework, we referred to several theories of organizational behavior, to engineering ethics, political ethics, and medical ethics. The "checklists" provide a middle ground for assessing individual and organizational ethics and, also, for understanding how they interact.

The Common Ground of Engineering Ethics

With this commitment to combining practical wisdom and analytic rigor, we return to Florman's objections to engineering ethics. Earlier, we defined ethics as critical reflection of what one does and why one does it. Our review of medical ethics indicated many ways to achieve this critical reflection. The two most common approaches are theoretical and principle-based or case-based and practical.

Florman was particularly disdainful of the new field of engineering ethics for several reasons. First, he observed that it was dominated by philosopher ethicists who had a tendency to simplify the engineer's moral obligation by stressing safety above all else.[38] Here, he was describing the philosopher's penchant for crafting generalized rules of behavior applicable to all. Philosopher Caroline Whitbeck, for example, argues that "there is an emerging consensus that safety is the foremost responsibility of the engineer."[39] In stressing the overreaching need for safety, Florman contends, ethicists ignore the "benefits of economy . . . (and) economy must be recognized as being practically and morally desirable."[40] In defense of Whitbeck, it should be noted that she does not cite philosophical theory solely as the basis of this obligation, but also includes the ethical codes and guidelines of engineering societies. Five such codes describe safety as the "foremost responsibility," while another seven combine the focus on safety with other obligations.[41] So safety is clearly an obligation recognized by engineers; but not all agree it is "foremost." At times, it may be given a lower priority than meeting a deadline or staying within budget, and this is what Florman relates to. He recognizes that moral dilemmas exist, but observes that they are more complex than ethicists admit.

Martin and Schinzinger's book, *Ethics in Engineering,* provides another example of the theory-based approach. The book introduces a summary of four major ethical theories: rights, utilitarian, deontological, and virtue ethics. It also cites newer critiques such as feminist ethics and reviews stages of moral development. A basic text in the field, it provides a comprehensive and thorough examination of cases, principles, theories, and professional

codes. Given that one of its goals is to demonstrate when theoretical application is useful, it illustrates Florman's point.[42] In the book, for example,
the case of the "DC-10" is pivotal in illustrating how ethical theories can
aid decision making in engineering practice. Specifically, Martin and
Schinzinger appeal to theories to justify "the general obligations of engineers to others involved in technological development." Basically, they
regard the primary focus of engineering ethics as "the promotion of safety
while bringing useful technological products to the public."[43] As a way of
explaining this point, they describe the DC-10 case:

> In 1974 the first crash of a fully loaded DC-10 jumbo jet occurred over
> the suburbs in Paris; 346 people were killed, a record for a single-plane
> crash. It was known in advance that the crash was bound to occur
> because of the jet's defective design.
>
>   The fuselage of the plane was developed by Convair, a sub-
> contractor for McDonnell-Douglas. Two years prior to the crash, Con-
> vair's senior engineer directing the project, Dan Applegate, had written
> a memo to the vice president of the company itemizing the dangers
> that could result from the design. He accurately detailed several ways
> the cargo doors could burst open during flight, depressurize the cargo
> space, and thereby collapse the floor of the passenger cabin above.
> Since the control lines ran along the cabin floor, this would mean a loss
> of control of the plane. Applegate recommended redesigning the doors
> and strengthening the cabin floor. Without such changes, he stated, it
> was inevitable that some DC-10 cargo doors would open in midair,
> resulting in crashes.[44]

Martin and Schinzinger report that Convair management did not dispute
technical facts or the predictions. It maintained, however, that the possible
financial liabilities Convair might incur if a redesign were made prohibited
them from passing the information on to McDonnell-Douglas; grounding
the plane to make safety improvements would place both companies at a
competitive disadvantage. In later sections of the book, the authors cite
both regulatory and organizational management voids existent at the time
of the crash.

    An examination of Applegate's obligations, as an engineer, is of prime
importance. For analysis, the authors rely on various theoretical approaches.
A utilitarian perspective, for example, would characterize the engineer's
dilemma in terms of personal versus public versus economic benefits. Con-

flicting duties to the public and one's employer would be noted by duty ethics. A rights perspective would speak to the rights of the public to be protected or the employer to not be questioned. The authors stress that while the theories have various perspectives, each supports the general position that the public's safety had a stake in the resolution – that is, that Applegate, Convair, and McDonnell-Douglas were not the only parties of import! Martin and Schinzinger concede that references to moral theories can help an engineer rank conflicting obligations. If this had been done in the DC-10 case, the protection of the lives of the public and its right to know the risks would have been justified and stressed. A valuable lesson in safety!

The authors use the case to demonstrate a range of ethical approaches and to engage the reader in exercises that challenge the moral imagination. They also provide justification, in this case, for redesign of the DC-10. Applegate's dilemma is cast as a common one engineers face. It occurs because of conflicting obligations at the personal, professional, organizational, and societal levels. In addition to citing different theories, Martin and Schinzinger encourage the reader to use various perspectives to think critically about the case. They cite, for example, the "social experiment" model of responsibility and query whether the public should have been informed about the DC-10's defect.[45] This model is based on a definition of engineering used in this book as an open-ended heuristic process, one that is always testing the limits of knowledge by trial and error. Within this view, the traditional goal of engineers – creating a risk-free environment – is impossible. Failures (accidents), in fact, are inevitable. Thus it follows that the individuals affected by the technology should be made aware of the limits of engineering design and given the opportunity, through informed consent, to be the subjects of an experiment.

The moral relationship between the engineer, the organization, and the user of the technology in the "social experiment" model becomes similar to that between physician and patient.[46] For example, when one flies in a commercial airplane, the specific risk of the aircraft could be listed along with air fares. The riskier crafts could have lower fares, and the traveler would explicitly accept the risk. If individuals do not have or are not capable of understanding the knowledge necessary to provide an informed consent, then a dilemma is again created for the engineer. Who should make risk–benefit decisions affecting an "incompetent"? These are challenging perspectives and important "thought experiments" for the engineer.

Looking to general trends, Martin and Schinzinger compare Applegate with Roger Boisjoly, an engineer at Morton Thiokol, who was frustrated and outraged by the launch of the *Challenger* (see Chapter 14). Applegate and Boisjoly, the authors conclude, did not feel they "had the kind of professional backing which would have allowed them to go beyond the organization directly to the affected public."[47]

Although critical of this theoretical approach, Florman might acknowledge that it does clarify where moral dilemmas occur in practice – that is, it sensitizes engineers to areas of moral conflict. The appeal to theory also offers a justification "outside the profession" for moral action to uphold the safety of the public. If economy is practically important, as Florman states, it is not morally justified in this case. Churchill clarifies further what a theory can and cannot do in medical ethics. Aside from identifying a morally significant dilemma, it can point to principles of action and guide their application. "Theories," however, "don't solve problems, people do." Here is where other aspects of ethics, such as judgment, perception, and insight into practice, become important. A theory, as a set of systematically linked rationales, is an aid to critical reflection. It is not the final arbiter. Given this perspective, the engineer as a moral agent could – along with other problem-solving skills – use an ethical theory to help solve a moral problem.[48]

Florman was suspect of the new engineering ethics because it relied on outside experts to resolve moral dilemmas. In his opinion, it was not moral expertise that was lacking in engineers, but a sophisticated sense of how politics, economics, and the legal system impinged on practice. Other engineering ethicists agree.[49] Pavolic observes that:

> If an engineer is troubled because he believes that the product that he is working on for his company poses a significant threat to the public health, safety, and welfare, he does not need to be told that he is to hold paramount the public health, safety, and welfare. . . . he presumably believes something like that already. He needs to be able to trace out the web of obligations, to know where to go and how to get the information he needs, what options are open and how to evaluate them. . . . doing this will quickly take the individual far beyond the boundaries of engineering and even of business.[50]

For Pavlovic, his remedy to Florman's objections was to instruct engineers to gain moral expertise by acquiring the analytic tools to sort out options – not sophisticated theories but practical tools like writing down possibilities and assessing consequences. He also encouraged them to imagine motives, actions, and consequences other than the ones traditionally used and, finally, to acquire knowledge and information outside one's area of professional competence.

Philosopher Michael Pritchard outlines goals for introducing the engineering student to ethical dimensions of practice. While he encourages a knowledge of moral principles and theory, he also focuses attention on the examination of ordinary "code-driven" cases or on commentaries from consensus reports. He attends to the common dilemmas engineers face so as to stimulate the moral imagination, help students recognize ethical issues, and analyze key ethical principles and concepts. Pritchard aims to elicit a sense of responsibility from students and to help them tolerate diverse viewpoints. He maintains that if engineers are taught ways to reduce disagreement, ambiguity, and vagueness, they will maintain critical and reflective thinking.[51] Even this brief sketch of the field of engineering ethics indicates that it is not the all-or-nothing polarized commentary that Florman suspects.

We present another version of the DC-10 case to convey Pavolic's "web of obligations." Note that this historical account introduces issues not included in Martin and Schinzinger's version. Consequently, moral issues other than safety are identified. As part of a project sponsored by the Center for the Study of Ethics in the Professions, Illinois Institute of Technology, Fay Sawyier chronicled and discussed, in outline form, the forty-year history that preceded this fatal incident. First, she presents an annotated time line that spans the years 1933–74. It documents the context of international competition and economic hardships that the aeronautical industry was experiencing prior to the crash, and a fundamental conflict of interest inherent in the Federal Aviation Administration, whose inspectors originally were tied to the companies of the aircraft being inspected. It reports that as early as 1968, RLDC (the Dutch equivalent of the FAA) cited known defects in the design of the compartment floors of these new jumbo jets. The warnings went unheeded even after repeated testing and a near-fatal incident, over Windsor, Ontario, in 1972. Sawyier identifies individual players in the case such as the heads of both the FAA and McDonnell-Douglas, who verbally agreed to work out an "Airworthiness Directive" for the DC-10 after the "Windsor" incident. Other specifics leading up to the crash are cited. Clearly, information outside the realm of engineering expertise is provided.

Central to the case is the memorandum, written in June 1972, by F. D. Applegate. Sawyier describes Applegate's memo as documenting "his own shock at the protracted disregard both of safety and even of routine procedures, culminating in the 'Agreement.'"[52] She includes four in-depth commentaries on the specifics of the case and asks pertinent questions that assess consequences and help to sort out options. For example, how should one evaluate the effect of the economic gamble on the decisions of those involved? Given the public nature of the aerospace industry, should outside checks be put into place? Should safety inspectors be protected from the influence of economic pressures? What of sales to foreign countries who have flight and maintenance standards lower than those that exist in the United States? The DC-10 that crashed had been sold to Turkey and all safety instructions were in English. Ultimately, it was a non-English-speaking technician who latched the cargo door before the plane took off. Did McDonnell-Douglas have an obligation to insure that all safety instructions were understood? Criticism is also made of the "old boys' net-work" that bolstered the agreement between McDonnell-Douglas and the FAA. Finally, why did Applegate write his memo and then file it away after management decided not to pass it on to McDonnell-Douglas? Should Congress or the press have been notified?

Sawyier's case is excellent in its historical detail. Technical diagrams of the plane and its defect are included as well as the Applegate memo in its entirety. She offers practical suggestions to remedy the situation. After reading the case, one is clearly alerted to the "web of obligations" and somewhat overwhelmed at the relatively insignificant impact that a single engineer, even one in a management position, could have. A political and economic revolution of sorts, including a redesign of the FAA and the drafting of new government regulations, seems appropriate. Applegate's memorandum indicated that he was aware of the complexity, but powerless to effect a change. From the perspective offered in this case, we understand why Applegate did not resort to whistleblowing. A leading FAA official was involved in a "gentleman's agreement," so the government was already involved. The press, though alerted to problems with the DC-10 after the Windsor incident, showed little interest. Should an engineer pursue what reporters decided not to? Furthermore, we questioned if there were general maxims to be learned from this case. Perhaps, but the wealth of details without an overall analytic framework makes generalizations diffi-cult to recognize.

With this goal of identifying a tacit ethic or maxim, we return to the DC-10 case one more time, to probe a bit deeper. We will specifically exam-ine the Applegate memo, with full knowledge of the context Sawyier has provided. We reprint the memo entitled "DC-10 Future Accident Liability" and dated June 27, 1972, in its entirety.

> The potential for long-term Convair liability on the DC-10 has caused me increasing concern for several reasons.
>
> 1. The fundamental safety of the cargo door latching system has been progressively degraded since the program began in 1968.
> 2. The airplane demonstrated an inherent susceptibility to catastrophic failure when exposed to explosive decompression of the cargo compartment in 1970 ground tests.
> 3. Douglas has taken an increasingly "hard-line" with regards to the relative division of design responsibility between Douglas and Convair during change cost negotiations.
> 4. The growing "consumerism" environment indicates increasing Convair exposure to accident liability claims in the years ahead.
>
> Let me expand my thoughts in more detail. At the beginning of the DC-10 program it was Douglas's declared intention to design the DC-10 cargo doors and door latch systems much like the DC-8s and -9s. Documentation in April 1968 said that they would be hydraulically operated. In October and November of 1968 they changed to electrical actuation which is fundamentally less positive.
>
> At that time we discussed internally the wisdom of this change and recognized the degradation of safety. However, we also recognized that it was Douglas's prerogative to make such conceptual system design decisions whereas it was our responsibility as a sub-contractor to carry out the detail design within the framework of their decision. It never occurred to us at that point that Douglas would attempt to shift the responsibility for these kinds of conceptual system decisions to Convair as they appear to be now doing in our change negotiations, since we did not then nor at any later date have any voice in such decisions. The lines of authority and responsibility between Douglas and Convair engineering were clearly defined and understood by both of us at that time.
>
> In July 1970 DC-10 Number Two was being pressure-tested in the "hangar" by Douglas, on the second shift, without electrical power in the airplane. This meant that the electrically powered cargo door

actuators and latch position warning switches were inoperative. The "green" second shift test crew manually cranked the latching system closed but failed to fully engage the latches on the forward door. They also failed to note that the external latch "lock" position indicator showed that the latches were not fully engaged. Subsequently, when the increasing cabin pressure reached about 3 psi (pounds per square inch) the forward door blew open. The resulting explosive decompression failed the cabin floor downward, rendering tail controls, plumbing, wiring, etc., which passed through the floor, inoperative. This inherent failure mode is catastrophic, since it results in the loss of control of the horizontal and vertical tail and the aft center engine. We *informally* studied and discussed with Douglas alternative corrective actions including blow out panels in the cabin floor which would provide a predictable cabin floor failure mode which would accommodate the "explosive" loss of cargo compartment pressure without loss of tail surface and aft center engine control. It seemed to us then prudent that such a change was indicated since "Murphy's Law" being what it is, cargo doors will come open sometime during the twenty years of use ahead for the DC-10.

Douglas concurrently studied alternative corrective actions, in house, and made a unilateral decision to incorporate vent doors in the cargo doors. This "bandaid fix" not only failed to correct the inherent DC-10 catastrophic failure mode of cabin floor collapse, but the detail design of the vent door change further degraded the safety of the original door latch system by replacing the direct, short-coupled and stiff latch "lock" indicator system with a complex and relatively flexible linkage. (This change was accomplished entirely by Douglas with the exception of the assistance of one Convair engineer who was sent to Long Beach at their request to help their vent door system design team.)

The progressive degradation of the fundamental safety of the cargo door latch system since 1968 has exposed us to increasing liability claims. On June 12, 1972, in Detroit, the cargo door latch electrical actuator system in DC-10 number 5 failed to fully engage the latches of the left rear cargo door and the complex and relatively flexible latch "lock" system failed to make it impossible to close the vent door. When the door blew open before the DC-10 reached 12,000 feet altitude the cabin floor collapsed disabling most of the control to the tail surfaces and aft center engine. It is only chance that the airplane was not lost. Douglas has again studied alternative corrective actions and appears to be applying more "bandaids." So far they have directed us to install small one-inch diameter, transparent inspection windows through which you can view latch "lock-pin" position, they are revising the rigging instructions to increase "lock-pin" engagement and they plan to reinforce and stiffen the flexible linkage.

It might well be asked why not make the cargo door latch system really "fool-proof" and leave the cabin floor alone. Assuming it is possible to make the latch "fool-proof" this doesn't solve the *fundamental deficiency in the airplane*. A cargo compartment can experience explosive decompression from a number of causes such as: sabotage, mid-air collision, explosion of combustibles in the compartment and perhaps others, any one of which may result in damage which would not be fatal to the DC-10 were it not for the tendency of the cabin floor to collapse. The responsibility for primary damage from these kinds of causes would clearly not be our responsibility, however, we might very well be held responsible for the secondary damage, that is the floor collapse which could cause the loss of the aircraft. It might be asked why we did not originally detail design the cabin floor to withstand the loads of cargo compartment explosive decompression or design blow out panels in the cabin floors to fail in a safe and predictable way.

I can only say that our contract with Douglas provided that Douglas would furnish all design criteria and loads (which in fact they did) and that we would design to satisfy these design criteria and loads (which in fact we did). There is nothing in our experience history which would have led us to expect that the DC-10 cabin floor would be inherently susceptible to catastrophic failure when exposed to explosive decompression of the cargo compartment, and I must presume that there is nothing in Douglas's experience history which would have led them to expect that the airplane would have this inherent characteristic or they would have provided for this in their loads and criteria which they furnished to us.

My only criticism of Douglas in this regard is that once this inherent weakness was demonstrated by the July 1970 test failure, they did not take immediate steps to correct it. It seems to me inevitable that, in the twenty years ahead of us, DC-10 cargo doors will come open and I would expect this to usually result in the loss of the airplane. This fundamental failure mode has been discussed in the past and is being discussed again in the bowels of both the Douglas and

Convair organizations. It appears however that Douglas is waiting and hoping for government direction or regulations in the hope of passing costs on to us or their customers.

If you can judge from Douglas's position during ongoing contract negotiations they may feel that any liability incurred in the meantime for loss of life, property and equipment may be legally passed on to us.

It is recommended that overtures be made at the highest management level to persuade Douglas to immediately make a decision to incorporate changes in the DC-10 which will correct the fundamental cabin floor catastrophic failure mode. Correction will take a good bit of time, hopefully there is time before the National Transportation Safety Board (NTSB) or the FAA grounds the airplane which could have disastrous effects upon sales and production both near and long term. This corrective action becomes more expensive than the cost of damages resulting from the loss of one plane load of people.

A consistent theme in Applegate's memo is, Who is responsible? His concern, however, is not with moral responsibility but with legal liability: Who will end up paying for the repairs? The history of Convair's and McDonnell-Douglas's involvement is documented – for example, Douglas changed the design of the cargo door from hydraulic to electric, which was a known higher risk. Applegate confesses:

At that time we discussed internally the wisdom of this change and recognized the degradation of safety. However, we also recognized that it was Douglas' prerogative to make such conceptual system design decisions whereas it was our responsibility as a sub-contractor to carry out the detail design within the framework of their decision.

It never occurred to us at that point that Douglas would attempt to shift the responsibility for these kinds of conceptual system decisions to Convair as they appear to be now doing. . . . we did not then nor at any later date have any voice in such decisions. The lines of authority and responsibility between Douglas and Convair engineering were clearly defined and understood by both of us at that time.

If Convair had *legal* responsibility, would it have proceeded with the subcontract? There seems to be a tacit ethic, a legal interpretation of responsibility, that a contractor builds to specification – even if that requires a known degradation to safety! The justification for this action is that the subcontractor is not responsible for the design. What options did Convair have at the time, short of refusing the contract? What option did Applegate have at the time, short of refusing to work on the project? Was the acceptance, based on the traditional subcontractor agreement, "morally" justified?

Without belaboring the point, one can construct a series of questions at this juncture that may have led to a preventive ethics posture. Applegate clearly demonstrates his competence as an engineer by explaining the high risk of the faulty design and outlining a range of specific corrective alternatives. He acts responsibly by communicating his concerns to management and clearly acts in accordance with Cicero's Creed II by stressing the risk and documenting both short- and long-term effects of the inherent defect in the DC-10 floor. The language of legal liability and engineering technique, not moral theory, characterizes the memo. One wonders if the framing of the potential disaster in this way was chosen because it was the only perspective that Applegate felt would be listened to. Would framing it in moral terms have had an effect? Does the legal language "insulate" Applegate psychologically from moral responsibility? The technical alternatives he posed seem to substitute for the explicit mention of the ethical dilemma. That is, at the time the memo was written, neither Convair, McDonnell-Douglas, nor the government was willing to bear the responsibilities for the costs of the repair. The memo indicates that McDonnell-Douglas was aware of the situation; certainly Convair was. The last three lines of Applegate's memo stress the *tragic choice* that he feared was being made: The last line in particular breaks with both the legal and the technical language and introduces the moral nature of the dilemma.

> It is recommended that overtures be made at the highest management level to persuade Douglas to immediately make a decision to incorporate changes in the DC-10 which will correct the fundamental cabin floor catastrophic failure mode. Correction will take a good bit of time, hopefully there is time before the National Transportation Safety Board (NTSB) or the FAA grounds the airplane which could have disastrous effects upon sales and production both near and long term. *This corrective action becomes more expensive than the cost of damages resulting from the loss of one plane load of people.* (Emphasis added)

The statement is difficult to interpret. That Applegate made it suggests that the high probability of the loss of unidentified lives was calculated in a quantitative risk–benefit fashion. In ethical terms, a utilitarian bargain was struck and the overall economics of the organization trumped safety of the public.

When one examines this case closely – and we limited our examination to the memo – two tacit ethics are discovered. One concerns a subcontractor's obligation to build to specification (even known high-risk ones). Deborah Johnson describes this as the "guns for hire" approach to engineering and finds it to be untenable.[53] When held up to reflective reasoning, using the framework of competence, responsibility, and Cicero's Creed II, the "subcontractor" ethic seems suspect in the face of a known faulty design. It is within the realm of the competence of both the engineer and the organization to take responsible moral action when this scenario occurs, namely, to construct alternative designs. Applegate infers that even if they built a foolproof latch, some other force could collapse the floor, but his examples ("sabotage, mid-air collision, explosion of combustibles") are less likely to occur than the cargo door opening. The risk of disaster would have been decreased if the latch had been improved. Acceptable risk, not risk-free, is the goal here. Would that compromise solution have produced an "acceptable risk"?

Engineering, as we define it, is a heuristic, not an applied science. Based on tradition, experience, scientific knowledge, and judgment, engineers are asked to "improve the human condition before all scientific facts are in."[54] Broome refers to this as the "engineer's imperative."[55] The situations they address are typically poorly understood. Consequently, the knowledge base from which decisions are made is often incomplete and marked by uncertainty. Note that Applegate concedes that McDonnell-Douglas did not knowingly design the cabin floor to collapse! After it failed tests, however, and proved to be a high risk in flight, he asserts that repairs should have been made. When engineers know how technology will affect those who use it, what are they obligated to disclose with reference to risks? Given the heuristic definition of engineering, it is recognized that while a risk-free environment is impossible to achieve, an acceptable level of risk can be attained. The moral question now becomes, who should determine what is acceptable?

Petroski[56] expands on the view of engineering as inherently risk laden, and constructs a nominal typology of four factors or "design errors" that inevitably lead to design failures. He stresses the engineer's role in the competent designing of technology and the legal system's responsibility for policing wrongdoing and meting out punishment. Design procedures, for example, can be conducted to prevent failures. The causes of failure include conditions that approach design limit states (e.g., overloads), random or unexpected hazards that have not been considered in design (e.g., extreme weather conditions), human-based errors (e.g., mistakes, carelessness), and attempts to economize in design solution or maintenance. As discussed, Petroski looks both to the engineering profession and to the legal system to control accidents. Engineers are responsible for prevention. The assignment of moral and legal responsibility for a failure by official bodies can act as a deterrent to future mistakes. Neither of these safeguards worked in the case of the DC-10.

The second tacit ethic we identify has to do with individual responsibility: writing one's concerns in a memo and passing it on to a person in the organization with decision-making power. This form of communication about high-risk and safety issues is common in engineering organizations. As we discuss extensively in Chapter 4, it is a way for an individual engineer to exercise his or her own moral courage, engineering competence, and responsibility. If the memo falls on deaf ears, at least one's own power or lack thereof in the organization has been identified. Memos also leave a paper trail that can be used to trace or document responsibility within the organization. Sometimes, the memos are "leaked" to the press. In Chapter 14, we document this with reference to the *Challenger* launch and the prophetic memo of NASA budget analyst Richard Cook. Other times, as in Applegate's case, it remained filed away until a federal investigation of the disastrous crash caused it to resurface.

Conclusion

Case studies in engineering practice like the DC-10 are both conceptually and technically difficult to construct. They are also difficult to analyze. Vivian Weil, who has contributed ably to the case-based approach, warns that documentation of design decisions must rely on evidence not usually cataloged or preserved.[57] Furthermore, Pritchard and others have explained that these cases of disaster portray a limited vision of the actual issues in engineering ethics.[58] "If taken as representative," Pritchard writes, "the cases suggest that ethics in engineering is mainly about the newsworthy. But engineers realize that only the tiniest minority of the approximately 1.5 million engineers will ever be involved in anything so newsworthy as these unfortunate events. If this is basically what engineering ethics is about, they might conclude, most engineers have little to worry about."[59] As we have maintained throughout this book, "the ordinary world of engineers presents ethical issues no less in need of careful attention."

A careful historical narrative even of the "disaster cases," however, portrays the context in which the dilemmas occurred. It thus includes many decision points and opportunities for a variety of individuals to intervene to avoid a collision course. Our use of historical narrative to chronicle the design, testing, and operation of the SSMEs includes a realistic overview of both successful and problematic situations. Our identification and explicit definition of the tacit ethical dimension in our cases provide an additional and unique contribution to the field of engineering ethics, a contribution that includes a language to begin to understand the common ground between "individual whim and legality."

Notes

1. For the limits to this autonomy and how it is defined with reference to the engineer within the organization, see Chapter 4.
2. Samuel C. Florman, "Moral Blueprints," in *Blaming Technology* (New York: St. Martin's Press, 1981), pp. 162–80. For a comparable view of medical ethics, see George J. Annas, "How We Lie," *Hastings Center Report* 25(6), 1995: 512–14.
3. Ibid., p. 172.
4. For a full discussion of this emerging consensus, see the subsequent discussion in this chapter.
5. Dan Brock and Alan Buchanan, *Deciding for Others: The Ethics of Surrogate Decision Making* (Cambridge: Cambridge University Press, 1989), p. 7.
6. J. W. Ross et al., *Health Care Ethics Committees: The Next Generation* (Chicago, IL: American Hospital Association, 1993).
7. Criticism of Ethics Committees. See Ross et al., *Health Care Ethics Committees: The Next Generation*; Bernard Lo, "Behind Closed Doors: Promises and Pitfalls of Ethics Committees," *New England Journal of Medicine* (317) 1987: 46–50; George J. Annas, "Ethics Committees: From Ethical Comfort to Ethical Corves," *Hastings Center Report* (21), 1991: 18–21; R. F. Wilson, M. Neff-Smith, D. Phillips, and J. C. Fletcher, "Hospital Ethics Committees: Are They Evaluating Their Performance?" *HEC Forum* 5, 1993: 5–34.
8. Joint Commission for the Accreditation of Health Care Organizations, *Accreditation Manual for Hospitals* (1992; Oakbrook Terrace, IL, 1995), standard R1.1.1.321.
9. Brock and Buchanan, *Deciding for Others*, pp. 6–9.
10. Susan D. Wolf, "Healthcare Reform and the Future of Physician Ethics," *Hastings Center Report* 24(2), March–April 1994: 28–41.
11. Joint Commission on Accreditation of Healthcare Organizations, *Accreditation Manual for Hospitals*, standards R1.4 – 4.2.
12. American Health Consultants, "Managed Care Brings a Demand for Institutional Ethics Policies," *Medical Ethics Advisor* 10(10), 1994: 1–3.

13. N. M. King, L. R. Churchill, and A. Cross, *The Physician as Captain of the Ship: A Critical Reappraisal* (Boston: Reidel Publishers, 1986).

14. Larry R. Churchill, "Reviving a Distinctive Medical Ethic," *Hastings Center Report* 19 (2), 1989: 28–34.

15. King et al., *The Physician as Captain of the Ship*, p. xi.

16. Christine K. Cassel, "Deciding to Forego Life-Sustaining Treatment: Implications for Policy," *Cardoso Law Review* 6(2), 1984: 268–87.

17. For engineering, see T. H. Broome, "Engineering Responsibility for Hazardous Technologies," *Journal of Professional Issues in Engineering* 113(?), 1987: 139–49. For medicine, see Albert R. Jonsen, *The New Medicine and the Old Ethics* (Cambridge, MA: Harvard University Press, 1990).

18. Thomas L. Beauchamp and James F. Childress, *Principles of Biomedical Ethics*, 3d ed. (New York: Oxford University Press, 1989), pp. 3–24.

19. For a review of the various theories, see Ronald Munson, ed., *Intervention and Reflection: Basic Issues in Medical Ethics*, 4th ed. (Belmont, CA: Wadsworth, 1992); for theories of justice, see Larry R. Churchill, *Rationing Healthcare in America: Perceptions and Principles of Justice* (Notre Dame, IN: University of Notre Dame Press, 1987).

20. Baruch Brody, *Life and Death Decisionmaking* (Oxford: Oxford University Press, 1988), pp. 1–94. Explains the general problems associated with each theory as well as the theories themselves.

21. T. Beauchamp and J. Childress, *Principles of Biomedical Ethics*, 1st ed., 1978; 2d ed., 1982; 3d ed., 1989; 4th ed., 1994.

22. Ibid., 3d ed., appendix, 400–54 (includes thirty-eight cases); 4th ed., see introduction and pp. 3–4, 40.

23. Brody, *Life and Death Decisionmaking*, pp.10–11, 72–99.

24. A. R. Jonsen, "The Birth of Bioethics," *Hastings Center Report*, special suppl., November–December 1993, pp. 51–4.

25. National Commission for the Protection of Human Subjects of Biomedical and Behavioral Research, *The Belmont Report: Ethical Guidelines for the Protection of Human Subjects of Research*, DHEW Publication No. (08)70-00 (Washington, DC: Department of Health Education and Welfare, 1978).

26. Albert J. Jonsen and Stephen Toulmin, *The Abuse of Casuistry: A History of Moral Reasoning* (Berkeley: University of California Press, 1988).

27. Ibid., pp. 23–90.

28. Edmund Pellegrino, "The Metamorphosis of Medical Ethics: A 30-Year Retrospective," *Journal of the American Medical Association* 269(9), 1993: 1159–62.

29. T. Beauchamp, "The Principles Approach," *Hastings Center Report*, special suppl., November–December 1993, p. 59.

30. This definition of morality appears in the third edition of Beauchamp and Childress's book. They were discussing three criteria necessary, from a philosophical viewpoint, for an action to be moral. Summing up the formal requirements, they commented on the academic dispute of whether the conditions were necessary or sufficient. They concluded that the important point to look for was whether the conditions were helpful in identifying what is moral, not if they were infallible or essential.

31. Jonsen, "The Birth of Bioethics," pp. 51–4.

32. K. Danner Clouser, "Bioethics and Philosophy," *Hastings Center Report*, special suppl., November–December 1993, pp. 510–11.

33. Ibid.
34. Larry R. Churchill, "Theories of Justice," in *Ethical Problems in Dialysis and Transplantation,* ed. C. M. Kjellstrand and J. B. Dossetor (London: Kluwer Academic Publishers, 1992), pp. 21–5.
35. We gauged our study a success when our engineering graduate student could patiently carry on and understand a discussion of cases and terms with our philosophy graduate student and "vice versa."
36. Beauchamp and Childress, *Principles of Biomedical Ethics,* 3d ed., p. 21.
37. K. Danner Clouser, and B. Gert, "A Critique of Principlism," *Journal of Medicine and Philosophy* 15, April 1990: 219–36.
38. Florman, "Moral Blueprints," p. 173.
39. C. Whitbeck, "The Engineer's Responsibility for Safety: Integrating Ethics Teaching into Courses in Engineering Design," paper presented at the ASME Winter Annual Meeting, Boston, 1987, p. 1.
40. Florman, "Moral Blueprints," pp. 173–4.
41. Whitbeck, "The Engineer's Responsibility for Safety," p. 4.
42. Mike W. Martin, and Roland Schinzinger, *Ethics in Engineering,* 2d ed. (New York: McGraw-Hill, 1989).
43. Ibid., p. 45.
44. Ibid., p. 43.
45. Ibid., pp. 43–50, 70, 84, 90, 165.
46. Ibid., p. 84.
47. Ibid.
48. Churchill, "Theories of Justice," pp. 21–5.
49. Karl Pavolic, "Autonomy and Obligation: Is There an Engineering Ethics?" in *Engineering Professionalism and Ethics.,* ed. James H. Schaub, and Karl Pavlovic (New York: Wiley, 1983), pp. 223–32.
50. Ibid., p. 230.
51. Michael S. Pritchard, "Teaching Engineering Ethics: A Case Study Approach," paper presented at the Illinois Institute of Technology Meeting, Chicago, June 12–13, 1990, pp. 295–318. Charles E. Harris Jr., Michael Pritchard, and Michael J. Rabins, *Engineering Ethics: Concepts and Cases* (Albany, NY: Wadsworth Press, 1995), pp. 5–12.
52. Faye Sawyier, "The Case of the DC-10 and Discussion," in Schaub and Pavolic, *Engineering Professionalism and Ethics,* p. 390.
53. Deborah Johnson, "The Social/Professional Responsibility of Engineers," *Annals of The New York Academy of Sciences,* 577, 1989: 106–14.
54. Broome, "Engineering Responsibility for Hazardous Technologies," p. 142.
55. Ibid., p. 143.
56. Henry Petroski, *To Engineer Is Human: The Role of Failure in Successful Design* (New York: St. Martin's Press, 1985).
57. Vivian Weil, ed., *Report of the Workshops on Ethical Issues in Engineering,* Proceedings of the Second National Conference on Ethics (Chicago: Center for the Study of Ethics in the Professions, Illinois Institute of Technology, 1979), p. 7.
58. Michael Pritchard, "Beyond Disaster Ethics," *Centennial Review,* 34(2), 1990: 295–318.
59. Ibid.

4

Satisficing, Organizational Theory, and Other Theoretical Contributions to Engineering Ethics

Decisions are something more than factual propositions. To be sure, they are descriptive of a future state of affairs, and this description can be true or false in a strictly empirical sense; but they possess, in addition, an imperative quality – they select one future state of affairs in preference to another and direct behavior toward the chosen alternative. In short, they have an ethical as well as factual content.

This ethical premise describes the objective of the organization in question. . . . In order for an ethical proposition to be useful for rational decision making (a) the values taken as organizational objectives must be definite, so that their degree of realization in any situation can be assessed, and (b) it must be possible to form judgments as to the probability that particular actions will implement these objectives.
– Herbert A. Simon, Administrative Behavior

Framework for Organizational Analysis

For over a century, scholars have studied and written about organizations.[1] We will briefly review a few of the more important models, highlighting the reason why we selected Herbert Simon's concepts to guide us in construct-ing our conceptual framework.

The human relations school, developed in the late 1920s and 1930s, focused on organizations as social phenomena and discovered that not all

69

activities in organizations are rational from the perspective of the goals of the organization. Human relations researchers first observed that informal organizations develop within the formal structures of organizations, and these informal organizations exhibit norms, values, and behaviors that are distinct from those prescribed by the formal organization. This tradition of research continues today and emphasizes topics such as motivation, leadership, and group processes. Although there are many aspects of NASA in the current study that pertain to these issues, social relations within NASA are not our primary interest. Ethical behavior, as we have defined it, is embedded in the formal structure of the organization, and in the formal behavior of its participants. Thus the human relations approach does not magnify the organizational issues with which we are most concerned.

Another school focusing on the study of bureaucracy developed prior to the human relations school. Max Weber studied and wrote about bureaucracies in the first decade of this century. However, it was not until after 1946 when his work was translated into English that it became widely known to English-speaking organizational writers.[2] After 1946, no organizational theorist has had more influence on the field than Weber, who focused on the formal structures of organizations. The division of labor (jobs, work groups, departments, divisions) and the formal hierarchy of authority are particularly important structures of bureaucracies. Weber also identified the importance of abstract rules, impersonal social relations, and technical expertise as key attributes of bureaucratic organizations. Bureaucratic theory is an excellent context within which to study NASA and its relations with the rest of the federal government. The structural parameters of hierarchical level and departmentalization are crucial in understanding much of what occurred during the development of the shuttle. However, although bureaucratic theory provides an excellent context, we need more to understand what happened. We need a framework that focuses on what goes on inside bureaucracies.

In 1976, Herbert Simon published the third edition of *Administrative Behavior*[3] (the first edition was published in 1945). Over a three-decade span, Simon and colleagues created what is now called the decision theory school of organizations, which focuses on the formal, rational behavior of organizations, with specific attention to how organizational members collectively make complex decisions. Perrow has argued that decision theory provides "the muscle and flesh for the Weberian skeleton [of bureaucratic theory], giving it more substance, complexity, and believability without reducing organization theory to propositions about individual behavior."[4] We agree, and draw heavily upon the definitions and propositions of decision theory in our analysis of engineering ethics.

Starting in the late 1960s and continuing to present times, researchers began considering organizations as open systems. This school believes that understanding what happens in the environments of organizations is critical to understanding what happens to and within organizations. This school has several variants. Contingency theory proposes that the structure of an organization depends on the structures in the organization's environment.[5] For example, the technology of an organization is often determined by the industry or larger collectivity within which it is embedded. Contingency theorists study the issues and problems of organizational change, such as how should an organization be restructured to perform more efficiently. A second variant, known as the population ecology approach, applies classic ecology theory to organizations. The unit of analysis for this approach is the population of organizations of similar types. A third variant is the institutional approach to organizational analysis. This approach is most concerned with the nontraditional dimensions of organization behavior, both within and outside organizations. Institutional analyses pursue topics like organizational politics, coalition formation, networking, and other modes of noneconomic interaction. In its most recent forms, these scholars examine organizational problems at the societal level. For example, DiMaggio and Powell argue that all bureaucracies are becoming structurally similar.[6]

The study of engineering ethics is most closely aligned with the decision theory school of organization theory. Other organizational frameworks will certainly add to our analytic insights, but they do not focus on what is central to our study.

Facts and Judgments

Our engineering ethics framework incorporates concepts pertaining to both the individual engineer and organization within which the engineer works. What are, however, the linkages between the individual and the organization? In this chapter we explain how we applied Herbert Simon's theory of organization to make these linkages explicit. The fit between Simon's model of organizational decision making and our framework extends and supports our analysis.

For Simon, a formal organization is "a description that, so far as possible, designates for each person in the organization what decisions that person makes and the influences to which he is subject in making each of these decisions."[7] In our study, we focused on two types of engineering decisions central to engineering ethics, the performance decision and the design decision. The first is the responsibility of the engineering manager who must make decisions about whether a technology, or proposed technology, will work, whereas the second concerns the design engineer and the decisions he or she makes in the creation of technology. An analysis of both with reference to Simon's model provides insights into how organizational structure and management style influence the fundamental character of engineering ethics.

Organizational decisions, according to Simon, have two or more alternatives, and each alternative has two components: factual and ethical. Factual propositions are statements about the observable world and the way in which it operates. In principle, factual propositions may be tested to determine whether they are true or false – whether what they say about the world actually occurs.[8] The factual component of each alternative is like a hypothesis in scientific inquiry. It is a statement about the future if the alternative is chosen. The set of alternatives of a decision describes the alternative futures faced by the decision maker. One of the alternatives is usually to continue the status quo; the decision maker assumes the future will be like the present. Decision makers generate factual components from four primary sources: facts, theory, empirical experience (e.g., data), and judgment. Facts can be physical constants but they can also include standard operating procedures that are part of any organization. An example of the application of theory might be the prediction of the response of the market to a price change. Another example is calculating the required increase in thrust of rockets to accommodate an increase in weight; an experienced engineer might know that a particular coating has worked in a pure oxygen environment in the past. Engineering test data are an important source for factual components.

Simon's model also includes what he calls a judgment as a source for a factual component of a decision. For Simon, a judgment is a statement about the world that has not been tested, but which in principle can be tested. It therefore satisfies his primary condition for being a factual component of a decision. Engineering judgment plays an important role in engineering practice. Thus the factual components of decision alternatives are a set of statements about possible future outcomes of a decision situation.

These components can be based on knowledge or can consist of an individual judgment. Knowledge consists of statements that have been verified and tested (i.e., facts) or have been derived from theory or the laws of nature (i.e., judgments).

A design engineer, for example, uses both facts and judgments when he or she selects materials for a particular part of a mechanical device. A range of materials is first identified by matching the performance specifications with engineering data. The specifications may require a metal part of certain strength, usable over a particular temperature range, and resistant to corrosion from specified agents. The design task is to select a material that meets these and other criteria and to evaluate its use within the constraints of schedule and cost. The engineer may not have all the engineering data for each optional material. Tests can be conducted to generate data to fill in gaps. This process is called characterizing the materials.

After characterizing the materials in this manner, the engineer states that materials A, B, and C will work in the design and, in Simon's model, become feasible alternatives. Materials D and E may have been considered, but are not included because, in the judgment of the engineer, current knowledge (i.e., testing, data, theory) indicates they will fail.

Such a judgment can be interpolative or extrapolative. That is, if an engineer assesses how a material performs under a new combination of conditions, where each condition by itself is within a tested range, an interpolative judgment is being made. An extrapolative judgment occurs when the engineer goes beyond known conditions, such as using a material at a temperature or pressure higher than prior experience. Aerospace engineers commonly use extrapolative judgments to extend performance specifications by up to 10 percent. The overall judgment of the individual engineer fills in the gaps in the knowledge base. These gaps exist because most engineering problems and decisions have so many parameters that it is impossible to test all potential situations.

The factual components in Simon's model of decision making are the realized instantiations of an engineer's competence. In the case of the design engineer, the alternatives generated and considered follow directly from his or her knowledge base, theoretical and empirical, as well as judgment and creativity. The more competent the engineer, the better and more extensive will be the factual components employed in engineering design decisions. The same relation holds for performance decisions made by engineering managers. Their factual components will incorporate results from testing, as well as assessments of the quality of technical work of other engineers. Engineering managers might generate alternatives that consider going forward and succeeding, going forward and failing, delaying the next step. Simon's model contains more, however, than alternatives and factual components.

Decisions are something more than factual propositions. To be sure, they are descriptive of a future state of affairs, and this description can be true or false in a strictly empirical sense; but they possess, in addition, an imperative quality – they select one future state of affairs in preference to another and direct behavior toward the chosen alternative. In short, they have an ethical as well as a factual content.[9]

This ethical premise describes the objective of the organization in question. In order for an ethical proposition to be useful for rational decision making (1) the values taken as organizational objectives must be definite, so that their degree of realization in any situation can be assessed, and (2) it must be possible to form judgments as to the probability that particular actions will implement these objectives.[10] The ethical component, in Simon's use of the term, is the set of weights that the decision maker applies to the alternatives in order to select the one that best meets the objective(s) of the organization.

For example, a design engineer might be considering two alternative materials. Material A will yield a component life of 15,000 seconds whereas material B will yield a component life of 3,000 seconds. Also, material A costs more than material B. These data on the longevities and costs form the factual components of the two alternatives. The engineering decision, selection of one of the materials, depends on the weights the decision maker applies to the two performance outcomes specified in the factual components of the decision, longevity and cost. A cost constrained design will select material B, whereas a longevity maximizing goal will select the other material. Both outcomes are valid engineering decisions, depending on the nature of the evaluation component.

Where do decision makers acquire these values? As with factual components, there are several sources. We identify four primary classes of values: personal, social, institutional, and organizational. Simon's analysis of organizational decision making acknowledges the first three sources, but focuses on the fourth. We will comment on the first two before delving into the values transmitted by institutional and organizational sources. For Simon, the organization and the institution wield the most influence in determining values. We diverge from Simon's framework to clarify how and when individual values and preferences come to play. This is one modification – or addition – we make to Simon's work.

People have personal preferences, and although these, in Simon's opinion, are not supposed to dominate organizational decision making, he does recognize they are always present. For example, people exhibit different tolerances for ambiguity. A person with a low tolerance would give less relative weight to "fuzzy" but important alternatives than to "clearer" ones. Furthermore, this same person might choose a conceptually elegant design alternative over a mundane design, even though conceptual elegance has little to do with the performance of the technology. Preferences are usually given less moral weight than religious, social, or cultural values for they are perceived as less consistent and change with reference to different situations.[11] For example, religious beliefs may consistently influence how individuals assess moral issues. Like Simon, we acknowledge that personal preferences and values are part of organizational decision making. They are responsible for ethical conflicts that may arise for the individual worker.

We rely on definitions in medical ethics to articulate the role these play in decision making. Basically, we assume that the engineer is a moral agent. Likewise, we assume that the organization also can act as a moral force. In both cases, it is the people or groups of persons who strive (or don't strive) to define, identify, and solve moral dilemmas. Although moral agents may rely on theories, principles, rules, professional codes, or federal regulations as action guides, they invoke their individual perceptions and insights as they resolve the problems. It is the engineer or engineer manager who does the striving – not the theory, organizational policy, or federal regulation. How persons use these tools depends upon their individual values and predilections. Moral decisions, according to moral philosopher Larry Churchill, "bear the mark" of the individual.[12] They can be compared with the "pots of a potter," or with how one drives a classic automobile. "I could own a Lamborghini," Churchill admits, "but drive it like a Toyota."

While the focus of our book is on defining a normative structure for ethical decision making in engineering practice, the point here is that we also acknowledge that individual engineers acting as moral agents will contribute their unique abilities to how decisions are made. There is much latitude within the professional ethic regarding how zealous or neglectful one can be. Zealousness may mark one as a moral hero and neglect as being morally irresponsible. How one assesses the varied gray choices in the middle is the concern here.

Davis has argued that a professional ethical code should be regarded as a "convention between professionals."[13] He defines a profession as a socially recognized "practice among its members" to cooperate and serve a certain ideal. "Engineers are clearly responsible for acting as their professional code of ethics requires." Davis also asserts that "[engineers'] professional responsibilities go beyond the code."[14] Not only are they obligated to encourage others to support it, but he contends they should "ostracize, criticize, chastise, those that do not. "[15]

From this perspective Davis analyzes "wrongdoing" and concludes that in the decision to launch the *Challenger*, if Robert Lund, an engineer and manager at Thiokol, in command the night the decision to launch was made, had adhered to an engineering code of ethics that called for safety, he would have avoided the disaster. Jerald Mason, Lund's boss (who was also an engineer), urged Lund to adopt a managerial systems approach to decision making. "Take off your engineering hat," he chided. Davis maintains that if Lund had held his professional engineering ethic as primary, he could have avoided disaster.[16] Again, while we recognize the importance of the professional codes, we contend that how decisions are made is far more complex than the "rule-oriented" approach Davis outlines. Why did Lund change his mind? Simon also acknowledges the complexity but stresses that the overriding ethic is provided by the organizational goals. If we interpret the decision to launch in those terms, the "systems approach," which attends to previous successful launches, economics, national prestige, and the longevity of the space program, as well as safety, then Mason's perspective is justified. The moral dilemma occurs only when one's personal value system conflicts with organizational goals. We provide some additional insight into that fateful evening's decision in Chapter 14 when we discuss the events that took place over a nine-year period prior to the loss of the *Challenger*.

We recognize that while engineers receive extensive formal training in the physical and engineering sciences and the methodologies of engineering design and analysis, they receive virtually no education in ethics. As they acquire experience in their areas of technological expertise, however, tacit or unsaid values of the profession are transmitted. Broome refers to this as engineering "folklore."[17] We have already mentioned the relevance of the heuristic "safety factors" in our discussion of the principle of Cicero's Creed II. Safety factors reflect professional engineering values about how a design should be done. Other important values include a reduction of uncertainty through research and testing. Not leaving avoidable risks to chance is a tenet to good engineering practice. Cicero's original dictum was, in short, an expression of a professional value. Thus, engineers as knowledge experts and institutionalized professional experts bring a set of values to their decision making. Although these professional values are important and provide the basis in part for our framework, we also acknowledge that how or if the engineer strives to act on these professional values is a matter of individual moral choice. We now return to Simon to examine his rationale for considering organizational decisions as paramount.

Organizations and Decisions

According to Simon, the most important influences on an organizational decision maker come from the organization. Organizational members have superiors, and superiors attempt to control the premises subordinates apply to their decisions. This relation between the superior and the subordinate is fundamental to Simon's model. It is the relation of *legitimate authority*. In accepting the position, the subordinate accepts (i.e., legitimates) the right of the superior to control how the subordinate performs the job. For Simon, the key activity of the job is decision making, and control is realized through the influence of the values the subordinate applies in making decisions.[18]

In the example of the value component given previously, longevity and cost were the values applied to the selection of material; it is likely that cost criteria are transmitted to a subordinate by a superior. Budgets are a normal part of the engineering design process. The longevity criteria may also have been influenced because the longevity of all the parts of a technological system are usually evaluated collectively. It makes little sense to either "overdesign" or "underdesign" parts. These types of organizational influences are incorporated into the technical specifications of a design, and these design specifications are transmitted from superior to subordinate. This observation leads to another dimension of Simon's model.

Given the core components of a decision, Simon extends his model to define rational organizational decision making by noting that:

> Fact and value . . . are related to means and ends. In the process of decision making those alternatives are chosen which are considered to be appropriate means for reaching desired ends. Ends themselves, however, are often merely instrumental to more final objectives. We are

thus led to the conception of a series, or hierarchy, of ends. Rationality
has to do with the construction of means–ends chains of this kind.[19]

We can use our technical specifications example to amplify Simon's point.
An engineering manager is given the task of overseeing the design of a
complex mechanical device. The device is made up of several subassemblies. The manager receives the technical specifications for the device, and
the functional requirements of the sub-assemblies. The manager then
translates these functional requirements into more refined technical specifications, which he or she assigns to the engineers in the design group.
Thus the engineering manager is the means of meeting his or her superior's
goal: the design of the complex mechanical device. In turn, the engineering
manager achieves his or her end by using subordinates as the means of
achieving the goal, with each subordinate assigned the task of designing a
subassembly. It is in this way that authority relations define means–ends
chains through organization. It is also through these means–ends chains
that organizations achieve rationality. Moreover, this rationality is greater
than what an unorganized group of individuals could attain.

    Simon connects this concept of means–ends chains with organizational
goals in the following way: "The fact that goals may be dependent for their
force on other more distant ends, leads to the arrangement of these goals in
a hierarchy – each level to be considered as an end relative to the levels
below it and as a means relative to the levels above it."[20] Thus the goals
(e.g., ends) at each level of the organization are aggregated as one ascends
the organizational hierarchy. Organizations, in Simon's view, are a form of
collective rationality, where this hierarchy of goals is coordinated toward
some set of larger organizational goals.

Is the Engineer Autonomous?

This hierarchical structure typically is invoked as limiting the autonomy of
engineers who have been traditionally defined as being constrained by the
organization. The nature of this constraint is important to define for it
bears relevance to our concept of the engineer as a moral agent. Both historical study of engineering and recent empirical sociological data confirm
that engineers do exercise an autonomous posture within the engineering
organizational context.

Historian Edwin Layton vividly discusses what he referred to as "a revolt of engineers," during the early 1920s.[21] Peter Meiskins reevaluated Layton's work and confirmed that:

> Mechanical and civil engineers in particular seemed to be searching for a redefinition of their place in society, to be seeking a more active role in solving the problems (both technical and social) of the day. Many of them seemed to have concluded that engineers by reason of their training, experience and social position, could develop a different and superior kind of leadership than that exercised by business.[22]

Meiskens concluded that the quest by engineers for a social conscience was joined (if not overpowered) by a quest for higher wages. Rank-and-file engineers, who thought material gains could be won by following a select group of idealists, lost interest in pursuing complex social issues when it became apparent that a "trade-union" approach would better serve their needs. Their revolt in short was brief. Capitalizing on the thriving "pro-business prosperity" of the 1920s, the professional engineer's modus operandi was defined in terms of achieving success *within* the organization.[23]

Although historians agree that engineers aligned themselves with business during the 1920s, there is no agreement as to *why* this alignment took place. David Noble is convinced that the alignment was inevitable. The revolt was, in his terms, an aberration, which was quickly eradicated amid a dominant business ethic.[24] Layton and Sinclair recognize that, while accommodation to business took place, engineers were not necessarily "domesticated to serve capitalist reason."[25] They document recurring debates especially among mechanical engineers regarding the technical and social aspects of their profession. Sinclair sums up these various interpretations by concluding that engineers are "an ambiguous and complex group; accommodated to the needs of American business but unsure and uneasy about the consistency of this position with their own professional standards." He attributes the current period of renewed interest in engineering ethics since the 1960s to the "uneasy coalition of engineering and business."[26] Thus, although the consensus is that engineers typically define their professional responsibilities with reference to the goals of an organization or company, it is also recognized that this definition is complex and not completely devoid of an element of social concern. How or when that social concern spills over into the public arena is a general topic in current engineering ethics.

Meiskins and Watson recently reexplored this debate regarding "Professional Autonomy and Organizational Constraint" with reference to engineers.[27] They focused on two questions. The first was whether engineers were able to maintain high levels of autonomy in an organizational context. The second concerned whether autonomy is central to job satisfaction. Having surveyed eight hundred engineers in Rochester, New York, they then classified autonomy into two broad categories: *ideological autonomy* was defined as both "the freedom to set goals and define problems"; *technical autonomy* was defined as "the freedom to pursue these goals and maintain control over the resources needed" to accomplish them.[28] Furthermore, Meiskins and Watson distinguished two types of technical autonomy: work schedule and work process. Engineers, they contend, are more constrained in terms of "schedule and resources" than process.

While engineers do value autonomy, they concluded, it is autonomy of a particular kind – "the ability to choose how to go about their own job."[29] They assume, in fact, that they will have this, given that they are valued as "trusted workers." Likewise, although they are constrained in the areas of project choice and control over resources, they accept these areas as falling within the authority of a managerial decision. As long as management is viewed as "facilitative and coordinating rather than coercive," engineers willingly accept constraints as being an inevitable trade-off for being able to do interesting work.[30] "When other professional values are not satisfied, however, the constraints are suspect." The theme of organizational coercion and professional values will be illuminated in several of our case studies.

The setting of organizational goals, then, is central to how individual engineers carry out their professional responsibilities. Simon's model can be used to conceptualize the complexity that surrounds the actual setting of organizational goals. We return to our extended example of NASA organizational structure to illustrate this. We can also assume that while the means–ends chain from top to bottom is a rational one, it may not always be perceived as such. Additionally, it may not be trusted.

We can examine these means–ends chains from the perspective of the top of the organization. Goals at the top of the organization are broad in scope. Each level in the hierarchy decomposes its received goals, and sets more specific goals and tasks for the next lower level of the organization. This process is repeated until the very general goals at the top are refined into specific tasks and jobs that employees at the "working" level of the organization perform.

For example, the Nixon administration made the decision to pursue the shuttle program. Thus the goal NASA received from the president might be stated: "Build a space shuttle for $5.5 billion dollars and have the program operational within six years." The top administration of NASA decomposed this goal for the next lower level of the organization in the following manner. For the Marshall Space Flight Center, the goal became: "Design and build the propulsion systems (i.e., the external tank, the shuttle main engines, and the solid rocket boosters) for a given budget and schedule." For the Johnson Space Flight Center the goals became: "Design and build the orbiter, and train the crew to fly it for a given budget and schedule." The Kennedy Space Flight Center goals were: "Design and build the launch facilities for the shuttle program for a given budget and schedule." Thus, the major centers of NASA are the means by which the whole of NASA meets the global goal of the shuttle program. Note further that the goals of the three centers are linked. The budgets of the three centers cannot exceed the total budget given to NASA by the president. The designs of the major components assigned to each center must be coordinated and interfaced. The linkages are numerous and complex.

Each NASA center, in turn, sets goals with organizational parameters of budget and schedule for its suborganizations. The process continues until all the tasks required to build and operate the shuttle have been assigned and carried out. This decomposition process extends through NASA to the contracting organizations. For example, Marshall Space Flight Center selects Morton Thiokol, Inc. to design and build the solid rocket boosters (SRB). Within Morton Thiokol, the task of designing and building the SRBs is decomposed into the major components: solid propellant, rocket engines, tubular structure of the rocket, control systems. Each of these systems is further broken down. The tubular structure of the rocket requires multiple sections, joints, and seals so that it can be manufactured in Utah, shipped to Florida, and assembled at the launch site. It is at this level that the details of engineering design work are carried out. Obviously, goals defined at the beginning of the process by the president and the top administrators of NASA are very different from the set of constraints and goals faced by the design engineer at the level of detailed engineering design. Yet, there is – or at least should be – a logical chain of means–ends relations from top to bottom that define the rationality of the organization through its decision-making processes.

At the national level, for example, the president makes a strategic decision within a national political context. Thus the complex goals involving prestige, power, or reelection may have motivated his original decision to build the space shuttle. For Richard Nixon, the decision to build the shuttle was motivated by a concern about large unemployment in the California aerospace industry. (President Bill Clinton had a very similar concern two decades later, when he addressed the closing of military bases in California during a period of downsizing of the defense industry.) In 1971, a general view that the space programs were prestigious for the United States was also apparently important. Nixon was not an engineer. Neither were most members of Congress. As described in Chapter 10 where we explored the Senate investigation of NASA's "all-up" testing procedure and other technical decisions made in the SSME project,[31] the different perspectives of the key decision makers clearly hindered a shared understanding of the problems and the solutions. President Carter supported the space program for different reasons. He believed a civilian space program was important for national scientific and technological leadership, particularly as the space program pertained to energy programs. He was also concerned with the military's dependence on the shuttle as its only means of launching spy and other satellites.

It is interesting that Carter made the critical decision to continue the shuttle program during the time period when cost overruns and earlier problems were first coming to light. He also assured that resources would be committed to the project. Carter made these decisions with the realization that he might not reap political benefit from future shuttle successes. Finally, Ronald Reagan sensed, like Jack Kennedy before him, that the space programs were popular politics. National prestige and pride were at stake. Reagan was largely unconcerned with the economics and technical aspects of the shuttle program, although he may have linked the shuttle to his more ambitious Strategic Defense Initiative program.

Satisficing: The Practical Decision-Making Tool

These changing, politically motivated goals help shape constraints of time and money, the universal parameters of organizational decision making. Technological goals and constraints are intricately woven into the list of goals that influence decision makers. By the time the tasks and goals have filtered into the operational part of the organization, the value or ethical components for individual decision makers are very complex and not transparent. About this situation, Simon, revealing his pragmatic philosophy, states that "in the decision making situations of real life, a course of action, to be acceptable, must satisfy a whole set of requirements, or constraints. Sometimes one of these requirements is singled out and referred to as the goal of the action. But the choice of one of the constraints from many, is to a large extent arbitrary."[32] Simon recognizes that complexity has implications for how decisions are really made in organizations. In actual organizational practice, according to Simon, decision makers do not attempt to find an "optimal" solution for the whole problem. Instead, the decision maker generates a limited set of "reasonable" (i.e., feasible) alternatives, and then evaluates them against the array of constraints and goals imposed by the organization. At one time, the "goal" of cost minimization may be paramount, but the constraints of the physical design specifications still must be applied in the evaluation of each alternative. At another time, the "goal" may be weight reduction, but engineers must still keep to their budgets and schedules. Decision makers may emphasize certain values in alternative evaluation, but they also attempt to satisfy their complex constraint spaces. In summary, they seek satisfactory solutions through their decision making behavior. Simon labeled this organizational phenomenon *satisficing*.[33] Given the complexity of the constraints, goals, and knowledge that goes into most organizational decisions, feasibility, not optimization, is achieved. Sometimes, as we will see, even feasibility is unattainable.

A complexity of goals and constraints characterized the shuttle program, which had to satisfy a wide range of technical, political, personal, professional, and economic goals. During negotiations seeking political support of the Air Force, NASA accepted the constraint of "cross-range" capability. This new constraint was added to an already extensive list of technical requirements. The Air Force wanted to launch the shuttle into polar orbit from its secure space launch facility at Vanderburg Air Force Base in California. This situation required the orbiter to have significantly greater flying capability than previous designs. To use Vanderburg, the shuttle had to be able to fly at least 1,000 miles "cross-range." The more costly and heavier delta wing design stemmed from this political and technical constraint. In an interesting side note, Logsdon has proposed that the cross-range requirement may have been artificial and, if NASA had objected, could easily have been modified.[34] The ramifications of NASA's not challenging this "constraint" are considerable and will be discussed later.

What do we mean when we say "NASA could have objected"? Does NASA have the same moral responsibility as a person? This is not a simple question. It invokes scholarly interpretation of Roman, British, and American law and a philosophical analysis of personhood. We merely provide a sketch of the relevant information here, and refer the reader elsewhere for an in–depth examination.[35] Ladd points out that a corporation is not a person and cannot be held responsible for its actions; only individuals or groups of individuals can be held responsible.[36] The question, as we have posed it elsewhere, is "responsible for what?" French, on the other hand, refers to the "Reality Theory" and defines corporations as "intentional systems . . . that manifest the ability to do such things as make decisions, act responsibly, enter into both contractual and noncontractual relationships with other persons and so forth. They may be blamed or credited for what they do in their own right. And, . . . they can be punished."[37] As intentional systems, French explains that corporations have their own "identity over time." They are not dependent on a "steady-state membership of human individuals." Despite radical changes in personnel, he writes, corporations remain the same entity over time. Given an external and internal structure, "corporate intentionality" cannot be reduced to the intentions of the individual persons in the corporation.[38]

French discusses the concept of corporate identity or personality and introduces a construct called a "Corporate Internal Decision Structure," which "contains the organizational relationships and lines of internal authority and responsibility as well as the rules by which corporate policy is recognized and implemented. . . . A corporation's CID structure provides its 'common end' and the organizational edifice in which its members 'conspire.'" One of French's points here is that "the Reality Theory of legal personhood states that there is a moral reality to the entities recognized as persons in the law." Corporations that have CID structures are moral persons and so warrant legal status.[39]

Simon's definition of a formal organization is not unlike the CID structure French defined. For Simon, a formal organization is "a description that, so far as possible, designates for each person in the organization what decisions that person makes and the influences to which he is subject in making these decisions."[40] French refers to these as the "fixed elements of the edifice" and views the individual employee as an "occupant" of a particular place or station:

> The structure exists independent of the particular humans who have the choice to contract their services where and when they desire.[41]

> Humans are now economically and sociologically freer than ever before in history. They are also "irrelevant in a fundamental sense."[42]

Anyone can be replaced at any time. The positions endure.[43]

By examining the concept of the individual's "station" within the organization, French provides a key to unlock the moral complexity of an individual's responsibility within a group.[44] "Under what conditions," asks French, "is a particular member of a group liable to be held responsible for what the group as a whole did or did not do?" For French, both "power" and "control" are intricately related to this concept of responsibility. "When a person has power with respect to a particular occurrence or event," he or she can intentionally choose to perform an action. This decision insures that the event will or will not occur. Control can be causal, as in the individual act of stepping on the brake of an automobile to adhere to a stop sign, or regulatory, as in creating a code of professional ethics. "In order to be held responsible," writes French, "a person must have . . . at least one of the two types of control."

For managers in a formal organizational structure, at least in Simon's terms, this recognition of power and control is defined. But for engineers at the bottom of the hierarchy, the power that their engineering knowledge endows them with may not be evident. French recognizes the possibility of this and asserts that "a person has a moral responsibility to understand the station he or she occupies in society at large and in more restricted groups."[45] He admits that "individuals may have no idea of their power in any particular group." In such cases, it may not make sense to say that they should have known that if they did X, the group would have responded in a desirable way. People may not know their individual powers and may assume that they have no way of finding out about them. "Any rational person," French asserts, "should recognize the reasonable course of action and try to communicate it to the group." This is the moral minimum required. We incorporate this idea into our framework. "If the person is unable to persuade the group or to create the critical mass to accomplish the task, at least the person has tried and, in the process, learned the limits of his or her power."[46]

In our cases, A. O. Tischler exercises this type of power, control, and responsibility.[47] Pratt and Whitney's formal protest regarding the choice of Rockwell to construct the main engines is an example of an organizational protest.[48] In each case, the protests are launched by either the individual or the group of individuals. The response is made also by the group. NASA is not an amorphous organization but can be considered in Simon's hierarchical structure to be comprised of various components, each bearing moral responsibility for the decisions it makes.

We can summarize our application of Simon's model of organizational decision making into seven points:

1. Decisions are composed of a set of alternatives, with factual and ethical components.
2. Factual components of alternatives project the future if that alternative is chosen.
3. The ethical components of a decision are influenced by the decision maker's organizational context and his or her professional values.
4. The organizational uses of authority relations to create means–ends chains of rational action.
5. These means–ends chains decompose the goals of the organization into tasks and jobs.
6. Because of the complexity of organizational tasks, decision makers *satisfice*.
7. Because organizational tasks are connected through the hierarchy, decisions made in one part of the organization can become constraints for decisions in other parts of the organization.

Organizational resources constrain the ability to generate knowledge relevant to a decision. However, factual components themselves are based on knowledge and judgment, which lie outside the scope of goals. Moreover, professionals hold values derived from their institutional affiliations. Obviously, conflict situations may arise when the factual component in a decision, particularly a judgment, or professional values do not support the goals desired by the hierarchy. This is a classic conflict situation for professionals working in organizations, and it is by no means unique to engineers.

Notes

1. Charles Perrow, *Complex Organizations: A Critical Essay*, 3d. ed. (New York: McGraw-Hill, 1986), chapt. 2.
2. H. H. Gerth and C. Wright Mills, eds., *Max Weber: Essays in Sociology* (New York: Oxford University Press, 1946); Max Weber, *Theory of Social and Economic Organization*, trans. A. M. Henderson and Talcott Parson (New York: Oxford University Press, 1947).
3. Herbert A. Simon, *Administrative Behavior*, 3d ed. (New York: Free Press, 1976).
4. Perrow, *Complex Organizations*, p. 119.

5. Two of the leading contributors to contingency theory are James Thompson, *Organizations in Action* (New York: McGraw-Hill, 1967), and Paul R. Lawrence and Jay W. Lorsch, *Organization and Environment* (Cambridge, MA: Harvard University Press, 1967).

6. Paul S. DiMaggio and Walter W. Powell, "The Iron Cage Revisited," in *The New Institutionalism in Organizational Analysis*, ed. Powell and DiMaggio (Chicago: University of Chicago Press, 1991).

7. Ibid., p. 37.

8. Ibid., pp. 45–6.

9. Ibid., p. 46.

10. Ibid., p. 50.

11. P. Churchill, "Public and Private Choice: A Philosophical Analysis," in *The Moral Dimensions of Public Policy Choice: Beyond the Market Paradigm*, ed. John M. Gilloy and Maurice Wade (Pittsburgh, PA: University of Pittsburgh Press, 1992), pp. 341—52.

12. Larry R. Churchill, "Theories of Justice," in *Ethical Problems in Dialysis and Transplantation*, ed. C. M. Kjellstrand and J. B. Dossetor (London: Kluwer Academic Publishers, 1992), pp. 21–31.

13. Michael Davis, "Thinking Like an Engineer: The Place of a Code of Ethics in the Practice of a Profession," *Philosophy and Public Affairs* 20(2), 1991: 150–67.

14. Ibid., p. 153.

15. Ibid., p. 166.

16. Ibid., pp. 153–66.

17. Taft H. Broome, Jr., "Engineering Responsibility for Hazardous Technologies," *Journal of Professional Responsibility in Engineering* 113(2), 1987:139–49.

18. Simon, *Administrative Behavior*, pp. 123ff.

19. Ibid., p. 62.

20. Ibid., p. 63.

21. Edwin T. Layton Jr., *The Revolt of the Engineers: Social Responsibility and the American Engineering Profession* (Baltimore: Johns Hopkins University Press, 1986).

22. Peter Meiskins, "The 'Revolt of the Engineers' Reconsidered," *Technology and Culture* 29(2), 1988: 219–46.

23. Ibid.

24. David Noble, *America by Design: Science, Technology and the Rise of Corporate Capitalism* (New York: Knopf, 1977), pp. 39–41. As quoted in Meiskins, "The Revolt of the Engineers Reconsidered."

25. Bruce Sinclair with the assistance of James P. Hull, *A Centennial History of the American Society of Mechanical Engineers, 1886–1980* (Toronto: University of Toronto Press, 1980), pp. 21–8.

26. Ibid.

27. Peter F. Meiskins and James M. Watson, "Professional Autonomy and Organizational Constraint: The Case of Engineers," *Sociological Quarterly* 30(4), 1989: 561–85.

28. Ibid., p. 563.

29. Ibid., p. 577.

30. Ibid., p. 578.

31. T. H. Johnson, "The Natural History of the Space Shuttle," *Technology in Society* 10, 1988: pp. 419–21.

32. Simon, *Administrative Behavior*, p. 262.

33. Ibid., p. 272.
34. John M. Logsdon, "The Decision to Develop the Space Shuttle," *Space Policy* 2, May 1986: 103–19.
35. Peter A. French, *Responsibility Matters* (Lawrence: University Press of Kansas, 1992), chaps. 7, 9, 13,16.
36. John Ladd, "Collective and Individual Moral Responsibility in Engineering: Some Questions," *IEEE Technology and Society* 1(2), 1982: 3–10.
37. French, "Laws Concept of Personhood: The Corporate and the Human Person," in *Responsibility Matters*, p. 138.
38. Ibid., p. 139.
39. Ibid., 141–43. This type of reasoning was introduced into the American legal system in 1886 and clearly has a distinguished history.
40. Simon, *Administrative Behavior*, p. 37
41. French, "Laws Concept of Personhood," p. 143.
42. Ibid., p. 39
43. Ibid., p. 143. French reviews a variety of philosophical views of "personhood" and its relation to the corporation – for example, Rousseau believed individual humans can exist outside the corporate union of the state, but that they could not achieve completeness as moral persons in its absence.
44. French, "Power, Control and Group Situations: And Then There Were None?" in *Responsibility Matters*, pp. 71–8.
45. Ibid., p. 78.
46. Ibid.
47. See Chapter 8 .
48. See Chapter 5, on the decision to build the space shuttle.

PART II

Political and Technical Background, Plus the Cases

5

The Decision to Build the Space Shuttle: Pay Me Now or Pay Me Later

The political compromises that led to the decision to build the space shuttle defined the parameters of cost and schedule within which the various design engineers and project managers had to work. It is generally recognized that NASA "sold" the shuttle to Congress[1] and, in so doing, drastically changed its overall plans for a space program. What has not been recognized was that in selling the shuttle, NASA placed its engineers and managers in an environment where ethical compromise would be the rule, not the exception. As engineers designed, built, and tested the space shuttle main engines (SSME), they were continually balancing safety issues with deadlines and a restricted budget. In the process, a variety of technical dilemmas were identified and resolved. In the second half of this book, those dilemmas are described and then examined with reference to our ethical framework. In this way, the tacit moral components included in the decision making process are made "explicit." As noted previously, the principles of competence, responsibility, and Cicero's Creed II provide a common language for discussing the ethical dilemmas.

This chapter clarifies the historical context in which the decision to build the space shuttle was made. Along with Chapter 6, which discusses how the SSMEs work, it describes how and why the dilemmas identified as having ethical import occurred. Both chapters also provide a context for understanding why individual engineers chose to resolve the dilemmas as they did.

The Decision to Build

In 1969, when Thomas Paine was appointed as the new NASA administrator, he described what he termed "the next logical step" to follow the Apollo moon landing: a manned mission to Mars. Richard M. Nixon was then president, and he appointed a space task group chaired by Vice-President Spiro Agnew to formally assess plans for NASA's future. In keeping with Paine's vision, the group's final proposal outlined the components of a mission to Mars. This would consist of a manned space station in a low earth orbit from which the Mars mission would be "launched"; a shuttle-type transportation system to supply the space station from Earth; and, finally, a space tug to relocate vehicles and other structures into high orbit once they got into space. Thus the building of the space shuttle project, as we know it, was actually one small piece of a much more ambitious plan.

Even as the Apollo mission achieved its stated goal, its awe began gradually to fade and serious questions surfaced regarding the social worth of exploring space. In 1969, the country focused on a range of intricate and disturbing domestic social problems. Protests against the Vietnam War, the Civil Rights movement, runaway inflation, and a general mistrust of authority affected the "mood" of Congress.[2] Top administrators at NASA, always attuned to the influence of politics, feared that this mood would result in the demise of the entire space program and opted for an approach that would satisfy a range of influential constituents. Clearly only an astute political decision would guarantee continuation of the space program. The space shuttle, conceived as "a flexible, economic, reusable space transportation system," served as this political expedient. Originally designed as one component of a larger space station project, the shuttle idea found support in the scientific community, which wanted a space station and a space telescope; in the Department of Defense, which wanted an alternative to expendable Saturn rockets for launching satellites, and in commercial users, who foresaw profits from launching a variety of payloads. To all, an "inexpensive, reusable space vehicle" seemed the solution.

Historian Alex Roland has pointed out that the idea of a reusable spacecraft originated in Germany during the 1920s. The X-series of experimental U.S. aircraft of the 1950s also capitalized on this concept. In the late 1960s, the expendable *Saturn V*, which was relied upon to launch into orbit payloads of approximately fifty tons, cost $185 million, or $500 to $1,000 per pound. (Overhead was estimated to add another $500 per pound.)[3] In contrast, George Mueller, the associate administrator of NASA, envisioned a reusable vehicle with a launch cost approaching \$5 per pound.[4] Mueller's concern for cutting costs was part of his overall systems approach to managing the development of a complex technology. Richard L. Brown, a member of the first shuttle task group at NASA headquarters and the Marshall Space Flight Center MSFC Special Projects Office with the SSME Project Definition, described George Mueller's philosophy as "the way you do something in the aerospace business at minimum cost is to do it in the minimum schedule . . . that's technically feasible." To Mueller, "time was money" and, according to Brown, the original time frame he constructed for building the shuttle was to "define it in '69, start it in '70, and fly it in '76." Envisioning the publicity possibilities associated with the U.S. bicentennial, "his . . . goal was to try to have the first flight on July 4, 1976."[5] Based on Mueller's cost projections, the design originally conceived for the shuttle was a fully reusable two-stage vehicle. The first stage was manned and, after assisting the second stage into orbit, was to fly back and land on an airstrip. The engines and fuel tanks were internal components of the manned second stage. The cargo bay of this sleek, straight-winged vehicle was fifteen feet in diameter and forty-five feet long. It would accommodate payloads of 45,000 pounds.

This original design, however economical, was altered in several ways as the negotiations to satisfy all political allies proceeded. Design changes were the most obvious. The Air Force requirements, for example, included a sixty-foot cargo bay to carry payloads of 65,000 pounds. This represented a 45 percent increase in payload capacity. They also requested a cross-range capability and, to achieve this, the shuttle's straight wings were changed to delta wings. Next the thermal protection system was doubled in weight in order to protect the second-stage vehicle as it reentered the earth's atmosphere. To compensate for this total increase in weight, escape rockets that could be used during lift-off, if necessary, were deleted, as were the auxiliary jet engines designed to provide a second chance for the crew to land the vehicle. These added requirements and their subsequent design changes, in turn, eventually translated into an increased demand on the engines, which would now have to deliver 109 percent of their original design thrust. (As noted in Chapter 6, these engines have yet to be tested at this level.) A foreshadowing of a decision making style that would characterize the entire space shuttle program was NASA's willingness to compromise its original plan to satisfy perceived political realities.

Public policy analyst John Logsdon has documented, for example, that the Defense Department's insistence on its requested changes was negotiable and that it was less concerned with the shuttle's design than NASA assumed. Nonetheless, the changes were readily made in the hopes of courting an ally to ensure that the shuttle would be built.[6] Indeed, this Air Force requirement drove the final shuttle configuration. According to Bob Marshall, then director of Marshall's Program Development Preliminary Design Office, "the reason the shuttle is 65 feet long and 15 feet in diameter and carries 65,000 pounds in orbit was to carry an upper stage and a specific payload into synchronous orbit, that sized the payload bay . . . that sized the whole system. So, the shuttle was basically designed to satisfy an Air Force payload [and cross-range] requirement."[7] While these design changes altered the appearance and capacity of the shuttle, it was the Office of Management and Budget (OMB) decision, in 1971, to only allot half of the expected $10 to $14 billion for development, which changed the "intrinsic character" of the craft as well. This caused compromises regarding cost, risk, and schedule to be made at every decision point during the shuttle's development, testing, and final construction. In approving the shuttle project, President Nixon overruled both his science advisor and his science advisory committee. Nixon was impressed with the Department of Defense support for the shuttle. He obviously was not aware of how "uninvolved" that support was. He did, however, appreciate that it was an election year, and that the space program employed large numbers of people in key states, particularly California. As Nixon put it, "space flight was here to stay and we'd best be part of it."[8] On January 6, 1972, a bargain basement version of the original $10 billion shuttle concept was approved for $5.5 billion with a $1 billion contingency. To Logsdon, "the system chosen for development was, from the start, unlikely ever to meet its announced objectives, but the gap between rhetoric and reality persisted for 14 years."[9]

Thus, the shuttle's original design was modified so as to win political support. Its weight was increased, a number of features were eliminated, and the engines were required to deliver 109 percent of their original design thrust. So as to keep production within the budget, the fuel tank was externalized and made expendable; the manned flyback first stage was replaced with boosters, which were to use solid rather than liquid fuel.

Given the final budget constraints, NASA further compromised its original plan to cut development costs and, following a plan implemented by George Mueller, relied on the risky practice of all-up testing. This practice was pioneered on the Titan II and Minuteman projects, and was dramatically used during the Apollo project to meet the goal of "placing a man on the moon by the end of the decade." This success-oriented strategy (discussed extensively in Chapter 7) downplayed component testing and was the subject of a full Senate investigation in 1978. While the decision to rely on all-up testing slashed the one-time development costs, as A. O. Tischler forecasted, it increased future ongoing operational costs.[10] In actuality, NASA kept the shuttle's development cost within its 1972 planned limits until 1978, when it became public that the risks required to do this threatened the success of the project (see Chapter 10).

Operational costs were a different story. The Congressional Budget Office estimated that the 1985 launch cost was $2,298 per pound in orbit – a far cry from the $150 per pound predicted thirteen years earlier (and more than the comparable 1985 estimated cost of $1,667 per pound for the *Saturn V*). Clearly, NASA's original concept of low operational costs had been sacrificed in order to achieve its budgeted developmental costs.[11] By 1977 when engine testing began in earnest, it was becoming apparent that the shortcuts in development may not have paid off. A series of major failures resulted in damaged engines, increased costs, and postponed schedules. Having passed Mueller's first launch date of July 4, 1976, it was increasingly clear that the new official target launch date of March 1, 1978, would not be met either. Further, there would not even be an acceptable engine by then; nor would a second critical problem with the protective tiles be resolved.

Having agreed halfheartedly to the shuttle as its sole launch vehicle, by the late 1970s, the Defense Department was increasingly concerned about the lack of progress. The Senate committee hearings conducted in 1978 had clarified that a focus on engine development was clearly a priority. Predictable first launch dates gave way to the concept of letting a "safe engine" be the guide to timing. Yet, a delay of more than three years would leave intelligence-gathering satellites without back-up rockets. With pressure from the Defense Department to develop alternative launch vehicles, President Carter, hardly a space enthusiast, sought to resolve the problem by asking Congress for more money for the shuttle in 1979. Because of the concern with assuring a military satellite launch capability and the $10 billion that was already expended, Congress approved. Alex Roland, notes that "for the first time since 1971, cost was no longer the main determinant in shuttle development."[12] Perhaps NASA's strategies paid off.

Finally, on April 12, 1981, the shuttle *Columbia* was successfully launched on its first flight. Not quite five years and twenty-five launches later, on January 28, 1986, the shuttle *Challenger* exploded after seventy-three seconds of flight (see Chapter 14). With it went both concepts of low-cost space transportation and having a sole launch vehicle for civilian and defense purposes. To Logsdon, the policy analyst, one conclusion was obvious: Given the original intent of a low cost, routine transportation system the shuttle "must be assessed as a policy failure."[13] Clearly, the budget cut, the decision to use all-up testing, and the subsequent design changes each had a major impact on the development of the main engines.

The Engine as the "Pacing Item"

As early as 1969, the engines were earmarked as the "pacing item" of the new space transportation system. The official development of the SSMEs was to be divided into four stages, or contract phases: initial feasibility investigations (phase A) of a reusable liquid oxygen–liquid hydrogen engine, design (phase B), development (phase C), and production (phase D). In spite of this traditional government–technology process, phase A feasibility studies were conducted for the space shuttle system but not officially for the SSME. According to a NASA press release on February 18, 1970, "This type of engine (referring especially to the XLR-129 experimental engine) [had] been under investigation by NASA and the Air Force for the past five years."[14] The press release announced that the request for proposals (RFP) for preliminary definition and planning (phase B) studies of the engine were issued to six corporations including North American Rockwell, Rocketdyne Division; United Aircraft, Pratt & Whitney Aircraft Division; Aeroject General; TRW Inc.; Bell Aerospace Systems; and Marquardt Corporation.[15] With "up to three" contractors to be selected, of the six contractors, Rocketdyne, Pratt & Whitney, and Aerojet General submitted proposals for phase B studies, while the other three declined to enter the competition. It was expected that the next step, phase C studies, would involve two contractors who would build competing prototype engines.[16] All three companies that submitted a proposal were awarded phase B contracts, for preliminary design studies of the SSME, and NASA received their final reports on April 21, 1971.

Controversy over the political process interfering with the building of the main engines surfaced as soon as the phase C contract was awarded. After reviewing the papers, NASA selected only one company, Rocketdyne, on July 13, to complete the development and production of the thirty-six engines to be used in both the orbiter and booster stages.[17] The $500 million contract proposed that Rocketdyne would deliver the engines by 1978.

For the shuttle configuration at that time, two of these engines would be installed in each orbiter and twelve in each booster.[18] *Space Business Daily* reported that "NASA hopes to complete negotiations with Rocketdyne by September 1."[19] These hopes were destroyed when, on August 5, Pratt & Whitney filed a formal protest with the General Accounting Office (GAO) over the selection of Rocketdyne for the phase C SSME award.[20] It charged that the selection of Rocketdyne was "illegal, arbitrary, and capricious, and based on unsound decisions."[21]

Earlier in July, Representative William R. Cotter preempted these allegations and bluntly charged that the SSME award was a political decision. He requested that the GAO investigate. To support his claim, Cotter pointed to the ten years and $47 million in government funds in addition to the $31 million of Pratt & Whitney's own money devoted to developing the main engine concept.[22] Senator Edward J. Gurney questioned NASA about the choice and James C. Fletcher, administrator of NASA, responded. Justifying the choice, Fletcher wrote, on July 30, "We believe that [the Source Evaluation Board's] assessment of the relative strengths and weaknesses of the competing designs provided an objective basis for the source selection made."[23] However, while Fletcher attempted to assure the senator, Pratt & Whitney was already planning its charges for a formal protest. Immediately following the protest, nine southern senators, including two from Florida where Pratt & Whitney was located, urged NASA to wait until the GAO reached a decision to finalize an engine contract.[24]

Just ten days after the protest letter by Pratt & Whitney, Bruce N. Torell, president of a Washington, D.C., law firm – Reavis, Pogue, Neal, and Rose – charged NASA with six contentions.[25]

1. NASA did not conduct meaningful negotiations with the offerors – Pratt & Whitney was not made aware of the deficiencies or permitted to make revisions prior to the final selection.
2. Rocketdyne's proposal "was plainly non responsive to the requirements of the RFP." Rocketdyne proposed the use of welds and INCO-718 in the engine components, an approach not acceptable for a reusable engine.
3. NASA had acted arbitrarily and capriciously in citing Pratt & Whitney proposal deficiencies.
4. By disregarding the XLR-129 experience of Pratt & Whitney, the NASA selection involved procedures and decisions that invited a cost overrun.

5. The NASA selection had the improper effect of "wasting the knowledge and the dollars expended in the Air Force XLR-129 program."

6. Rocketdyne was given the unfair competitive advantage "of carrying on a technical effort germane to this project" with funds from the NASA Saturn launch support contract.[26]

Robert Anderson, president of North American Rockwell, upon hearing the allegations, was quoted as saying "I have full confidence that NASA's decision will be upheld by the General Accounting Office."[27]

Meanwhile, a few days after the charges were issued, Cotter reiterated that the decision was a political move by the Nixon administration. Although he said that he had no proof of this allegation, he claimed that there were indications that the decision criteria were changed during the judging of the contract proposals.[28] Then, on August 26, Cotter continued his crusade by asking NASA about the role of Dale Myers, the deputy administrator of NASA and director of the manned space flight programs, in awarding the contract. Before joining NASA, Myers was the vice-president of North American Rockwell. Cotter accused Myers of playing a key role in cutting off Air Force funding for the Pratt & Whitney XLR-129 experimental engine program.[29] In response to Cotter's accusations, NASA's assistant administrator for industry affairs, Daniel J. Harnett, said that the agency's records for the contract award were open for inspection. He further explained that, although Myers was responsible for selecting members of the Source Evaluation Board, the members were chosen before bids were even issued.[30]

In September, NASA awarded Rocketdyne a level-of-effort type contract for $1 million for four months to investigate engineering issues critical to the design of the engine. Anderson was quoted as saying, "it is most unfortunate that this protest has the effect of delaying full implementation of the program which also will mean a delay in the return to work of many of our laid off employees."[31] While the investigation continued, Rocketdyne was optimistic the award would be reaffirmed by the GAO and planned the first firing of the main engine by June 1973. The work being done under the interim contract was, according to Paul D. Castenholz, vice-president of Rocketdyne, necessary regardless of the award. It included providing interface data to the vehicle competitors, analyzing and reviewing design specifications, providing design support for the engine, studying fabrication methods applicable to the engine, investigating materials, analyzing the engine controller, and developing program plans. Several additional key technical areas, however, such as combustion stability, electronics, and valve fabrication methods, were also being studied by the engineers.[32]

On October 12, Congressman Jerome R. Waldie of California informed the House of Representatives of his "own confidence in the ability of Rocketdyne to perform excellently in every aspect of the program." For support, Waldie submitted a copy of NASA's official selection statement by Fletcher. After an initial evaluation, before any written or oral discussions, Rocketdyne was rated first and Pratt & Whitney second in the order of merit. Following the written discussions and oral interviews, the ratings, according to Fletcher, did not change. Rocketdyne had proposed the lowest cost originally, but after the committee adjusted the cost estimates from the weaknesses and the strengths in the companies' proposals, Pratt & Whitney's cost estimate was lowest. In spite of this, Fletcher claimed, "the design of the space shuttle main engine proposed by Rocketdyne was superior to the design of the other competitors."[33]

Later that month, Cotter accused NASA of using a $4 million interim contract as an award for Rocketdyne to learn how to build the engines, saying that GAO documents would support his claim.[34] In answer to Cotter's and Pratt & Whitney's protests, North American Rockwell filed a briefing with the U.S. comptroller general and challenged Pratt & Whitney's claims. Rockwell claimed that it had "more experience" in liquid rocket engine production then any other company. Also, it declared that NASA viewed the Pratt & Whitney design as "weak," especially in the area of solving the stability problem. In response to the accusation of using Saturn funds for SSME studies, Rockwell responded, "Of course work under one Government contract does not have to stop because the information acquired may be helpful in a competition for another contract. Indeed, during the SSME competition, Pratt & Whitney was funded under seven different Government contracts for doing work which it admits provided information helpful on its SSME proposal."[35]

Finally, on April 1, 1972, the GAO rejected the protest made by Pratt & Whitney and disclaimed each of the filed accusations. First, "Some of the major Pratt & Whitney deficiencies involved comparative weaknesses," and their discussion of these would have likely involved "leveling and technical transfusion." This type of discussion is normally avoided. Second, the GAO simply disagreed with the accusation that the proposed welding and the use of a specific metal alloy subject to hydrogen embrittlement did not meet the reusability requirement. Third, after thorough review of the volumes, the GAO concluded, "the evaluations were not arbitrary or capricious . . . but were comprehensive and objective and provided a sound basis for selecting the most advantageous proposal." Fourth, knowledge gained by Rocketdyne under the Saturn contract was not, the GAO believed, "of substantial benefit or that such advantage was unfair." Fifth, it felt that each company's experience was weighed fairly. And finally, cost aspects "were given comprehensive and objective consideration in the evaluation process."[36] Within a week of the GAO decision, NASA was negotiating a $450 million cost-plus-award-fee contract with Rocketdyne for twenty-five engines for the orbiter of the final shuttle configuration.[37]

Reusability: Politically Correct but an Engineering Red Flag

In developing the SSME, reusability became a crucial factor. While it satisfied the economic political expedient, it pushed the engines beyond the state of the art. In a 1971 article, Jerry Thomson, chief engineer of the SSME program at the Marshall Space Flight Center (MSFC), commented that even in 1969 "it became clear that the Shuttle would demand major advancements in propulsion technology beyond those applied in Apollo."[38] The engine would need to perform at a higher level of power and for longer times than the J-2, the liquid hydrogen–liquid oxygen engine used for the *Saturn V* launch vehicle.[39] However, according to Frank Stewart of the Engine Program Office at the time, the 1969 Pratt & Whitney XLR-129 reusable, experimental engine provided the technology to advance the state of the art of the J-2 to the technology necessary for the SSME.[40] Extending the life of the engine for the reusability aspect required increased combustion pressure, which in turn created problems of materials, high temperatures, high speed pumps, and a sufficient cooling system.[41] Therefore, as Tischler explained in a 1971 article, "Reusability poses serious technological problems. . . . These requirements, in turn necessitate strengthening and verifying our technological knowledge in materials, structures, (and) propulsion."[42] Tischler, however, felt these requirements could be met if a step-by-step evolutionary process were used in the engine's development.[43]

In order to achieve extended life of the engine components, given the requirements of compact design and low weight in combination with high combustion temperatures and pressures, materials such as heat-resistant superalloys, nickel and iron-based Iconel alloys, and stainless steels were used throughout the SSME.[44] In 1971, Dale Myers, NASA associate administrator in charge of shuttle development, had assured Congress that "The high-temperature materials required for use in this engine are already available. . . . Although significant innovation is required, principally to keep development costs low, the basis for the technology required is solidly in hand."[45] Yet, two years later, MSFC liquid propulsion experts John McCarty and Joseph Lombardo wrote that "The barrier to economic reusability which must be overcome is the damage to the materials of which the engine is constructed. . . . The effects of high operating pressures are also felt by bearings and seals in the pressure producing pumps. Such components are subject to high cycle fatigue failure and wearout."[46]

Whose assessment was correct? In 1977, major testing began and four costly engine failures occurred, due respectively to a ruptured seal, overheated bearings, and cracked turbine blades (twice), all in either the high-pressure oxygen or hydrogen pumps. It was clearly apparent that this was going to be a much more difficult task than NASA had optimistically anticipated, and the original March 1978 launch date could not be met.[47]

Nor would the engine's technical problems be quickly solved. Nearly a decade later, in examining 1,400 unsatisfactory condition reports (UCRs) filed by Rocketdyne engineers between 1980 to 1983, Glover, Kelley, and Tischer found that almost 600 were attributable to cracking and erosion. The cracking was usually caused by vibration or thermally induced fatigue. Erosion was primarily a problem in the high-temperature areas. In particular, erosion and fatigue cracking in the turbine blades, the turbine sheet metal, and the preburner (to the turbine joint area) of the high-pressure fuel turbopump (HPFTP) were the subject of many of these UCRs. The high-pressure oxidizer turbopump (HPOTP) experienced similar problems, but to a lesser degree. Here the major complexity was bearing wear, which first occurred in 1977. Erosion and cracking were also observed in the fuel preburner, main injector, hot gas manifold, and main combustion chamber.[48] These difficulties limited the life of the components and, in turn, the life of the entire engine. NASA engineers McCarty and Lombardo were correct: "Expensive materials evaluation . . . must be addressed to arrive at workable solutions."[49] As two of our cases illustrate, while this was a known engineering problem, materials evaluation was deemphasized in favor of trying to meet cost and time factors.

Political Ethics: An Oxymoron?

All engineering decisions involve trade-offs between performance (risk versus safety), cost, and schedules. In this regard, decision making that took place when the shuttle was built is similar to any engineering decision. Deciding what the nature of these trade-offs will be comprises the tacit value judgments all engineers are taught to make during the years of their professional education. Every project an engineer works on, in fact, is framed within these parameters. Sylvia Fries, former director of the NASA History Office, explained this within the context of building a space vehicle:

> Safety, reliability, and quality assurance engineering (SRQA) . . . are involved at each phase of a manned spacecraft's development and operations: developing specifications, evaluation of bidders' proposals, definition and preliminary design, production and operations. . . . Throughout these phases tradeoffs have necessarily had to be made between performance, cost, and schedule, on the one hand; and optimum standards of safety, performance, and quality on the other. These tradeoffs constitute the measures of "risk" for each program's many elements.[50]

Because some degree of risk is inherent in all engineering, an engineer is constantly striving to minimize it and maximize performance, while meeting the proposed schedule and cost. When a higher than minimal risk factor is accepted, defensible reasons are required to justify this action.

When an engineer offers defensible reasons to accept a higher than minimal risk in a given project, he or she is making an ethical decision. This chapter chronicles the political context that influenced NASA's decisions regarding how the space shuttle would be built and suggests that the cut-rate budget introduced pressures that led to an acceptance of a high-risk testing strategy into the overall program. These decisions had ramifications on organizational goals and, in turn, on individual engineers.

The politics of NASA have been a focus of study for several years.[51] We review three representative interpretations of the politics of space policy. Robert W. Smith has documented the fascinating interaction between Congress, NASA, and the scientific community in *The Space Telescope: A Study of NASA, Science, Technology and Politics*. Not unlike the narrative described in this book, Smith provides us with a commentary regarding the overall importance of looking critically at the political process within which enormously expensive, federally funded long-term projects are "born." We previously cited Smith's analysis of how the telescope's scientific capabilities were shaped by what was "politically feasible" and how this process contributed to NASA's inclusion of a higher risk than had been intended.[52] Smith also described the consequences that playing the political game had.

> Capital Hill rhetoric and the claims made during the selling, and sometimes overselling, of the space telescope a few years earlier provided a sharp contrast with the hard reality of what NASA now was publicly claiming was a taxing design and development effort. During the selling campaigns, the argument had been made repeatedly by NASA, the contractors, and the astronomers that developing the technology for the telescope would not pose major problems. Yet, three years into the program, they faced a sizable cost increase, combined with an extra year to fifteen months more on the schedule, as well as warnings about the magnitude of the technical tasks.[53]

The House Subcommittee on Space Science and Applications acted as the liaison between Congress and NASA. It held monthly hearings to keep current on progress and problems, and also served as a technical advisory committee. Early in 1981, after a series of reported difficulties with several major aspects of the telescope's development, the committee queried NASA about the problems. "We really thought that NASA and the contractors pretty well understood that project and what it involved. What went wrong?"[54] To which a NASA official replied: "It is a question of optimism at the outset, not fully appreciating the complexities of the program."[55]

Smith's careful examination of the long-term development of the telescope, including its initial funding process, forced him to conclude that "optimism is surely too simple and easy an answer. The problems being encountered indicated that the phase B studies had not been extensive enough, and not nearly enough work had been done to prove the concepts that were to be put into practice in phase C/D, in part because NASA had not secured sufficient planning funds."[56]

The space telescope project was funded under the same allocation as the shuttle. Trying to compensate for what was clearly an inadequate budget, NASA cut corners in development. Its strategy, which is discussed in Chapters 7 and 10, was to encourage contractors to adopt a high-risk, success-oriented approach. Referred to as all-up testing this concept was used successfully in the Apollo project. If the strategy worked, savings would be substantial. If not, more funds would have to be procured at a later date to maintain the ongoing progress of the program. While this was never explicitly stated by NASA, it is a common strategy in securing federal funds. Whether it was ethically defensible to apply it to the development of both the space telescope and the SSME is a dilemma we pose in this book. Clearly, it had long-term ramifications for both of these complex technologies.

Thomas H. Johnson, a political scientist who examined the historical development of the space shuttle program from its inception to the present day, reached a conclusion slightly different from Smith's.[57] To Johnson, neither the development process nor the technology was the culprit. In his terms, "scientific rationality" was simply overridden by one president who ignored professional advisors, and by a second president who overruled the Defense Department's protest of the shuttle's monopoly on launches. Managers in both the booster and engine programs willfully ignored their engineers who understood that the systems were being run beyond design limitations. "Mixing explicit technical and economic goals with subjective criteria of politics and prestige," warns Johnson, is a very risky business and results in a confusion of goals. The stress on competition promoted what he labeled "irrational elements in decision-making." That is, cautious judgment and detailed analysis took a back seat to glamour, national prestige, and the political realities of the funding process.

John McCurdy, an organizational theorist, studied NASA's evolution as a federal agency. He concluded that NASA's cultural progression was a classic case study of how government agencies change with age.[58] NASA's original organizational culture, based on "civilian" characteristics, incorporated norms that valued both "technical discretion and tight management control."[59] Further, it inherited a tradition of open communication and flexibility to handle new demands. "With a strong sense of mission," McCurdy writes, NASA flourished. "Government," according to McCurdy, with its "volatile political environment and increasing bureaucracy, works against such cultures and forces them into the 'bureaucratic mold.'"[60] One result of this process was for NASA to adopt a more casual attitude toward risk. According to McCurdy, the original NASA culture embraced the notion that space flight was inherently risky, that failure was a normal part of the learning process, that trouble was a constant companion, and that each space flight had to be treated as a discrete event. The nature of space travel during the 1960s supported this perspective, when most flights were short and daring. By the 1980s, changes had occurred. NASA officials made plans to engage in missions that varied from frequent to continuous. They encouraged the radical notion that much of what they did in space could be made safe, routine, and familiar. Such notions were totally contrary to the norms of NASA's original culture (see charts in McCurdy's text). While we explain in this book that experiences within NASA promoted a "success-oriented" philosophy and that this also added confidence that the higher risk was a "good bet," McCurdy's general theory also explains NASA's response to the political pressures.

The interplay between national politics, NASA management, and individual engineering decisions has been ably described. Smith fully blames the political process for the shortcuts incorporated into the feasibility, planning, and development phases of the Hubble Telescope project. Johnson is more explicit in his remarks, naming and blaming individual politicians for allowing politics to influence technical decisions at all. McCurdy interprets NASA's political involvement from an evolutionary approach, seeing the events as predictable. He actually describes this as "organizational aging."[61] These views contribute to our understanding of the complex decision-making processes involved in building federally funded technology. They provide hindsight to identify the dilemmas that occurred. They do not, however, explore if or why an individual, organization, or government should or could be held responsible for the problems.

Dennis F. Thompson, a philosopher, does just that. He has created a conceptual scheme to analyze political decisions ethically.[62] Two aspects of political life, according to Thompson, create ethical dilemmas. One centers on the reality that "officials act for us" and the other on the given that officials do not act alone but "with others." Thompson explains it in this way:

> For the sake of those for whom they act, officials may take their duties to permit or even require them to lie, break promises, and manipulate citizens. These and worse violations of our shared moral principles create what is known as the *problem of dirty hands*. It concerns the political leader who for the sake of public purposes violates moral principles.[63]

A classic public health dilemma, being played out currently, serves to illustrate this dilemma. Laws that require the reporting of HIV-positive persons and partners have been proposed to track the spread of the AIDS pandemic. Clearly, the laws are constructed to promote the welfare of all citizens. In insuring this "greater good," the confidentiality and privacy rights of some will be violated. The social costs of such violations can be devastating to the individual. Another dilemma for public health officials involved "carrying out experiments for the social good," or using citizens as "ends."[64] Recent reports about radiation exposure and testing in the late 1960s or the forty-year history of the Tuskegee syphilis study are both heinous examples of this problem. In the case of "dirty hands" Thompson contends that one must explicitly identify which moral principle should apply and then provide justification for its application. This process of justification is standard fare in ethics but, as Thompson contends, it takes on special importance in politics. There is rarely a public statement justifying why certain citizens' rights are violated. In funding the shuttle, the political compromises were made to satisfy all interest groups. In doing so, each group's interests may also have been compromised. The political importance of bolstering the aerospace industry in California by awarding Rocketdyne the SSME contract may have been the event that elected Nixon as president. The chain of events – from Watergate to the explosion of the *Challenger* – would clearly be a study of the politics of "dirty hands." Was Nixon "up to his elbows"?

What is the problem of "many hands"? It is often assumed that because officials act with many others, moral responsibility is diffuse and no one can take responsibility for decisions and policies. Thompson describes it this way: "Many kinds of practices and policies of government . . . especially those that extend overtime and produce unintended consequences, reveal no fingerprints."[65] Although it may be difficult to point to individuals as blameworthy, or to elicit arguments justifying one action over another, it is the task of political ethics to define criteria for doing so. Thompson "resists the growing tendency to deny responsibility to persons, and the complementary tendency to attribute it to collectivities of various kinds." His aim, in analyzing political ethics, "is to preserve the essentials of the traditional idea of personal responsibility against the pressures of organizational life." In so doing, he recognizes that traditional notions of personal responsibility are altered. Yet they provide "a link between the actions of individuals and the structures of organizations."[66]

Thompson formulates the criteria for an ethics of political judgment much in the way we framed our principles in engineering ethics. He proceeds by examining real-life cases and developing criteria from them. "Political ethics cannot proceed without examples. Examples draw our attention to one of the most difficult but least examined steps in political judgment – identifying the issues themselves. Because political ethics must function in non-ideal circumstances, the issues do not usually come with labels announcing themselves as moral dilemmas."[67] One can view the Senate investigation described in Chapter 10 as a political ethics exercise. Prior to the investigation, the House of Representatives provided the liaison committee responsible for monitoring NASA's technical decisions and basically accepted what it was told. We document how trust was eroded between the committee and NASA and how the Senate investigation set new guidelines to insure that a safe shuttle would be built and that NASA would remain accountable for its actions. A year later, Carter allocated more funds to the shuttle project – insuring the safety of the vehicle. This new oversight process brought the government back into the critical role of investigating NASA's decisions and, in so doing, accepted responsibility for the technology.

Thompson's method and his concepts are included here to aid in identifying the issues of responsibility at the political level. In the previous chapter we discussed this by examining Herbert A. Simon's organizational theory and insights offered by philosopher Peter French.

Conclusion

We do not answer the question of who, in the political arena, was responsible for creating the environment that ultimately placed individual engineers in a position to compromise their professional judgments. We do conclude that some level of responsibility does reside with political officials involved in the funding and in the continuation of the space shuttle project. Philosopher David Bella is generous in his identification of this problem. His generosity, however, speaks to "intent" and not to "ends."

> It may be extremely difficult to identify who is responsible, to what degree, and in what respect, when things go wrong. There need not be malicious intent, deliberate deception, or the like in order for very bad things to happen. An organization of basically decent individuals can cause serious harm to others or even to the overall quality of the environment.[68]

Robert S. Ryan, a senior NASA engineer who has been involved with the design of a number of major systems, as well as participating in the investigation of approximately 150 failures and problems over a thirty-five-year career, uses experienced engineering judgment and places us back to the specifics of the building of the SSMEs. He succinctly states:[69]

> Engineering is a process based on principles of technical disciplines, management, etc. Its application involves risks. Reduction of risks requires both knowledge of the process and the inherent subprocesses, the technical disciplines, and the ability to judge merits of each. The system focus is fundamental to lowering these risks through decision making involving the system from front to end, concept to operations, and considering total cost. Total cost is, in all probability, the hardest item to define, certainly it is the hardest to calculate. The great challenge is how one balances total cost with risks to arrive at an acceptable answer.

Difficult as this balancing act is, Ryan's engineering experience has evolved into a general principle, a tacit ethic to good engineering:

> What is done up front in a project – requirements, constraints, etc. – determines to a large extent, the quality and performance of the design. What is done in production and operations tends to be fine tuning of this predetermined configuration and is very costly if robustness is not present. . . . Or, pay me now, or pay me later.

Perhaps public officials involved in deciding the building of complex technologies should memorize Ryan's principle and include it in their assessment of proposed budgets.

Notes

1. Sylvia Fries, "2001 to 1994: Political Environment and the Design of NASA's Space Station System," *Technology and Culture* 29, July 1988: 568–93; Alex Roland, "Priorities in Space for the USA," *Space Policy*, May 1987: 104–11; Alex Roland, ed., *A Spacefaring People: Perspectives on Early Spaceflight*, NASA History Series (Washington, DC: NASA, 1985), pp. 81–107.
2. Alex Roland, "The Shuttle: Triumph or Turkey?" *Discover*, November 1985: 29–49.
3. Ibid.
4. George E. Mueller, "The New Future for Manned Spacecraft Development," *Astronautics and Aeronautics* 7, March 1969: 24–32.
5. Richard L. Brown, taped interview, in Management Operations Office, MSFC, *Oral Interviews: Space Shuttle History Project*, Transcript Collection, December 1988, pp. 240–1.

6. John M. Logsdon, "The Space Shuttle Program: A Policy Failure?" *Science* 232, 1986: 1099–105.
7. Robert Marshall, taped interview in Management Operations Office, MSFC, *Oral Interviews: Space Shuttle History Project,* Transcript Collection, December 1988, p. 119.
8. Thomas H. Johnson, "The Natural History of the Space Shuttle," *Technology and Society* 10, 1988: 417–24.
9. Adelbart O. Tischler, "A Commentary on Low-Cost Space Transportation," *Astronautics and Aeronautics* 7, August 1969: 50–64.
10. Ibid.
11. Roland, *A Spacefaring People.*
12. Ibid.
13. Logsdon, "The Space Shuttle Program."
14. "Space Shuttle Engine RFP," NASA News Release No. 70-26, February 18, 1970.
15. Ibid.
16. "Aerojet/NR/UAC Bid for Shuttle Main Engine," *Space Business Daily* 49 (20), March 27, 1970: 129.
17. "Space Shuttle Engine Negotiations," NASA News Release No. 71-131, July 13, 1971.
18. "NR's Rocketdyne Picked for Shuttle Main Engine," *Space Business Daily* 57(8), July 14, 1971: 52.
19. "Shuttle Engine Deliveries to Begin in 1975," *Space Business Daily* 57(9), July 15, 1971: 60.
20. U.S. Congress, House of Representatives, *Extensions of Remarks,* Hon. Louis Frey Jr., *Congressional Record,* E9211, August 5, 1971.
21. George C. Marshall Space Flight Center, *Chronology: MSFC Space Shuttle Program Development, Assembly, and Testing Major Events* (Huntsville, AL, 1988), p. 14.
22. "GAO Asked to Investigate Space Shuttle Engine Contract," *Space Business Daily* 57(16), July 26, 1971: 106.
23. James C. Fletcher, letter to Hon. Edward J. Gurney, July 30, 1971, NASA History Office.
24. "Space Shuttle Struggle: NASA Award to Rocketdyne Is Challenged," *Washington Post,* August 8, 1971, p. F1.
25. "P & W Files Formal Protest on Shuttle Engine Award," *Space Business Daily* 57(34), August 19, 1971: 230.
26. Ibid., p. 231.
27. Lyons, "Propriety of Space Shuttle Contract Is Questioned," *New York Times,* August 22, 1971, sect. 5, p. 24.
28. Ibid.
29. Philip Warden, "Ask Report on Space Shuttle Pact," *Chicago Tribune,* August 27, 1971.
30. "Limited Shuttle Engine Effort Set Pending GAO Review," *Space Business Daily* 58(1), September 9, 1971: 2.
31. Ibid., p. 1.
32. "Shuttle Engine Still on Schedule," *Aviation Week and Space Technology* 95, September 13, 1971: 21.
33. U.S. Congress, House of Representatives, *Extensions of Remarks,* "Rocketdyne's Space Shuttle Contract," Hon. Jerome R. Waldie, *Congressional Record,* E10689, October 12, 1971.
34. "Congressman Challenges Interim Shuttle Engine Contract," *Space Business Daily* 58, October, 26, 1971, p. 196.

35. "NR Challenges P & W Engine Experience Claim," *Space Daily* 59, November 3, 1971: 16–17.
36. Katherine Johnson, "Shuttle Engine Speeded by GAO Decision," *Aviation Week and Space Technology* 96, April 10, 1972: 14–15.
37. Ibid.
38. Jerry Thomson, "Shuttle: The Approach to Propulsion Technology," *Astronautics and Aeronautics* 9, February 1971: 64.
39. Malcolm McConnel, *Challenger: A Major Malfunction* (Garden City, NY: Doubleday, 1987), p. 42.
40. Taped interview in Management Operations Office, Frank Stewart, MSFC, *Oral Interviews: Space Shuttle History Project*, Transcripts Collection, December 1988, p. 64.
41. "Space Shuttle Main Engine – Interactive Design Challenges," NASA Contract Publication 2342, pt. 2, 1985, p. 600.
42. Adelbart O. Tischler, "Space Shuttle," *Astronautics and Aeronautics* 9, February 1971: 24.
43. In a private communication, he stated: "Let me say immediately that such an engine, although advanced over the J-2 and F-1 engines (for which I wrote the original specifications) was *not* beyond the state-of-the-art in 1971. In many ways it is conservatively specified. For example, the turbine inlet temperatures are more than 1000F below those of commercial aircraft engines at rated power. And aircraft engines run for 20,000 hours, not all of that at rated power, of course, but in marked contrast to the 2,000-second life of the SSME turbines. The beyond-the-state-of-the-art story was fabricated by those who were assigned to manage it for obvious reasons, and it served Rocketdyne's purpose of extracting over eight billion dollars (thus far) from the Treasury."
44. Greg Arnold and Clyde Jones, "Welding the Space Shuttle Engine," *Robotics Today*, October 1986: 13–16.
45. As reported in Richard S. Lewis, *The Voyages of Columbia: The First True Spaceship* (New York: Columbia University Press, 1984), p. 60.
46. John P. McCarty and Joseph A. Lombardo, "Chemical Propulsion: The Old and the New Challengers," *AIAA Student Journal* 11, December 1973: 24–5.
47. Ibid., pp. 66–7; also see M. H. Taniguchi, *Failure Control Techniques for the SSME, NAS836305, Phase I Final Report*, Rockwell International, Canoga Park, CA, RI/RD86-165 (revised), undated.
48. R. C. Glover, B. A. Kelley, and A. E. Tischer, *SSME Failure Data Review, Diagnostic Survey and SSME Studies and Analyses of the Space Shuttle Main Engine: Diagnostic Evaluation*, BCD-SSME-TR-86-1, Contract No. NASA 3737 (Columbus, OH: Battelle, December 15, 1986), pp. 13–6.
49. McCarty and Lombardo, "Chemical Propulsion: The Old and the New Challengers," p. 25.
50. Sylvia Fries, "NASA: Safety Organization and Procedures in Manned Space Flight Programs," unpublished manuscript, NASA History Office, Washington, D.C.
51. See, for example, Roland, "Priorities in Space for the USA"; Roland, *A Spacefaring People: Perspectives on Early Spaceflight;* Roland, "The Shuttle: Triumph or Turkey?"; Mueller, "The New Future for Manned Spacecraft Development"; Brown, taped interview, pp. 240–1; Logsdon, "The Space Shuttle Program."
52. Robert W. Smith, *The Space Telescope: A Study of NASA, Science, Technology and Politics* (Cambridge: Cambridge University Press, 1989).

53. Ibid., p. 288.
54. Ibid., pp. 288–9.
55. Ibid.
56. Ibid.
57. Johnson, "The Natural History of the Space Shuttle," pp. 417–24.
58. Howard E. McCurdy, *Inside NASA: High Technology and Organizational Change* (Baltimore: Johns Hopkins University Press, 1993).
59. McCurdy, *Inside NASA*, p. 159.
60. Ibid., p. 172.
61. Ibid., p. 170.
62. Dennis F. Thompson, *Political Ethics and Public Office* (Cambridge, MA: Harvard University Press, 1987).
63. Ibid., p. 4.
64. Ibid., p. 5.
65. Ibid.
66. Ibid., p. 40.
67. Ibid., p. 41.
68. David Bella, as quoted in Michael Pritchard, "Beyond Disaster Ethics," *Centennial Review* 34(2), 1990: 295–318.
69. Robert S. Ryan, "The Role of Failure/Problems in Engineering: A Commentary on Failures Experienced – Lessons Learned," NASA Technical Paper 3213, George C. Marshall Flight Center, Huntsville, AL, March 1992.

6

The Space Shuttle Main Engine: An Overview and Analysis

Early in the space shuttle's planning, NASA engineers realized that its engines would require "major advancements in propulsion technology beyond those applied in Apollo."[1] In particular, lightweight, high-performance, reusable liquid oxygen and hydrogen engines with sophisticated propellant-feed systems would have to be developed. Further, the engine would be the long-lead-time element in the shuttle's developmental time frame.

The main engine would be the first reusable, computer-controlled liquid hydrogen–liquid oxygen rocket engine of the 500,000-pound-thrust class. Clustered together in the back end of the orbiter and supplied with propellants from the external fuel tank, the three main engines would have to operate for approximately 8.5 minutes, the first two of which would be in parallel with solid rocket boosters, in order to launch the shuttle into orbit. According to the design, any contained engine failure would be safely overcome by employing one of several abort modes that possibly could require the remaining engines to operate for up to fourteen minutes.[2]

In recognition of the need for an innovative, complex engine, prior to the final approval of the shuttle project, three preliminary design contracts (phases A and B) for the space shuttle main engine (SSME) were let to Aerojet-General, Pratt & Whitney Aircraft, and the Rocketdyne Division of the then North American Rockwell Company.

Rocketdyne's proposal was based on its experience in developing the J-2 engine used for the upper two stages of the *Saturn V* rocket. As part of its phase B contract, Rocketdyne tested a full-scale model of its proposed engine with all elements except for the turbopumps. Pratt & Whitney's design was based on an experimental, reusable rocket engine, the XLR 129, which it was developing for the Air Force. Aerojet's proposal utilized development work on a NASA–Atomic Energy Commission nuclear engine, particularly in the area of pump technology.[3]

NASA had initially planned to select two of these three contractors to proceed into phase C (development). Frank Stewart, SSME deputy manager of the Shuttle Project Office, MSFC, explained the rationale for this approach:

> We had $25 million in the first year of Phase C for our concept to keep two contractors on board. In view of the uncertainties and the fact that contractors like to propose more than they can handle – a "promise you anything kind of thing," to some extent. They are optimistic, let me say, in their promises. This "fly before you buy" concept promoted by David Packard, former Secretary of Defense, looked pretty attractive to us, because we knew there was some unknown ground to plow. We had structured it so we'd go up to a few engine level test firings with two contractors of the three. And then we make a final selection on which one is going to do the SSME.[4]

In spite of this apparently sound reasoning, NASA reconsidered its position and chose to make only one award. Foreshadowing a type of decision making that would become a recurrent theme in the development, testing, and manufacturing of the shuttle, Stewart recalled the circumstance surrounding this decision:

> Just before we got to that point [of selecting two contractors] . . . headquarters decided they needed approximately $25 million. I made a plea to keep it, to no avail. They took the $25 million away and we had one contractor to select Phase C. In view of what turned out with the engine, I think in the long run NASA would have been ahead if we could have kept the two contractors in for a competitive bid.[5]

During the development of the Apollo, Congress gratuitously gave NASA an additional $4 million for advanced propulsion work! A. O. Tischler clarified that this allowed the agency to look at a number of new liquid fuel possibilities, including all the amines and select hydrocarbons. It also insured that designers could spend sufficient effort on materials technology and cooling techniques.[6]

The original shuttle design called for a piloted, reusable first-stage booster with twelve engines, and an orbiter with two additional engines of similar design. NASA planned for the engines to throttle from 50 to 109 percent of the rated power level, with the above-100-percent operation reserved for abort modes. Just eleven months after the three competing contractors had submitted their phase B design studies, the same political compromises that insured continuation of the space program also severely restrained the design of the shuttle (see Chapter 5). These resulted in a configuration of three engines mounted at the base of the orbiter, and assisted at launch by two solid-fuel boosters. The engines' liquid oxygen and liquid hydrogen fuel would be carried in an expendable, external tank. Minimum throttling requirements were increased to 65 percent from 50 percent for dynamic pressure and acceleration control. The original 100-mission-life SSME design was reduced to 55 missions, after the performance requirements were increased, under the assumption that the lifetime of the engines could be met with the same basic design. In hindsight, this assumption did not hold.[7] In order for the three-engine combination to carry a 65,000-pound cargo, it would have to operate occasionally at 109 percent rated thrust. As the shuttle's development proceeded and the design weight of the vehicle increased, it became apparent that the engines would have to operate normally at 109 percent. Thus, the margin of error originally incorporated into the design was used up. The engines have yet to deliver the desired 109 percent of capacity, and consequently the heaviest payload that has been launched has been 47,000 pounds.[8]

These decisions were made in light of Congress's allocation and the resultant need to reduce the development cost of the project. As noted in the previous chapter, a consequence of reduced developmental costs was the greatly increased operating expenditures. While this was recognized early on by A. O. Tischler,[9] it was officially ignored as the push to begin the project took on urgency.

After NASA awarded $500 million phase C/D contracts to design and build the engine to Rocketdyne, it then had to contend with a formal protest lodged by Pratt & Whitney based on several issues. According to Pratt & Whitney, Rocketdyne proposed to use materials and techniques not permitted under the terms of the NASA proposal request. Specifically, by welding rather than bolting, Rocketdyne was compromising reusability according to the losing bidder. Further, Rocketdyne engineers had proposed using INCO-718, which Pratt & Whitney charged was unacceptable because that particular metal is subject to hydrogen embrittlement. Pratt &

Whitney also charged that NASA based its decision on "design analysis" rather than on Pratt & Whitney's test results and experience (with the reusable XLR 129 engine). This, it claimed, increased the risk of cost overruns. Pratt & Whitney also claimed that Rocketdyne was permitted to finalize part of its technical effort with funds provided under a NASA Saturn launch support contract.[10] Following an eight-month delay to investigate these charges, the Government Accounting Office upheld the award to Rocketdyne. During the delay period, Rocketdyne was awarded a series of interim contracts totaling $8 million in order to begin work on the engine.[11] As the case studies show, Pratt & Whitney had a solid basis for its charges as problems arose due to both hydrogen embrittlement and the large number of welds.

Officially NASA (and Rocketdyne) viewed the SSME as an advanced version of the J-2 engine. That claim is part of what has since been called "the selling of the shuttle." It enabled NASA to justify its "success-oriented" approach, assuming that the previous experience of both contractor and agency would result in an economical engine to build with a minimum of separate component testing.[12] In reality, the SSME called for a new concept in rocket engines, "line-replaceable units" (LRU). For the first time a rocket engine was to be built with components that could be replaced either in the factory or in the field. A decade later, key NASA engineers would refer to the "SSME as the greatest challenge ever imposed on rocket-design engineers."[13] Alex Roland is less charitable:

> Development of a new engine was a curious risk for NASA and it was probably taken mainly to give the Marshall Space Flight Center something to do. NASA compounded the risk by betting that its new engine would deliver 109 percent of rated thrust – in a bargain-basement development program this gamble never had a chance. The engine was simply too advanced to work full capacity the first time around.[14]

How a Liquid-Fuel Rocket Engine Works: A Short Overview of the Fluid Mechanics[15]

In order to better understand some of the design issues that the engineers faced, we have provided an overview of how a liquid-fuel rocket engine operates. Although some aspects are quite complex, the liquid-fuel rocket engine is one of the easier modern-day power plants to understand. The

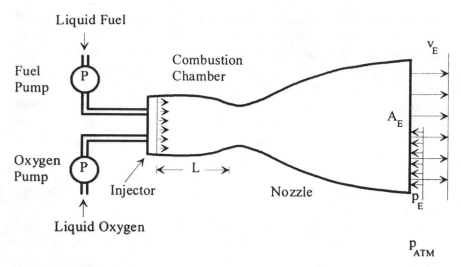

Figure 6.1 Simple liquid–fuel rocket engine.

nonmathematically inclined reader may wish to skim through this section and move on to the next. Those who can follow the mathematics are encouraged to do so, for it will add to their understanding of the dilemmas the engineers faced.

In order to create thrust, fuel (hydrogen in the case of the SSME) and oxygen are injected into a combustion chamber and burned. The enormous expansion of the newly combined gases creates high pressure. This high pressure accelerates the gases into and then out of the rocket nozzle. As shown in Figure 6.1, the exit area of the nozzle is A_E; gases exit the nozzle at a high velocity, V_E.

It is possible that pressure in the exit area, p_E, is different from the atmospheric pressure, p_{ATM}. In this case, the thrust, T, of the rocket engine is:

$$T = \dot{m}V_E + \left(p_E - p_{ATM}\right)A_E$$

where m is total mass flow rate of the gases from the engine.

Combustion takes place in the combustion chamber, which is of length L. In this length, droplets of fuel and oxygen must be mixed, vaporized, and finally burned. Although there is some underlying theory which can be applied in order to determine the length L, for the most part experimental methods are used to find L.

A schematic framework is given in Figure 6.2 which shows only one pump for each fluid. Modern liquid rocket engines have two pumps in series for the oxygen and two in series for the hydrogen (the fuel). This is necessary in order to avoid cavitation and obtain maximum pressure rise for minimum input (i.e., to operate at peak efficiency). The low-pressure pumps for both the liquid oxygen and liquid hydrogen are axial flow pumps and run at rotational speeds that are much less than the two high-pressure centrifugal pumps. The low-pressure pumps are driven by bleeding off the fluids from their respective high-pressure pumps and running them through turbines.

The high-pressure pumps are driven by axial gas turbines. The gases that drive the turbines come from separate combustors that burn the oxygen and hydrogen bled off the high-pressure pumps. Although much detail has been left out, it is important to mention that the liquid hydrogen is first run through tubes surrounding the main engine nozzle and combustion chamber before being burned. This substantially helps these parts to withstand the high combustion temperatures.

Now consider the flow of gases from the combustion chamber to the nozzle exit area. Assume that ρ (gas density), A (area), and V (velocity) vary only in the x-direction (not in the r or radial direction) as shown in Figure 6.3.

The conservation of mass for steady flow is $\rho AV = \dot{m}$ where $\dot{m}$ is a constant value. Taking the derivative,

$$\frac{d}{dx}(\rho A V) = 0$$

or

$$\frac{1}{V}\frac{dV}{dx} + \frac{1}{A}\frac{dA}{dX} + \frac{1}{\rho}\frac{d\rho}{dx} = 0.$$

The momentum equation (from $F = ma$) is:

$$-\frac{dp}{dx} = \rho V \frac{dV}{dx}$$

where p is the gas pressure at the same point x as ρ, A, and V.

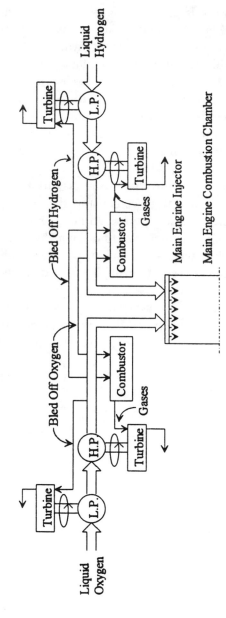

Figure 6.2 Schematic overview of liquid–fuel rocket engine.

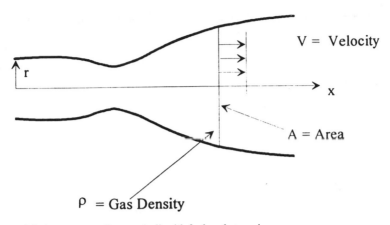

Figure 6.3 Area versus distance in liquid-fuel rocket engine.

The speed of sound is

$$c = \sqrt{\frac{dp}{d\rho}}; \quad \text{thus} \quad -\frac{dp}{dx} = -\frac{dp}{d\rho}\frac{d\rho}{dx} = -c^2\frac{d\rho}{dx}.$$

The momentum equation now becomes

$$-c^2\frac{d\rho}{dx} = \rho V \frac{dV}{dx}$$

or

$$-\frac{1}{\rho}\frac{d\rho}{dx} = \frac{V}{c^2}\frac{dV}{dx}.$$

Eliminate $\dfrac{1}{\rho}\dfrac{d\rho}{dx}$ from the conservation of mass using this last equation or

$$\frac{1}{V}\frac{dV}{dx} + \frac{1}{A}\frac{dA}{dx} = \frac{V}{c^2}\frac{dV}{dx}.$$

Since $M = $ mach number $= \dfrac{V}{c}$ (speed of sound for given pressure and density); therefore:

$$\frac{1}{A}\frac{dA}{dx} = \left(M^2 - 1\right)\frac{1}{V}\frac{dV}{dx}$$

or

$$\frac{dV}{dx} = \frac{V}{\left(M^2 - 1\right)} \cdot \frac{1}{A} \cdot \frac{dA}{dx}.$$

A plot of A versus x for the rocket engine is shown in Figure 6.4.

We now need to consider the relationship between velocity and distance for subsonic and supersonic flows in both the converging (before the throat) and diverging (after the throat) areas of the engine. For subsonic flow ($M < 1$) and within the converging area of the nozzle (i.e., $\dfrac{dA}{dx}$ is a negative), the preceding equation then implies that $\dfrac{dV}{dx}$ will be a positive value. That is, V (velocity) will increase with increasing values of x – that is, as the gases are forced toward the nozzle. Since A (area) is always positive and V is in the positive x direction, for this situation V will always be positive. In contrast, for the same for subsonic flow ($M < 1$) and

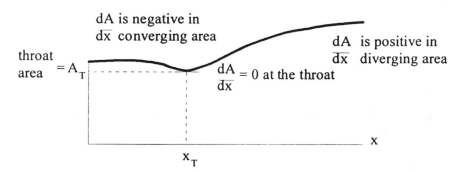

Figure 6.4 Throat area versus distance; note converging and then diverging areas.

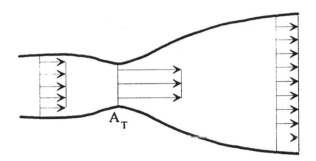

Figure 6.5 Velocities for subsonic rocket engine flow.

within the diverging area of the nozzle $\left(\text{i.e., } \dfrac{dA}{dx} \text{ is positive} \right)$, it implies

that $\dfrac{dV}{dx}$ will be negative, which means that V will then be decreasing with increasing values of x. (See Figure 6.5.)

In a similar manner, for supersonic flow (M > 1) within the converging area of nozzle implies that $\dfrac{dV}{dx}$ is negative. That is, V will decrease with increasing x. Again, in contrast, for M > 1 within the diverging area, $\dfrac{dV}{dx}$ will be positive, which means that V will increase with increasing x. (See Figure 6.6.)

For sonic velocity (M = 1), the previous equation implies that $\dfrac{dV}{dx}$ will be infinite, which means that an infinite change in V will result from an infinitesimal change in x. Clearly this would require an infinite force that cannot exist. The only explanation is that $\dfrac{dV}{dx}$ must be zero whenever M = 1, which means M = 1 can only exist at a throat as shown in Figure 6.4. Note that the inverse argument does not hold; that is, if there is a throat, the equation does not imply that M must be one.

In an actual rocket engine, the velocity is very small at the injector. Thus the flow is subsonic in the combustion chamber. To accelerate this flow, a

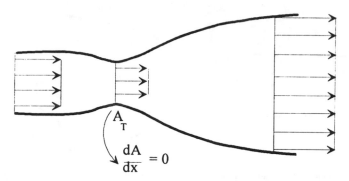

Figure 6.6 Velocities for supersonic flow at all points in a rocket engine.

converging area is needed. As soon as velocity equals sonic speed ($M = 1$), there must be a throat. If the velocity is to continue to increase, then the increasing flow must be supersonic, and a diverging area is required to maintain this acceleration. This explains the shape of the rocket engine.

In order to create this desired flow, an exact pressure in the combustion chamber must be achieved. This pressure is called the combustion design pressure, $p_{c_{DESIGN}}$. To understand what happens if p_c is not equal to $p_{c_{DESIGN}}$, consider the following situations where p_c is increasing from a low value.

For p_c very low, the flow is subsonic everywhere, as illustrated in Figure 6.7. As p_c is increased (by burning more fuel), $M = 1$ occurs at the throat, as shown in Figure 6.8.

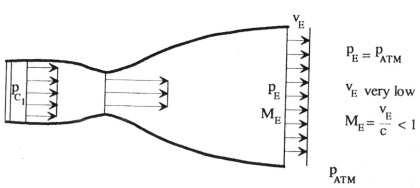

Figure 6.7 Low combustion pressure.

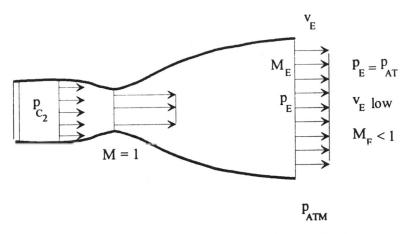

Figure 6.8 Increased combustion pressure with sonic velocity reached at throat.

Increasing p_c above p_{c_2} causes a normal shock wave to move down the diverging portion of the nozzle. A fixed p_{c_3} causes the shock wave to stand at a fixed location in the nozzle, which is illustrated in Figure 6.9.

As p_c becomes larger, the normal shock will be blown out of the nozzle exit plane and oblique shocks will form external to the nozzle. This situation is called overexpansion as p_E (exit pressure) is less than p_{ATM} (atmospheric pressure). This condition is shown in Figure 6.10.

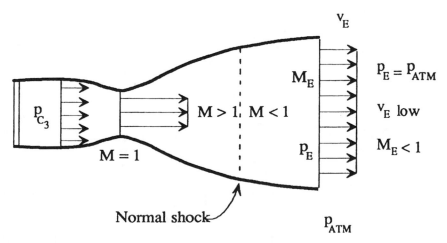

Figure 6.9 Standing shock wave in diverging section of nozzle.

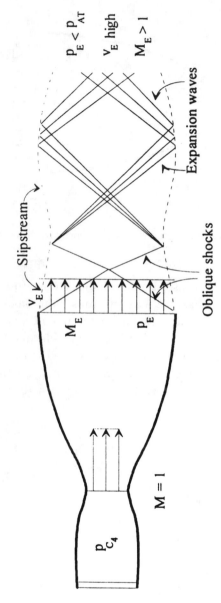

$p_E < p_{AT}$

v_E high

$M_E > 1$

Expansion waves

Slipstream

Oblique shocks

v_E

M_E

p_E

$M = 1$

p_{C_4}

Figure 6.10 Oblique shocks external to nozzle.

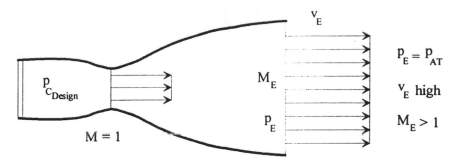

Figure 6.11 Design combustion chamber pressure.

Finally as p_c is increased further, the external shocks will disappear and $p_E = p_{ATM}$. This is the correct (or design) combustion chamber pressure, $p_{c_{DESIGN}}$. It gives the same V_E (exit velocity) as p_{c_4}, but with no thrust penalty due to the pressure differential – that is, $(p_E - p_{ATM})A_E$ The flow in the nozzle is completely supersonic. (See Figure 6.11.)

If p_c is increased above $p_{c_{DESIGN}}$, then external expansion and compression waves appear external to the nozzle. This condition is called underexpansion as p_E is greater than p_{ATM}. In this case, the $(p_E - p_{ATM})A_E$ term now adds to the thrust. This is shown in Figure 6.12.

For a fixed p_c these different flow patterns as described in the previous figures can also be achieved as p_{ATM} is decreased. This condition actually occurs as the rocket ascends in the atmosphere.

The Working of the Space Shuttle Main Engine

The constraints of cost and reusability drove the parameters within which design criteria for the SSME were required to operate. The resulting parameters represented unprecedented regimes of pressures, temperatures, and rotating machinery speeds.[16] System pressures as high as 7,900 pounds per square inch (psi) would be a factor of three greater than those of earlier rocket engines. The pressure level and allowable system weight resulted in turbomachinery with an unprecedented horsepower-to-weight ratio as high as 100 horsepower per pound. Further, since 1 second of specific impulse (thrust multiplied by burn time per total mass of fuel) is equivalent to 1,000 pounds of payload, the early planners calculated that each 1 second increase in specific impulse would be worth $162,000 of payload per flight.[17] The

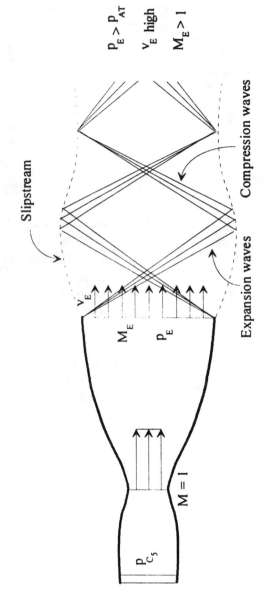

Figure 6.12 Underexpansion condition; exit pressure exceeds atmospheric pressure.

weight constraints also resulted in designs that relied on welded instead of bolted joints, forgings instead of castings to produce complex parts such as manifolds and volutes, and the use of high-strength materials in the structure of the engine. Unfortunately, as Pratt & Whitney claimed, certain materials chosen were sensitive to hydrogen and required the use of protective treatment like coatings, gold plating, or weld overlays. This led to a complex engine design which is difficult to manufacture and maintain.[18]

The design criteria called for a maximum specific impulse of 455 seconds, a significant increase over the J-2's 426 seconds. The engine had to produce 375,000 pounds of thrust at sea level and, as noted, be throttleable between 65 to 109 percent of rated power. To achieve the desired specific impulse, the velocity by which the combustion gases (steam) are expelled had to be increased, thus increasing the engine's thrust. This was accomplished by significantly increasing the engine's combustion chamber pressure to over 3,000 psi compared with 780 psi for the J-2 engine and maximizing, within a tight envelope, the expansion ratio of the exhaust nozzle exit to its throat area. Table 6.1 presents a comparison among the SSME, J-2, and the experimental XLR 129.

Raising the pressure that amount was not a simple matter. A two-stage combustion cycle that utilized separate sets of pumps for the liquid oxygen and hydrogen fuel was designed. Each set consisted of low- and high-pressure units. (See Figure 6.13 for a schematic overview and Figure 6.14 for an illustration with the major components.) During the first cycle the propellants are partially burned in preburners, producing hydrogen-rich gas to power the high-pressure turbopumps. In the second cycle, this fuel-rich steam is routed through the hot-gas manifold to the main injector where it is injected, along with additional oxidizer and fuel into the main combustion chamber at high pressure and a 6:1 (oxygen to hydrogen) mixture ratio.

Specifically, the low-pressure pumps located at the propellant inlets raise the pressure to several hundred pounds per square inch, providing the proper head for the two high-pressure pumps. Fuel (hydrogen) at the high-pressure fuel pump (HPFP) discharge is also used to cool the main combustion chamber (MCC), nozzle, and other hot-gas flow path areas of the engine. In turn, the chamber coolant discharge is utilized to drive the low-pressure fuel turbine (LPFT). The nozzle coolant discharge is mixed with a bypass flow and then fed to the two preburners. Oxygen is routed from the high-pressure oxygen pump (HPOP) to drive the low-pressure oxidizer turbine (LPOT) and feed the main injector. It is also fed to a preburner boost pump (PBP), which is utilized to provide oxidizer to the two preburners.

Table 6.1. *Engine comparison chart*

	J-2	XLR-129	SSME
Company	Rocketdyne	Pratt & Whitney	Rocketdyne
Vehicle	Saturn V	Experimental	Space shuttle
Type	Expendable liquid	Reusable liquid	Reusable liquid
Time period	1960–73	1966–70	1970–present
Weight	2,000 lbs.		6,700 lbs.
Size	10 ft. long, 6.5 ft. wide		14 ft. long, 7.5 ft. diameter
Thrust	200,000 lbs.	250,000 lbs.	375,000 lbs.
Specific impulse	426 sec.	382 sec., 446 sec.	363.2 sec., 455.2 sec.
Combustion pressure	780 psi	2,740 psi	3,126 psi
Mixture ratio	5.5:1	5.7:1	6:1
Maximum pump rotation speed	25,000 rpm	48,000 rpm	35,000 rpm
Turbopumps	2 (1 Ox/1 H_2) two-stage pumps	Low-speed inducers; high-speed turbo (1 Ox/1 H_2 each)	4 pumps (2 Ox high/low) (2 H_2 high/low)
Materials	Stainless steel (thrust chamber tubes)	Nickel base alloy (primary material)	Nickel base superalloy (primary material)
		Pure copper (main combustion chamber liner); titanium alloy (fuel pump impeller)	Copper alloy (main combustion chamber liner); nickel base alloy (blade material)
Connection	Welding/seals	Bolts	Welding

Source: Data from David A. Atherton, "J-2, M-1 Engine Design Details Reported," *Aviation Week and Space Technology* 7, May 6, 1963: 56–9; Walter Dankhoff, Paul Herr, and Melvin C. McIlwain, "Space Shuttle Main Engine (SSME) – The Maturing Process," *Astronautics and Aeronautics*, January 1983: 6–32, 49; and *Space and Shuttle News Reference*, NASA (Washington, DC: U.S. Government Printing Office, 1981), pp.2-4–15.

Each of the preburners provides fuel-rich combustion gases to their respective high-pressure turbines. At the turbine discharges, the gas is then routed through the hot-gas manifold (HGM) and fed to the main injector. The final

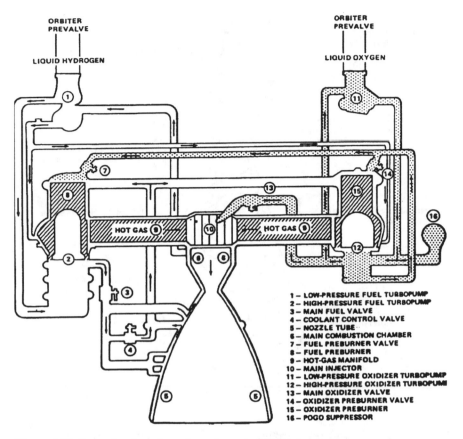

ORBITER PREVALVE

LIQUID HYDROGEN

ORBITER PREVALVE

LIQUID OXYGEN

HOT GAS

HOT GAS

1 – LOW-PRESSURE FUEL TURBOPUMP
2 – HIGH-PRESSURE FUEL TURBOPUMP
3 – MAIN FUEL VALVE
4 – COOLANT CONTROL VALVE
5 – NOZZLE TUBE
6 – MAIN COMBUSTION CHAMBER
7 – FUEL PREBURNER VALVE
8 – FUEL PREBURNER
9 – HOT-GAS MANIFOLD
10 – MAIN INJECTOR
11 – LOW-PRESSURE OXIDIZER TURBOPUMP
12 – HIGH-PRESSURE OXIDIZER TURBOPUMP
13 – MAIN OXIDIZER VALVE
14 – OXIDIZER PREBURNER VALVE
15 – OXIDIZER PREBURNER
16 – POGO SUPPRESSOR

Figure 6.13 Space shuttle main engine schematic (NASA drawing).

combustion occurs in the MCC and the resulting gases are expanded through the supersonic nozzle. A heat exchanger located on the oxidizer side of the hot-gas manifold converts liquid oxygen to gaseous oxygen for the orbiter oxygen tank and the pogo suppresser system. The latter's purpose is to absorb vibrations and oscillations caused by the engine structure and the combustion cycle.[19]

The high-pressure oxidizer turbopump (HPOTP) actually consists of two centrifugal pumps (HPOP and the PBP) on a common shaft driven directly by the two-stage, hot-gas turbine. The shaft is to spin at 29,194 revolutions per minute (rpm) at 109 percent rated power. The high-pressure fuel turbopump (HPFTP) is a three-stage, centrifugal flow pump, directly driven by a two-stage, hot-gas turbine. As described, it provides fuel for cooling the main combustion chamber, nozzle, and hot-gas manifold, driving the

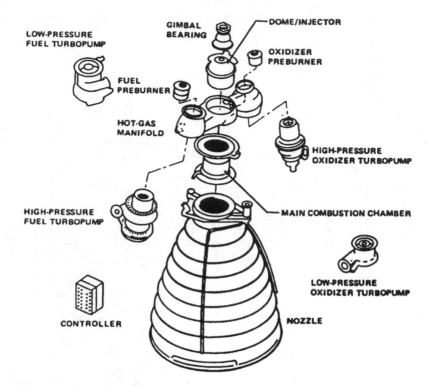

Figure 6.14 The space shuttle main engine's major components.

LPFPT turbine, and pressurizing the vehicle fuel tank. At 109 percent of rated power, it will spin at 36,595 rpm.

The fluid flow for the engine operation is controlled by five main valves: main oxidizer valve (oxidizer flow to the main injector and the main chamber augmented spark igniter); main fuel valve (fuel flow to the thrust chamber coolant circuits, low-pressure fuel turbopump turbine, hot-gas manifold coolant circuit, both preburners, and the three augmented spark igniters); oxidizer preburner oxidizer valve (oxidizer flow to the oxidizer preburner and preburner augmented spark igniter responsible for controlling engine thrust level); fuel preburner oxidizer valve (oxidizer flow to fuel preburner and fuel preburner augmented spark igniter, which maintains desired mixture ratio); and chamber coolant valve (controls fuel flow to the main combustion chamber and the nozzle coolant circuits).

An electronic controller operating in conjunction with the engine's sensors, valves, actuators, and spark igniters provides a self-contained system for control, checkout, and monitoring. Control of thrust and mixture ratio are updated fifty times per second. The control system is redundant; normal operation will continue after the first failure with fail-safe shutdown after the second failure of any control system component.

As noted, welding was extensively used during the final assembly of the engine in order to further reduce weight. This required a special computer controlled electron-beam welding system and joint designs to detect defects better. To build the main combustion chamber, a special electroforming process on a scale never before attempted was required. The materials used for certain sections of the turbine could not handle the high thermal transients and high temperatures, requiring a change in the engine valve sequencing and the use of more hydrogen for cooling.[20] Throughout the development and testing phase, materials problems resulted in cracked turbine blades, cracked sheet metal in turbine sections, and limited life of the turbopump bearings.

In short, the planned missions for the shuttle dictated a unique and technology-extending rocket engine, rather than the use of "off-the-shelf" components or a straightforward extension of technology. Ryan, who analyzed the structural design of the shuttle following the *Challenger* accident, noted that the high specific impulse requirements, in conjunction with a 55-mission life (scaled down from the extremely optimistic 100) and the volume and the weight constraints, combined to produce unique, demanding structural design, manufacturing, and verification requirements. As noted earlier, the two-stage combustion system resulted in unprecedented operating temperatures, pressures, and rotating machinery speeds. In turn, these high rotary speeds and the combustion processes created mechanical, acoustical, and fluctuating pressure environments.[21]

Ryan adds that the rotary dynamics problems were "compounded by several orders of magnitude in the high-pressure fuel and oxidizer pumps due to the high-energy concentration (energy density) and speed ranges. As a result, supposedly small changes (causes) are greatly magnified in the responses." For example, the bearing life of these pumps is very limited and is to a large degree a function of the dynamic response.[22]

NASA's Testing Philosophy: Is the Whole Greater Than the Sum of Its Parts?

The engine development program had a history marked with problems, not unlike other rocket engine development programs. According to the development contract, Rocketdyne was to employ the design verification specification approach, which required tests to be performed at the lowest possible assembly level (e.g., component, subsystem, system) in order to demonstrate that the design specifications were met. In late 1973 and early 1974 numerous problems such as component facility construction problems, weight-driven design changes, and changes to achieve needed structural strength delayed component fabrication and testing. For example, engine-level testing of the high-pressure turbopumps began only three months before component-level tests commenced. Subsequently, a number of component test facility failures occurred, which, coupled with a very high rate of hardware attrition in the test program, led to the decision to cancel the comprehensive turbomachinery-level test program. The plan to conduct turbopump component-level development tests was abandoned because of difficulty in manufacturing components on schedule and major failures of component test facilities.[23]

Among the requirements placed on engine flight components by NASA were that they be test-fired and that there would be an equivalent component within the ground test program that had at least twice the firing time as the highest expected flight firing time. The ground test program as designed required six types of engine tests:

- Development tests to solve problems and investigate improvements.
- Certification tests to expose the "all-up" improved configuration to a simulated flight series of firings in order to qualify design improvements for flight.
- "Green run" tests to expose new or newly overhauled components to a first firing.
- Acceptance tests to verify that new hardware meets requirements for admission to fleet.
- Main propulsion test article (MPTA) to provide cluster firings in order to qualify the three-engine cluster and external tank as a system.
- Flight readiness firings (FRF) to qualify new vehicles along with their main engines on launch pad.

Hot-fire testing of the main engine subsystems began in 1974, with the first engine test in 1975. Extensive testing at Rocketdyne's Santa Susana Field Laboratory in California and the National Space Technology Laboratories (NSTL) in Mississippi followed. Six development engines were used on two test stands in Mississippi and one in California.[24] During the period between

1977 and early 1986, the main engine had undergone over 1,200 static firings with accumulated total engine operating time in excess of 270,000 seconds.

In order to detect engine anomalies and prevent major damage, a multi-faceted detection-shutdown system had been designed by Rocketdyne. It consisted of sensors, automatic redline and other limit logic, redundant sensors and controller voting logic, conditional decision logic, and human monitoring. Approximately 300 to 500 measurements were sensed and recorded during each test; 100 of these were used for control and monitoring.[25]

The engine-level test program required fifty tests, eleven turbopump replacements, and over three months to develop acceptable start and shutdown sequences and reach 50 percent of the rated thrust level. After reaching that point, all efforts were directed toward meeting the specified 109 percent power level and fifty-five-flight system life. However, a large number of problems were encountered due to the combination of more severe internal conditions (compared with previous engine developments), manufacturing defects, operational errors, and the consequences of design assumptions that proved incorrect. The more significant of these problems included a high-pressure fuel turbopump subsynchronous whirl and turbine blade failures; the high-pressure oxidizer turbopump explosions caused by seal and bearing failures; and problems associated with the preburner and main combustion chamber injectors. The Williams Committee noted that "the elimination of many instruments and their ports and bosses to reduce weight led to difficulty in determining failure causes because data on internal engine conditions were not available. Also, in uncontained failures, the hardware was consumed by resulting fire, and conclusive evidence of cause could not be obtained."[26]

During the first three years of testing (1977 through 1979), the test success rate increased from 49 to 71 percent; for tests greater than 300 seconds in duration the success ratio went from 53 to 66 percent. Both the number of premature test cutoffs and new failure modes were considerably less than that experienced for the J-2 at similar points in its development program.[27] However, between 1977 and 1986, there were 40 tests with significant anomalies including 27 major failures. These resulted in program schedule delays ranging from three weeks to six months and engine and facility damage costs ranging between $1 million to $26 million per test. In addition the extensive failure analyses that resulted were also costly (see Table 6.2).[28] Four failures occurred during launches or launch attempts (pre-*Challenger*), all of which were relatively minor in nature.[29] Of particular concern were a series of turbopump failures and resultant damaged engines which occurred in 1977–8. Cikanek summarized the chronology of testing

Table 6.2. *SSME test incidents*

Incident damage	Date	Damage	Damage cost	Delay
Injector failure				
Test 901-173 (LOX post fractures, erosion–MCC)	3/31/78	12B, 13C, 14B	UA	UA
Test 901-331 (LOX post fractures, erosion–MCC)	7/15/81	7C, 12A, 13B, 14B	$4.1M	24 weeks
Test 750-148 (LOX post fractures, erosion–MCC)	9/2/81	12B, 13C, 14B	$7.0M	8 weeks
Test 901-183 (LOX post fractures, erosion–MCC)	6/5/78	12C, 13C, 14C	UA	UA
Test 902-198 (LOX post fractures, erosion–MCC)	7/23/80	12C, 13C, 14C	$1.0M	12 weeks
Test 901-307 (LOX post fractures, erosion–FPB)	1/28/81	2C, 3B	UA	UA
Test 902-244 (LOX post fractures, erosion–DPB)		See Test 901-317		
Test SF10-01 (FPB anomalies)	7/12/80	2C, 3C, 5C, 14C, 15C, 17C	$1.5M	16 weeks
STS-8 (localized: FPB damage, PWC failure)		No additional information		
Test 750-160 (fuel blockage: water left in FPB injector by EDM process)	2/12/82	2B, 3B, 7B, 11A, 12A, 13B, 14A	$15.0M	12 weeks
Control failure				
Test 901-284 (erroneous sensor, lee jet, PWC failure)	7/30/80	1C, 2C, 3C, 4C, 5C, 7A, 9B, 10A, 14B, 15C, 17B	$9.2M	16 weeks
Test 902-132 (main oxidizer valve misindexed)	10/3/78	2B, 7C, 11B, 12B	UA	UA
Duct, manifold, or heat exchanger failure				
Test 750-041 (steerhorn anomaly, fuel leak)	5/14/79	2B, 3B, 5C, 7B, 11B, 12B, 13C, 14B	$8.5M	8 weeks
Test SF6-03 (steerhorn anomaly, fuel leak)	11/4/79	1C, 2B, 7B, 11B, 12A, 13A, 14B	UA	6 weeks

Table 6.2. *SSME test incidents (Continued)*

Incident damage	Date	Damage	Damage cost	Delay
Test 750-259 (MCC outlet manifold neck, fuel leak)	3/27/85	3B, 4A, 5A, 6A, 7B, 8C, 9A, 10A, 12A, 13A, 14B, 15C, 17A	UA	UA
FRF-2 (MCC outlet manifold neck, fuel leak)		No additional information		
Test 901-485 (nozzle tube rupture, fuel leak)	7/24/85	14C	UA	UA
Test 750-175 (catastrophic structural: HCF in high-pressure oxidizer duct)	8/27/82	1C, 4B, 7B, 9B, 10A, 11C, 13B, 14B, 16C	$11.0M	8 weeks
Test 901-222 (heat exchanger, weld failure, HAL)	12/5/78	7B, 11B, 12B	UA	UA
Test 902-112 (fuel block: solidified nitrogen blockage of fuel pump inlet duct)	6/10/78	1B, 2B, 7B, 12C	UA	UA
Test 901-345 (localized: MCC cavity burst diaphragm leak)		No additional information		
Valve failure				
Test SF6-01 (main fuel valve: structural, fuel leak)	7/2/79	2C, 4B, 14C, 16C, 17C	$8.3M	14 weeks
Test 901-225 (main oxidizer valve: HAL)	12/27/78	1C, 4A, 5C, 6B, 7B, 9A, 10A, 11A, 12B, 13B, 14C, 15B, 16B, 17C	$10.0M	4–6 weeks
Test 750-168 (OPOV component: seal failure)	5/15/82	7B, 11B, 12B, 13C	$4.4M	8 weeks
High-pressure oxidizer turbopump failure				
Test 901-110 (rotor/seal support, HAL)	3/24/77	4B, 5C, 6C, 7A, 8C, 9B, 10A, 12C, 13C, 14C, 15B, 17B	$3.3M	4 weeks
Test 901-136 (rotor/seal support)	9/8/77	4A, 5C, 6C, 7A, 8C, 9B, 10A, 12C, 13C, 14C, 15B, 17B	$2.4M	4 weeks

Table 6.2. *SSME test incidents (Continued)*

Incident damage	Date	Damage	Damage cost	Delay
Test 902-120 (heat addition to LOX)	7/18/78	4B, 6B, 7B, 9C, 10A, 13C, 14C, 15C, 17B	$1.65M	5 weeks
High-pressure fuel turbopump failure				
Test 901-340 (turn around duct cracked/torn)	10/15/81	2A, 3C, 12C, 14C	UA	UA
Test 901-363 (turn around duct cracked/torn)	3/30/82	2C	UA	UA
Test 902-118 (turn around duct cracked/torn)	7/21/78	2B, 13B	UA	UA
Test 902-383 (localized: turn around duct cracked/torn)		No additional information		
Test 901-436 (coolant liner buckle)	2/14/84	2A, 5A (engine was totally gutted and retired)	UA	UA
Test 901-364 (hot-gas intrusion to rotor cooling)	4/7/82	2A, 5A (engine was totally gutted and retired)	$26.0M	8 weeks
Test 902-209 (hot-gas intrusion to rotor cooling)	11/16/80	2C, 3C	UA	UA
Test 750-165 (hot-gas intrusion to rotor cooling)	5/15/82	7B, 11B, 12B, 13C	$4.4M	8 weeks
Test 901-147 (power transfer failure, turbine blades)	12/1/77	NA	UA	UA
Test 902-249 (power transfer failure, turbine blades)	9/21/81	2A, 7B, 11A, 12A, 13A, 14A	$15.1M	3 weeks
Test 902-095 (power transfer failure, turbine blades)	11/17/77	2A, 12B, 13B	UA	UA
Test 901-346 (localized: turbine blades)	11/19/81	2B	UA	UA
Test 901-362 (power transfer failure)	3/27/82	2C, 7C, 12C, 13C	UA	UA
Test 901-410 (power transfer failure)	5/20/83	2B	UA	UA

Table 6.2. *SSME test incidents (Continued)*

Abbreviations/notations		Damage nomenclature key:	
		ID No.	*Component*
UA	Item unavailable	1	Low-pressure fuel turbopump
NA	Not applicable	2	High-pressure fuel turbopump
MCC	Main combustion chamber	2	Fuel preburner
OPOV	Oxidizer preburner oxidizer	3	Fuel side valves
	valve	4	Fuel side ducts
LOX	Liquid oxygen	5	Fuel side ducts
EDM	Electrical discharge machining	6	Low-pressure oxidizer turbopump
HCF	High cycle fatigue	7	High-pressure oxidizer turbopump
HAL	Heat addition to LOX		and heat exchanger
PWC	Pressure wall containment	8	Oxidizer preburner
		9	Oxidizer side valves
Suffix to I.D. no.		10	Oxidizer ducts and POGO
A	Component destroyed	11	Hot-gas manifold
B	Component heavily damaged	12	Main injector
C	Component lightly damaged	13	Main combustion chamber
		14	Nozzle
		15	Stand equipment
		16	Stand structure
		17	Controller

Source: Data from M.H. Taniguchi, "Failure Control Techniques for the SSME, NAS836305, Phase I Final Report," Rockwell International, Canoga Park, CA, RI/RD86-165 (revised), undated.

and failures between 1977 and 1986 as follows: 1977, 4 failures out of 115 tests; 1978, 6 of 144; 1979, 3 of 136; 1980, 3 of 80; 1981, 5 of 132; 1982, 4 of 128; 1983, 0 of 96; 1984, 1 of 29; 1985, 1 of 33; and 1986, 0 of 6.

Of the twenty-seven major failures, six were initiated by the high-pressure fuel turbopump; the main burner was responsible for five; the preburners and the high-pressure oxidizer turbopump were responsible for three each. The most common cause of these failures has been attributed to high-cycle fatigue, which had an initiating role in at least eight incidents. Manufacturing or maintenance process deficiencies were a factor in at least four, and design deficiencies were responsible for causing at least four of the failures and contributed to at least twenty of the twenty-five failures for which a formal board investigation report was available.

For example, the various structures within the SSME are very sensitive to flow and acoustic–structural interaction problems. In particular, the liquid-oxygen (LOX) inlet has a two-blade splitter in order to create a more uniform LOX flow distribution into the combustion chamber dome. Any type of vane which protrudes into this high-flow environment is susceptible to flow-induced oscillations and, thus, fatigue cracking, which, in fact, occurred. Based on the follow-up analysis and testing, certain cracking is believed to have occurred at the vane's natural frequency of 4,000 hertz (Hz). Its cause was eventually attributed to small manufacturing differences in the bluntness of the vane trailing edge and possible small vane angle–offset differences or slight modal shifts. To Ryan, "the message is clear, high performance systems are very sensitive to small changes, whether in manufacturing or environments, many times leading to failure or strict operational requirements."[30]

According to both Taniguchi and Cikanek, a review of the major failures indicated a number of instances in which more extensive analysis and testing were warranted. They observed that most of the engine failures were a result of insufficient knowledge of the environment and loads imposed on the components, particularly dynamic loads. Cikanek notes that design changes (11) and environment definition improvement (10) were recommended the most often following the failure analyses.[31] These recommendations have underlined the need during engine development to improve environment definition. They also suggest that, in certain cases, the severity of failures may have been reduced or eliminated through better monitoring.

Although the engine was given preliminary certification in March 1980, it did not achieve the desired 109 percent rating. Initially it was limited to rated power level (100 percent) and severely restricted reusability. However, three improvement programs followed, ultimately aimed at achieving the original objective of certification at full power level and mission life. After various design changes were incorporated, the engine achieved certification for operation at 104 percent of rated power level, although its operation was constrained by precautionary controls and restrictions such as special inspections and severe service life limitations on many parts. A substantial number of major design changes were proposed to improve safety margins and service life and achieve full-power-level (FPL) certification for normal operation. However, budget constraints limited the scope of the improvement program to achieving certification at FPL, reducing many inspection and maintenance requirements primarily imposed on the high-pressure turbo pumps and extending the engine service life to 5,000 seconds.[32]

Problems were not limited to testing; on the first fourteen missions between April 1981 and November 1984, there were delays for unexpected engine work on all but the maiden flight of *Columbia*. These delays required an average of 633 man-days for engine repairs. Seven of the delays involved the high-pressure fuel turbopump and one for the high-pressure oxidizer turbopump. Much of the damage was attributable to vibration and the high turbine-inlet temperatures, which caused cracking. For three other flights, at least one engine had to be replaced or removed for modification. Prior to the initial *Challenger* launch, all three engines were removed, repaired, or replaced due to leakage in the hydrogen lines.[33]

A major concern of ours has focused on the debate over "all-up" versus component testing. For the SSME, three levels of testing needed to be addressed: the individual components; a single SSME; and the complete system, which consists of the three SSME clusters. As amplified in Chapter 10, which reviews the workings of the "Covert Committee," NASA relied more heavily on computer models to reduce its need to evaluate new designs through testing and, consequently, to reduce the need for redesign work than was the case for all previous engines. Thus, "if or when malfunctions occurred during the testing of the operational engine, new hardware may need to be designed, constructed, and retrofitted, causing delays in the program."[34]

While the Covert Committee noted that the combustion chamber, nozzle, preburner, valves, and computer controller were tested as components and were performing relatively satisfactorily, significant difficulties were encountered in trying to test the high-pressure turbopumps as components. Using a facility constructed at Rocketdyne for this purpose, testing began in May 1975 and concluded in September 1977, having achieved less than three minutes of cumulative testing each on the oxygen and hydrogen (fuel) pump systems. This lack of progress was caused by the very high system pressures, which, in turn, required heavy hardware that was difficult to manage. As changes were made, long delays resulted.[35]

Rocketdyne and NASA finally concluded that component testing of the high-pressure turbopumps could not be carried out at a rate that would be useful in the development program. Therefore the decision was made to conduct the turbopump development tests by running the pumps as part of the total engine system. The Covert Committee found this decision to have great bearing on any examination of the problems in the turbopumps, since it precluded testing the pumps at conditions other than the normal operating conditions. The committee concluded that the lack of an adequate test rig to conduct component tests outside of the engine expeditiously appears to be the primary reason that the turbopump problems were not discovered before initial engine testing and the resultant four test failures due to the turbopumps. Further, an additional thirteen failures, five of which were considered to be major incidents, were to occur after the Covert report was submitted. The Covert Committee pointedly reported that no rocket engine using turbopumps had ever been produced without extensive development testing of the turbopumps first as components.[36]

Ryan has pointed out that "a part or component responds differently when it is separate than it does when it is a part of the whole. As a result, we must understand interaction and sensitivities to provide the proper design. Verification becomes even more complex in that it must account for these extremes, including failures and their redundant paths, etc."[37]

The Engines Post-*Challenger*

Following the *Challenger* Flight 51-L incident, a number of modifications were made to the engine, primarily improvements in the turbine blades, temperature sensors, and cooling system. Robert Marshall, Preliminary Design Office Director for MSFC, noted that:

> After 51-L [NASA] changed the design of blades on the turbine side, changed the damper configuration between blades on LOX pump; also changed the way we design or treat the blade. Made some changes in way we cool the bearings, some changes in the sheet metal which guides flow into the pumps, changes to some valve time configurations, how they open and close. Some 50 changes to engine. Also recertified the engines.[38]

Ryan, who had defended the all-up testing policy, nevertheless identified currency as a major problem with the SSME program. His analysis following the *Challenger* accident found that reliance on inadequate models, including erroneous assumptions concerning the operating environments, were prevalent. For example, although the SSME structural audit concluded that the engine had been adequately analyzed and verified, it also found that certain parts had more limited life than specified: indeed, in some cases, the required discrepancy approval requests (DARs) had not been filed. The cause for the gap in reporting varied from actual mistakes in assumptions to simple "oversight" on the part of individual engineers. To Ryan, "the final responsibility (on reporting) rests with the engineer, technician, etc., doing the job. He must not just do the work but communicate it."[39]

Following the *Challenger* accident, a reexamination of the safety and operating margins resulted in approximately forty additional changes. The power-level objective was formally changed to 104 percent with the 109 percent level reserved for the abort contingency. Full-power-level capability was demonstrated during the certification test program as well as by a short-duration run during each flight engine acceptance test. However, even though the engine accumulated some 90,000 seconds of test time, it still required frequent removal, disassembly, and overhaul of both high-pressure turbopumps (one to three flights). In addition, a large array of design approval requests remained for the entire engine.[40]

In somewhat of an understatement of the difficulties ahead for the shuttle fleet, Bob Marshall concluded that we "still have a couple of problems with pumps that we want to better understand before we commit them to two flights, or three flights or four flights."[41] The original objectives for the engine were 55 starts and 27,000-second run time. Two years after the *Challenger* accident, the objective of a 55 mission life without a major overhaul remained elusive; as the engines remain certified for only 10 to 15 missions.[42] Changes in specifications required 30 starts and 15,000-second run time for all components other than high-pressure turbopumps and the flexible ducts; the latter must be capable of 20 starts and 10,000 seconds. Bell and Esch noted that serious risks remained with the SSME, including welds that cannot be inspected; turbine blades subject to cracking, and main bearings not capable of handling torque if one blade is broken.[43] Indeed, from late April 1990 through the end of September, the entire shuttle fleet was grounded due to pervasive hydrogen leaks.

The Williams SSME Assessment Team found that several components still did not meet all contract requirements. "To ensure that operating margins exist, life limits, special inspections, and other criteria and limits are imposed."[44] The Assessment Team found that "the continuing occurrence of hardware problems indicates that the SSME is not as rugged as desired for such a machine. Also, the manpower consumed in executing all the precautions and controls certainly adds substantially to the recurring costs of using the engine." However, the team indicated that several improvements in varying states of development, including the single-tube heat exchanger, alternate high-pressure turbopump, large-throat main combustion chamber, and two-duct powerhead, if they achieve their objectives, will greatly enhance engine reliability and safety. Although each component has largely overcome the development problems encountered to date, because of the budgetary and other problems encountered over the years, improvements have been undertaken serially, and current plans for development and certification are not as efficient and coherent as they might be.[45]

Clearly, a major concern has been the high-pressure turbo pumps. In 1985, prior to the *Challenger* accident, "it was concluded that, within the constraints presented by the existing designs, no group of physically possible modifications could produce the more rugged, reproducible and reliable machines needed." Consequently, a decision was made to design and develop a new set of high-pressure turbopumps. These new pumps were not to be burdened with weight restrictions on the existing machines. The designs were to be responsive to lessons learned from the more than ten years' experiences with the current turbopumps.[46] The major design differences on the new pumps would be:

- Incorporate lessons learned with emphasis on increased safety margins.
- Increase performance and structural margins.
- Utilize 115 percent RPL (rated power level) as design point.
- Use single-crystal turbine blades.
- Eliminate welds and sheet metal by using advanced precision castings.
- Eliminate thermal and hydrogen environment coatings by use of improved materials.
- Reduce number of rotating parts by 50 percent.
- Provide safe shutdown in event of turbine blade failure.
- Design for inspectability, producibility, and operability.

Pratt & Whitney (which, as noted, had protested the 1971 contract award to Rocketdyne) informed NASA that, to correct the problems with the engines, a major design change was needed, which would require three years of additional testing followed by another four years before the new machinery would be ready. The estimated cost could be as high as $1 billion, or twice the 1971 award for the complete engine.[47] Eleven months after the *Challenger* accident, in December 1986, Pratt & Whitney received a $198 million contract to develop new high-pressure turbopumps that would be interchangeable with the current ones. These pumps were to achieve the 55 mission life including 7.5 running hours at 109 percent of rated power. Rocketdyne also received funding for improvements to meet the same objective.[48]

Table 6.3. *Comparison of current and alternative HPOTP*

Objective	Alternate HPOTP	Current HPOTP
Minimum welds through fine-grain investment castings	7	300
Eliminate uninspectable welds	None	250
Provide subcritical rotordynamic operation Stiffen rotor system	Integral Tiebolt/Disk	Shaft coupled to 1st disk; bolted to 2d disk
Minimize rotating elements	28	50
Provide significant suction (NPSP) margin	40%	Marginal
Minimize LOX cooled bearings	1	4
Eliminate coatings/closeouts required for hydrogen embrittlement protection	None	Gold coating/weld closeouts
Reduce shaft RPM	22,400	28,000

The high-pressure oxidizer turbopump (HPOTP) has been the most troublesome of the two on SSME (see Table 6.3). Testing at engine level began in 1991 but ran into a number of problems – turbine inlet cracking, turbine bellows failure, turbine bearing outer race cracking, and high synchronous vibration of rotor assembly. Corrective actions were devised and implemented for the first three problems, two of which were attributable to manufacturing problems. However, it took one and a half years, several attempts, and a team of experts to solve the high synchronous vibration problem.[49]

The high-pressure fuel turbopump (HPFTP) began engine-level testing in May 1991. The ability to operate at 109 percent of rated power was demonstrated and the redesigned HPFTP accumulated 2,200 seconds of operation during twenty-three tests. The testing revealed several design deficiencies – cracking of turbine inlet associated with thermal transients, lift-off seal leakage, ball bearing inner race cracks, a high-cycle fatigue crack at the corner of a second-stage turbine blade. All but the blade-cracking problem were corrected and solutions implemented. However, in December 1991, in order to concentrate resources on more critical and difficult problems of HPOTP, the HPFTP program was placed on hold for two years because of budgetary constraints.[50] To the Aerospace Safety Advisory

Panel, "using the new HPTOP with the existing HPFTP while feasible, will not achieve the operating margin increases sought for the engine system."[51]

Also discontinued in 1991 were the development of large-throat main combustion chamber and advanced fabrication processes for the SSME due to budget limitations. After learning of this, the Aerospace Safety Advisory Panel stated that "both of these efforts eventually would have led to significantly enhanced safety and reliability of the SSME." Hence, they recommended that these important safety-related programs be restored.[52]

On an optimistic note, the Safety Advisory Panel reported that the in-flight performance of the SSMEs

> has been very consistent and without significant anomalies since the return-to-flight after Challenger. There are now [in 1991] sufficient engines to provide four shipsets plus three spares. The practice of removing all three engines after each flight and conducting the post- and pre-flight tests in the "engine room" has proved beneficial and effective. Except for the HPTPs, the major components of the engines have demonstrated service lifetimes in excess of the specified 55 equivalent flights.[53]

The following year, the panel found that individual major component improvement programs continued to make progress, although the total engine upgrade was being delayed because the HPFTP part of the advanced turbopump program was on hold. In addition, it indicated that the "highly effective" large-throat main combustion chamber[54] has finally been made a formal part of the SSME program by NASA but has been denied appropriations by Congress. The panel again recommended that the HPFTP development be restarted and that NASA continue to press for approval of the large-throat main combustion chamber development.[55] In particular, the panel noted that its use provides significant increases in the operating margins of most of the SSME components, especially the HPTPs.[56]

However, while the SSME performed well in flight in 1993, the causes of launch delays and on-pad launch aborts were primarily attributable to manufacturing control problems. A result was a series of reinspections of delivered hardware that required at least partial disassembly of major engine components, particularly the turbomachinery. This, in turn, caused a shortage of usable turbopumps. The advisory panel found that "'sheetmetal' cracks in the current HPFTP have become more frequent and are larger than previously experienced. This has led to the imposition of a 4,250 second operating time limit and a reduction of allowable crack size by a factor of four." Of greatest concern to the panel was the generation of fragments that could, if they struck a turbine blade, cause blade failure and lead to a catastrophic engine failure. The panel indicated that "the delayed alternative HPFTP program should eliminate the cracking problem," and for the third straight year recommended that NASA restart the development and testing of the alternative HPFTP immediately.[57]

Two other problems were noted by the panel. They observed that engine sensor failures had become more frequent and were a source of increased risk of launch delays, on-pad aborts, or potential unwarranted engine shutdown in flight. They also pointed out that "the SSME health monitoring system comprising the engine controller and its algorithms, software and sensors is old technology. The controller's limited computational capacity precludes incorporation of more state-of-the-art algorithms and decision rules. As a result, the probabilities of either shutting down a healthy engine or failing to detect an engine anomaly are higher than necessary."

Into the Twenty-first Century

On March 8, 1995, NASA outlined its plan to develop a new launch system,[58] the first to be developed since the advent of the space shuttle. Touting an innovative "fast track" procurement process, the plan features partnerships between government and industry to design, develop, and produce "workhorse, resusable launch systems." Designated as "X-33" and "X-34," the development would proceed in two phases. Total government funding would be $24 and $70 million for the X-33 and X-34 respectively during the fifteen-month phase I period in which the selected contractors – Lockheed Advanced Development Company, McDonnell-Douglas Aerospace, Rockwell International Corporation (X-33), and Orbital Sciences Corporation (X-34) – would propose detailed investment strategies, operations planning, and vehicle design to permit competitive selection of an industrial partner or partners for NASA. Then, an administrative decision on whether to proceed with phase II to design, build, and flight test the vehicles would be made. That phase would last through the end of the decade. The Marshall Space Flight Center would be the NASA host site for these projects.

"The goal of NASA's Reusable Launch Vehicle (RLV) technology program is to enable significant reductions in the cost of access to space to promote the creation and delivery of new space services and other activities that will improve U.S. economic competitiveness." Administrator Daniel S. Golden is quoted as saying that the procurement process, which was initiated and completed within a two-month period, "is a true harbinger of how the twenty-first-century 'faster, better, cheaper' NASA intends to conduct business."

The primary objectives of the X-33 program are to mature the technologies required for a single-stage-to-orbit (SSTO) rocket, reduce the technical risk, and encourage private investment in the commercial development and operation of the next-generation system. It is the SSTO rocket that A. O. Tischler strongly supported twenty-five years earlier in lieu of the shuttle (see Appendix B). Have we now come full circle?

It is also worth noting that selling the SSTO, in particular, and the new NASA, in general, to Congress and to the public, and clarifying that "faster, better, cheaper" is also actually safer indicate just how difficult it is to maintain the type of space program the United States has taken for granted. While joint government–industry partnerships may reduce the financial risk to both parties, the risk of harm to human astronauts, and a spectrum of vexing dilemmas, still exist. Clearly, the question remains whether even this "reduced" high price is worth the effort. While the nineteenth century's Jules Verne thought so, it is no longer clear that the U.S. Congress, as it approaches the twenty-first century, will continue to do so.

Notes

1. Jerry Thomson, "Shuttle: The Approach to Propulsion Technology," *Astronautics and Aeronautics* 9, February 1971: 64–7.
2. Walter C. Williams, *Report of the SSME Assessment Team* (Washington, DC: NASA, January 1993), p. 3.
3. William S. Hieronymus, "$1-Billion Shuttle Engine Program Seen," *Aviation Week and Space Technology* 94, June 21, 1971: 60–3; Michael L. Yaffee, "P&W Shuttle Engine Based on XLR129," *Aviation Week and Space Technology* 94, June 14, 1971: 51–7.
4. Frank Stewart, taped interview in Management Operations Office, MSFC, *Oral Interviews: Space Shuttle History Project*, Transcript Collection, December 1988, p. 65.
5. Ibid., pp. 65-6.
6. News article, M&R, May 18, 1964, p. 9, NASA History Office.
7. Robert S. Ryan, "The Role of Failure/Problems in Engineering: A Commentary on Failures Experienced – Lessons Learned," NASA Technical Paper 3213, George C. Marshall Flight Center, Huntsville, AL, March 1992, p. 11.

8. Thomas H. Johnson, "The Natural History of the Space Shuttle," *Technology and Society* 10, 1988: 417–24.
9. Adelbart O. Tischler, "A Commentary on Low-Cost Space Transportation," *Astronautics and Aeronautics* 7, August 1969: 50–64.
10. "Shuttle Engineer Protest Detailed," *Aviation Week and Space Technology 95*, August 23, 1971: 23.
11. Katherine Johnson, "Shuttle Engine Speeded by GAO Decision," *Aviation Week and Space Technology* 96, April 10, 1972: 14–5.
12. Richard S. Lewis, *The Voyages of Columbia: The First True Spaceship* (New York: Columbia University Press, 1984), p. 63.
13. Walter Dankhoff, Paul Herr, and Melvin C. McIlwain, "Space Shuttle Main Engine (SSME) – The Maturing Process," *Astronautics and Aeronautics*, January 1983: 26–32, 49.
14. Alex Roland, "The Shuttle: Triumph or Turkey?" *Discover*, November 1985: 30–49.
15. This section has been provided by Joel Peterson, associate professor of mechanical engineering, School of Engineering, University of Pittsburgh.
16. Dankhoff et al., "Space Shuttle Main Engine."
17. Dominick J. Sanchini, *The Space Shuttle Main Engine*, A74-42369 (New York: American Institute of Aeronautics and Astronautics, 1974), pp. 193–200.
18. Williams, *Report of the SSME Assessment Team*, p. 4.
19. Harry A. Cikanek III, MSFC,"Characteristics of SSME Failures," paper presented at the AIAA/SAE/ASME/ASEE 23d Joint Propulsion Conference, San Diego, June 29–July 2, 1987.
20. Dankhoff et al., "Space Shuttle Main Engine."
21. Robert S. Ryan, "Practices in Adequate Structural Design," NASA Technical Paper 2893, George C. Marshall Flight Center, Huntsville, AL, 1989, p. 63.
22. Ibid., p. 64.
23. Williams, *Report of the SSME Assessment Team*, p. 4.
24. H. I. Colbo, "Development of the Space Shuttle Main Engine," paper presented at the AIAA/SAE/ASME 15th Joint Propulsion Conference, New York, June 18–20, 1979.
25. M. H.Taniguchi, "Failure Control Techniques for the SSME, NAS836305, Phase I Final Report," Rockwell International, RI/RD86–165 (revised), Canoga Park, CA, undated.
26. Williams, *Report of the SSME Assessment Team*, pp. 4–5.
27. Colbo, "Development of the SSME."
28. Taniguchi, *Failure Control Techniques for the SSME.*
29. Cikanek, "Characteristics of SSME Failures."
30. Ryan, "Practices in Adequate Structural Design," pp. 15–18.
31. Cikanek, "Characteristics of SSME Failures."
32. Williams, *Report of the SSME Assessment Team*, p. 5.
33. Mike Toner, "It's Pay Off or Perish for the Shuttle," *Science Digest*, May 1985: 64–7, 87–8.
34. Eugene E. Covert, *Technical Status of the Space Shuttle Main Engine: A Report of the Ad Hoc Committee for Review of the Space Shuttle Main Engine Development Program* (Washington, DC: Assembly of Engineering, National Research Council, National Academy of Sciences, March 1978), p. 2.
35. Ibid., p. 11.
36. Ibid., pp. 11–2; Taniguchi, *Failure Control Techniques for the SSME*, p. 21.
37. Ryan, "Practices in Adequate Structural Design," p. 21.

38. Bob Marshall, taped interview in Management Operations Office, MSFC, *Oral Interviews: Space Shuttle History Project,* Transcript Collection, December, 1988, pp. 132–3.
39. Ryan, "Practices in Adequate Structural Design," 1989, p. 63.
40. Williams, *Report of the SSME Development Team,* p. 6.
41. Marshall, taped interview, p. 135.
42. Harald Kranzel, "Shuttle Main Engine Story," *Spaceflight* 30, 1988: 378–80.
43. Trudy E. Bell and Karl Esch, "The Space Shuttle: A Case of Subjective Engineering," *IEEE Spectrum* 26, June 1989: 42–5.
44. Williams, *Report of the SSME Assessment Team,* p. 12.
45. Ibid., pp. ii–iii.
46. Ibid., pp. 19–20.
47. Toner, "It's Pay Off or Perish for the Shuttle."
48. Edward H. Kolcum, "Pratt Reviving Propulsion Plants to Attract Future Engine Contracts," *Aviation Week and Space Technology,* June 6, 1988, pp. 46–7.
49. Williams, *Report of the SSME Assessment Team,* pp. 20–1.
50. Ibid.
51. Aerospace Safety Advisory Panel, *Annual Report* (Washington, DC: NASA, March 1992), pp. 38–9.
52. Ibid., p. 11.
53. Ibid., p. 35.
54. The large-throat main combustion chamber design has the throat area increased by 11 percent and the operating pressure decreased by 9 percent.
55. Aerospace Safety Advisory Panel, *Annual Report* (Washington, DC: NASA, March 1993), p. 10.
56. Ibid., p. 28.
57. Aerospace Safety Advisory Panel, *Annual Report* (Washinton, DC: NASA, March 1994), pp. 9–10, 26.
58. This press release is an example of the material that NASA has placed on the Internet. The interested reader is strongly encouraged to "browse" the World Wide Web and examine the large volume of material the different NASA centers have made available.

7

All-Up Testing versus Component Testing: Normative Practice or Flawed Judgment?

As mentioned earlier, innovative engineering endeavors, like the development of the space shuttle main engines, involve uncertainty and risk. The most effective way to deal with uncertainty and risk is at the base, an engineering judgment call concerning the nature, amount, and scope of testing. Although risk assessment inherently is ethical in nature – a high-risk factor could cause harm – acceptance of a high-risk factor in and of itself does not constitute a moral wrong. There are at least two common defensible reasons for accepting a high risk. The first is to assure that those persons most immediately affected by that risk are informed of it and agree to take it. The second has to do with the long-term success of the project. Does the acceptance of the risk threaten this success? Does its acceptance compromise the organizational obligation of "stewardship" principles, that is, the equitable use of public funds?

NASA adopted an all-up and high-risk testing approach to the development of the shuttle. This case examines whether or not its reasons for doing so were "defensible." Specifically, NASA cited the complexity of the engine, the fact that it was "beyond state of the art" and built like a "Swiss clock." Because of this complexity, they reasoned that the engine bed itself was the only way the components could be tested.

Two Testing Philosophies

Two very different engineering design philosophies have been utilized in the development of highly complex flight vehicles. They differ significantly

149

in their approach to testing. The traditional, conservative approach is a bottom-up or component design philosophy. In this case, the components are individually designed, materials to be used are characterized, purchased parts are specified, and the emerging component is tested and redesigned as necessary. Functional subassemblies are constructed and tested to ensure that component coordination works as planned and, finally, the end product is constructed and proved to be flight worthy. This approach allows each component, large and small, to be tested under conditions that actually exceed in-flight demands. Consequently, it is possible to develop contingency plans, which may become necessary in later flight testing activities. Generally, alternate designs of critical components are developed in parallel until one can be proved, through testing, to be superior. Component testing in this relatively conservative manner is a lengthy and costly exercise.[1] However, advocates of this method argue that in the long run it is easier to locate root problems and less expensive to make modifications while the end product is still in its early developmental stages.[2]

An alternative design philosophy which has been used in the development of complex flight vehicles has been referred to as top-down, system, or all-up testing. "Up" implies that a component or subassembly is a flight-ready piece of hardware, not just a mock-up simulating its actual weight and space requirements. "All-up" says that *every* component and subassembly in the *first* launch (i.e., the first test) of the final flight vehicle configuration is an actual flight-worthy and functioning unit.[3] This seemingly all-or-nothing approach is in sharp contrast to the component design method where progress is accomplished in stages.

The all-up philosophy is the nontraditional approach and its extensive adoption is unique to NASA. Historically, it was developed by the Department of Defense during the Titan II and Minuteman missile programs, before Apollo took shape.[4] Its original intent was to provide a realistic alternative to the methodical component testing approach in which design engineers seemed to want to test and then overtest each item. Although this alternative all-up philosophy was described as "reckless" and "high risk,"[5] it was later implemented in Apollo as well. Its application in Apollo was a major extension from the earlier programs, which were comparable in neither size nor complexity to Apollo, nor were these missiles manned. Furthermore, there were significant differences in the motivation of the programs; the Minuteman was considered a matter of national security[6] whereas Apollo originated as a matter of national pride.[7] A look at the historical context in which NASA first used the all-up testing approach and the complex organizational structure in which it was applied will lay the foundation needed to examine its "defensibility" as applied in the development of the shuttle.

Apollo's Testing Philosophy

In 1961, the Apollo program to land a man on the moon and return was initiated by President John F. Kennedy[8] in response to the Soviet Union's recent success with Sputnik, the first vehicle to be placed into orbit and to return. In these early years, the space program was a high national priority, and the American public, not wanting to concede technological superiority to the Russians, enthusiastically supported "the race." However, by the middle of 1963, public concern with cancer research and foreign aid challenged the wisdom of spending billions of dollars on space.[9] Although Apollo had originally managed to obtain more than its required critical economic support, its financial security was not maintained. In November 1963, Congress cut $612 million[10] from NASA's budget request. Voicing a refrain which was virtually to undercut long-term NASA development, the questionable public and financial support hinted at future problems.

Technical problems were also present in the Apollo program, especially in the development of the F-1 engine for the first stage of the *Saturn 5* liquid rocket which was to take the astronauts to the moon. The F-1 was to outsize the current largest operational engine, the H-1, by 800 percent, which equated to an increase of 1.32 million pounds of thrust. Perhaps the most difficult task in developing a liquid rocket is ensuring that the propellants, an oxidizer and a fuel, burn smoothly. In more technical terms, the engines must exhibit combustion stability. The F-1's enormous size was complicating this already difficult task to an extraordinary degree. The problem was so serious that, in mid-1962, the top engineers of NASA's and Rocketdyne's Liquid Fuel Engines groups were pulled from their management positions and assigned to the newly formed Combustion Devices Team. Within Rocketdyne, this was the absolute highest priority in the company.[11]

A year later, while the combustion instability problem was still unsolved, unsettling organizational politics within NASA began to surface and add to the growing list of problems. Brainerd Holmes, the highly respected head of the Office of Manned Space Flight (OMSF), resigned his position. Holmes was popular among his colleagues and had received good press as well. He had appeared on *Time*'s cover for his managerial success with the BMEWS (ballistic missile early-warning system) early-warning radar network, and, referring to his resignation, *Missiles and Rockets* editorialized "an American tragedy." Administrator Webb repeatedly refused to grant Holmes's request, one that had been made and denied by his predecessor, that the NASA center directors report directly to him. Holmes wanted maximum authority over the Apollo program. There were other disputes with Webb and, eventually, Holmes left NASA "to return to private life for financial reasons."[12]

George Mueller was chosen to replace Holmes as the head of the OMSF. Aware of the pressures from "without" and the technical problems "within," Mueller enlisted the aid of John Disher and A. O. Tischler, two experienced NASA headquarters men, to conduct an objective and confidential assessment of Apollo's schedule for landing a man on the moon by the end of the decade. In two weeks, Disher and Tischler reported to Mueller that "lunar landing cannot likely be attained within the decade with acceptable risk." Their personal estimate was that the probability of accomplishing Kennedy's goal was about 10 percent, not very high considering the stakes involved and the money already spent. Mueller, having obtained the ammunition he was apparently searching for, asked the two men to report these findings directly to Bob Seamans, NASA's associate administrator, in order to make clear the realities as he took over.[13]

The two design philosophies, all-up versus component, were in competition at NASA, with each corresponding to a different group's professional perspectives. The systems engineers of that era built *unmanned* missile systems and early-warning radar systems for the Department of Defense. Their job was to provide the type of managerial leadership needed to coordinate effectively large-scale, highly complex technical systems. Systems engineers interpreted the flight-testing phase of development as a trade-off between reliability and cost. Their flight vehicles were unmanned, hence, they were not necessarily aiming to make a perfect piece of hardware. The second group of engineers included both the German rocket team, housed at the Marshall Space Flight Center (MSFC) and led by Werner von Braun, and the NACA (National Advisory Committee for Aeronautics, and NASA's forerunner) engineers whose experiences were in-flight testing. Because the engineers in this group saw their failures materialize in grand explosions and/or the loss of human lives, they were accustomed to conservative, meticulously planned advances in their hardware capabilities.[14]

Mueller was in the first group, a systems engineer who took a manage-rial-like "total-picture" view of the program. He was known for being impeccably logical when dealing with the people from the centers. He insisted, without exception, that all decisions were to be unquestionably supported by data and he appeared to offer no consideration to arguments based on emotion or the human element. Mueller was not a disliked man; he was admired by most and especially by the people who worked closely with him.[15] He was, however, immovable in his logical analysis of risk. Mueller's passion for purely logical reasoning played a definitive role in the decision to adopt the all-up testing philosophy for Apollo. His analysis con-cluded that NASA could never fly the *Saturn V* enough times to statistically boost its confidence to the point of man-rating it. He also felt that the ground and support tasks were sufficiently different when stages were added one by one that you really did not learn anything from the exercises anyway. His third point was that, in a step-by-step approach where multi-ple launches were planned to accomplish one thing, NASA was not in the position to "take advantage of success"; the schedule would actually be dis-rupted if success came "early," that is, before the last try.[16] Werner von Braun said of Mueller's idea, "it sounded reckless" but "Mueller's reason-ing was impeccable."[17] No one was able to prove, quantitatively, that the idea would fail, and so Mueller persisted. All of these factors – the Disher and Tischler's grim private assessment, the technical gridlock of combus-tion instability, the systems engineer's style of management, and the pro-fessional reputation of Mueller – combined to make a reality out of what Charles Murray, in hindsight, has labeled the single most significant idea which contributed to Apollo's eventual 1969 moon landing. The decision was to use all-up testing.

Mueller introduced his idea in October 1963, only two months after assuming his position as head of the OMSF, at a meeting of the Manned Space Flight Management Council. There were representatives of MSFC and Houston's Johnson Space Center present and all met the idea with immediate disbelief.[18] But Mueller, because of his authority and respect within the organization and his penchant for flawless logic, implemented the new philosophy as the only possible chance of reaching the moon on time. In retrospect, Apollo's success was probably due to the solid engi-neering foundation from which development proceeded, the generous resources initially provided by Congress which allowed for parallel devel-opment as needed, and luck.[19]

Testing Philosophy for the Shuttle's Main Engines

Many of the NASA personnel who experienced Apollo's success remained with the organization through the shuttle's inception; Administrator Webb is one example. As already documented, the shuttle program, from beginning to end, suffered from a severe lack of funds and, in a global sense, this affliction provoked the eventual rejuvenation of the all-up testing philosophy. In contrast to Apollo, the schedule was not the predominant constraint, except for the obvious connection between time and money. But financial support was not all that the shuttle lacked. It never enjoyed the public and Congressional support that Apollo, at least initially, experienced. The Vietnam War and other social issues consumed the public's attention; there was no time left to think about space.

For the entire shuttle project, not only for the main engines, NASA adopted a success-oriented strategy, which assumed that components and tests would perform as expected on the first try. The objective of this strategy was to save time and money by eliminating, supposedly, unnecessary parallel development efforts and hardware testing.[20]

William H. Sneed, Marshall's program planning office director during the time of the shuttle development, compared that project to the Apollo. To Sneed, the technical challenges

> were both equally great. The scheduling . . . Apollo might have had a little tougher job; there was a declaration that we would do something before the end of the decade. . . . the one thing that we had with Apollo that we didn't have in the Shuttle program . . . was . . . ample money. . . . Any time we got into difficulties with the Apollo program we had the money to . . . "buy our way out of it." . . . we could initiate a backup development if we had a problem with a component not working well. . . .
>
> The Shuttle was a little more difficult in that we had a fixed budget. . . . That seemed to really drive everything that we did. . . . it turned out that the schedule was kind of a variable. . . . But to vary the schedule is a demand for more money. So there was a conflict really built in from the outset. . . . you were trying to force fit everything into a fixed technical set of requirements, a fixed budget, and really a fixed schedule. . . . it forced . . . both engineering types and management types to be very frugal in developing a test program; they had to be very selective in going into backup developments. In fact, I'm not even sure we had

backup developments. . . . And as a result of that, I think it introduced an element of risk far beyond what we had on the Apollo program.[21]

Although the effects of this design strategy were felt throughout the entire engine development program, its impact on the development of the main engine's high-pressure turbopumps (HPTPs) undoubtedly caused the most controversy. While some engine components – for example, the combustion chamber and the computer controller – were tested as components and performed relatively well, the high-pressure turbopumps, as noted in Chapter 6, had severe problems.[22] A component testing facility had been constructed for the turbopumps and testing was initiated. The program was later dropped when NASA claimed that progress was not being made fast enough to be useful in the rest of the development program. Indeed, progress was very slow. Lee F. Webster, a member of the Ad Hoc Committee for Review of the Space Shuttle Main Engine Development Program, estimated that in the twenty-seven months of turbopump component testing, ten times more testing would normally have been accumulated, based on experience with past rocket engine development programs.[23]

Initially, the Coca 1A and 1B test stands in Santa Susana, California, were used until September 1977 in an effort to test the engine's complex turbopumps. During the congressional hearings of the Ad Hoc Committee, NASA outlined the following reasons for closing the stands:

1. The initial performance mapping had been achieved for the turbopumps and full engine testing was now more efficient.
2. The testing facilities were limited in terms of test rate and duration. (One consequence was that the effects of reusability could not be adequately studied without extensive facility modifications.)
3. Coca 1A and 1B would have to compete with the engine test program for the limited number of turbopumps.

In reality, only 161 seconds of testing had been completed in the oxidizer pump system and 111 seconds on the hydrogen pump system.[24] The Ad Hoc Committee understood and appreciated the difficulties associated with developing a test rig of the complexity necessary to understand fully the turbopump technology; developing these facilities can possibly take years.[25] However, in March 1978, during the first of the two Senate hearings, the Ad Hoc group strongly recommended that NASA again acquire a component development test rig to explore the problems facing the turbopumps. They claimed that the lack of such a rig was the main reason for the delay in discovering the problems.[26]

In response, NASA instead chose to reactivate the Coca 1C stand and test the turbopumps not as components but rather as part of the entire engine system, using the complete engine as the test bed. Dr. Frosch claimed that building a suitable test stand was not any easier than getting the engine itself to operate properly and, therefore, a component-development test stand was not necessarily a profitable option.[27] This plan was carried out during the time between the two hearings held before the Senate subcommittee (i.e., March 1978 to February 1979). If the SSME may be considered a final end-product, and its technological complexity certainly allows it to be called this, then all-up testing was again being implemented as the alternative testing philosophy in the space program.

James B. Odom, the external tank manager within Marshall's Shuttle Project Office, defended this decision, recalling

> that engine was extremely advanced for the time. It made it very difficult. . . . one of the problems that we had in developing the engine was, normally you will go off and develop the pumps in one test stand, . . . the turbines in another test stand, and . . . the combustion chamber in another test stand. . . . you . . . use a facility that will provide the gases and liquids that will run each one of these pieces. In the case of the main engine, because of the sheer velocities and the sheer quantities of hot gas that it takes to drive the turbine, you can't build a facility to drive it. So we had to build an engine and put it all together, because that was the only way that you could get high enough temperatures and high enough pressures and high enough flow rates . . . out of an actual pump. . . . every time you want to test a pump you have to build a whole engine. And when you lose a pump, you lose a whole engine. It's a very difficult way, a very expensive way, to really push technology as far as we did in that engine program.[28]

However, during the second hearing in February 1979, Dr. Eugene Covert strongly criticized NASA's all-up approach and its decision not to accept the committee's original recommendation. He again recommended that component test facilities be developed for the turbopumps even though the investment in time and money would be significant. The committee's reasons were as follows:[29]

1. Major failures could result in long delays if the engine itself is used as the test bed.
2. A continuing engineering program would be needed to support the engines throughout their operational life and this would most economically and efficiently be carried out through the use of a component test stand.
3. The spare parts and engines may not be available to sustain the testing at Coca 1C.

As an example, the committee pointed to the source of a major failure that occurred on the recently reactivated engine test stand. A fire broke out in the main oxidizer valve, resulting in extensive damage to the engine and test stand. In their opinion, this incident may have been prevented had the proposed component test stand been available to isolate the problem before exposing the entire engine to the untested components.[30] Nevertheless, for the second time, and although Dr. Covert described the approach as "far from ideal,"[31] NASA defended its philosophy and refused to change.

Analysis

From an ethicist's point of view, the central question regarding the adoption of all-up testing is, Was it ethically defensible? In other words, did NASA management make a competent, responsible decision that satisfied Cicero's Creed II? The economic and political contexts and technical challenges faced when the decision to use all-up testing was made in the Apollo project differed considerably from those existing when the shuttle was developed. By comparing and contrasting decision making in these two projects, some idea of what constitutes an ethically defensible reason can be appreciated.

Recall that all-up testing was initially developed as an alternative to a growing pedantic and seemingly "overtesting" philosophy of engineers designing unmanned vehicles. Applying it to the Apollo program was an unprecedented leap in risk taking. It was one thing to risk losing costly technology; another to risk the lives of the astronauts. From the previous review, however, the combination of the following "facts" makes the decision understandable. Since Apollo had always had public and congressional support – hence the needed funds – it had been developed successfully, using traditional component techniques, until the combustion instability problem was encountered. This technological roadblock, combined with the seemingly flawless logic of Mueller and his ability to mobilize support among a doubtful management team, appeared to account for the "reckless" notion of all-up testing. The purpose, however, was to solve a fairly circum- scribed problem, and the efficacy of Mueller's reasoning was appreciated by the engineers who had been frustrated and unsuccessful by using traditional means. In sum, the all-up approach was not adopted to save money. It was adopted relatively late in the Apollo program and with the recognition within the organization that a high risk was being taken. Although all may not have agreed that the risk was worth taking, the rea- sons for adopting all-up in the Apollo program satisfied our criteria for competent and responsible decision making, both individually and for the organization.

The shuttle project was "born amid" an impoverished budget. It has been shown that the rhetoric that sold the concept of an economical, reus- able space craft, developed by off-the-shelf technology, was disputed and articulated by at least one manager, A. O. Tischler. Scholarly studies evalu- ating the decision also conclude that it was a "policy failure." The all-up approach to testing was, nonetheless, adopted shuttlewide. The Covert Committee hearings again voiced concern. NASA, however, was convinced that the promise of cost effectiveness and its track record for previous suc- cess outweighed concern for risk.

The testing philosophy was an attempt to control costs by forcing engi- neering design into an almost production-line mode. Proponents of this approach in NASA, particularly Associate Administrator Mueller, argued that component testing wasted both time and money. Major components or elements cannot be designed until subcomponents and subsystems are designed and tested. Schedules are fairly flexible because of incremental development of the design. It was possible, in all-up testing using com- puter models and other modern techniques, to design without exhaustively testing every piece of a technology. Components are tested simultaneously as part of an entire system, such as the SSME. It was argued that all-up testing was therefore more efficient and provided engineers with the same information as the more tedious and costly component testing. "All-up" testing saved both time and money, and was obviously the way to meet the shuttle's stringent budget constraints.

Was All-Up "Economical"?

Did all-up save the shuttle project in terms of dollars? Consider the following "major incidents" which occurred during twenty-seven test failures of the main engines:[32]

Test 901-110 (March 24, 1977)

A fire occurred in the seal region of the high-pressure oxidizer turbopump, which separated the cryogenic oxidizer pump from the hot combustion gas-driven turbine. The fire led to the failure of both the pump and the surrounding structures, a second external fire, and further damage. Review board recommendations included provision of "gradual buildup in test duration with thorough review for untested components." The cost for damages was $3.3 million; the length of the delay was four weeks.

Test 901-136 (September 8, 1977)

A pump end-bearing failure in the high-pressure oxidizer turbopump was precipitated by unequal load sharing due to coolant flow forces. This led to an internal fire resulting in a breached structure and external fire. Among the review board's recommendations were redesign bearings for better load sharing, increase bearing stiffness and rotor balance, implement a bearing test program, improve understanding of rotor dynamics, and utilize internal monitoring instrumentation. The damage cost was $2.4 million with a four-week delay.

Test 902-120 (July 18, 1978)

A failure of the high-pressure oxidizer turbopump was centered on the first-time use of a capacitance device designed to measure its shaft, bearing, and bearing cartridge movement. Rubbing between the device's pads and a speed nut ignited a fire, which burned into the turbine end bearings and the main pump. An estimated $1.65 million in damages and a five-week delay resulted.

Test SF10-01 (July 12, 1980)

A distorted preburner faceplate produced a locally high fuel mixture ratio, resulting in thermal overload which breached the liner. This led to structural-wall hot-gas impingement and eventually burned through, initiating a localized external fire. The review board's recommendations were to provide a full understanding of the flow in the end cap region and the failure mechanism, establish maximum value of allowable faceplate distortion, and develop component testing to understand the engine start transient effects on the preburner. The incident resulted in $1.5 million in damages and a delay of sixteen weeks.

Test 902-198 (July 23, 1980)

High-cycle fatigue led to failure in a main injector liquid–oxygen post, allowing oxygen to leak into the hot-gas regions. The ensuing fire burned many adjacent posts and led to primary faceplate damage. Review board recommendations included establishing a life prediction criteria, using the criteria to predict life for flight; analyze post materials and consider a material change if an advantage is shown; continue study to understand the injector environment. Damage cost was $1 million; delay was twelve weeks.

Test 750-160 (February 12, 1982)

Water utilized for cooling drained into the chamber coolant valve discharge elbow. The pretest drying purges and dew point measurements were not sufficient to remove or detect the presence of water. During the engine start, cryogenic hydrogen introduced into the line froze the water and created a 25 percent flow blockage in the fuel preburner. The resulting fuel mixture ratio was too high, severely eroding the fuel turbine and downstream hot-gas flow-path hardware. The review board tartly recommended that the established methods for contamination removal be tested in order to prove that they work. The cost of this incident was $15 million with a twelve-week delay.

Test 901-364 (April 7, 1982)

A "Kaiser Hat" failure resulted from the first-time testing of a new design for the high-pressure fuel turbine end cap, and provided a lesson on the need for more analysis and component testing. From NASA's incident report: "During the investigation, it was established that all changes, including the nut which caused this failure, (were) reviewed formally both by Rocketdyne and NASA. Late changes to a design, such as the undercut feature of this nut, may not have had the thorough evaluation that the original design had been given. The undercut was made for structural consideration and its significance as a potential flow path cause apparently was overlooked."[33] Damages totaled $26 million; an eight-week delay occurred.

Test 750-175 (August 27, 1982)

A catastrophic failure of the high-pressure oxidizer duct was initiated by a high-cycle fatigue (HCF) crack adjacent to a specially developed ultrasonic flow transducer. The high-cycle fatigue was caused by a combination of thinning the duct wall to install the transducer blocks, physically adding the block masses to the duct, and increasing the local stresses brought about by brazing the blocks to the duct wall. Rocketdyne's incident report stated that "it is clear that brazed joints are not to be relied upon for HCF application without extensive analysis and testing. The HCF properties of Rocketdyne braze alloys do not exist, but should be presumed to be lower than parent metal properties. Braze fillet geometry is difficult to control, and the surfaces of braze fillets inherently have shrinkage voids. Therefore, relying on braze fillets to reduce stress concentration is not conservative."[34] The estimated damage cost was $11 million; the resultant delay lasted eight weeks.

These incidents caused program schedule delays, resulted in engine and test facility damage, and typically required an extensive failure analysis before testing could continue. Most of these engine failures have been attributed to insufficient knowledge of the environment and loads imposed on the components, particularly dynamic loads. The most frequently cited mechanism for failure was high-cycle fatigue, initiating at least eight of the incidents. Manufacturing or maintenance process deficiencies were factors in at least four of the cases. Recommendations included design changes (in eleven of the incidents) and improved environment definition (for ten incidents).[35]

Although the threat of such explosions occurring "in flight" was clearly a possibility, it was not until NASA lost "six identifiable lives" that remedial design of the engines took place. The following comment from Bill Sneed (Marshall's program planning office director during much of the

SSME design and development phase) made after the *Challenger* accident
is a compelling commentary on "experience" and risk–taking behavior.

> I don't know how many billions of dollars we've spent in the last two and
> a half years [since the *Challenger* accident], but had that kind of money, a
> small portion of that money, been put back into the R&D program ini-
> tially, even into sustaining the engineering program after we started to
> fly, you can see it would have been good return on the investment.[36]

It is hard to argue in this particular case – the development of the main
engines – that the all–up philosophy was economical. In Chapter 9, we will
discuss a case in which a test stand was being developed. Even in relatively
minor aspects of engine development, problems associated with the all–up
approach were documented.

Did All-Up Provide Needed Information?

Did the all–up approach provide engineers with data comparable to that
obtained by the more traditional method? After reviewing the facts presented in this chapter, the answer to this question is apparent. Not only did
it inhibit learning from the failure, it depleted the project of spare parts.
Richard Feynman, Nobel Prize–winning physicist and member of the
Rogers Commission, conducted an individual investigation of the engineers working on the main engines. His observations confirm this conclusion. In Feynman's words, "most airplanes are designed 'from bottom up'
with parts that have already been extensively tested. The shuttle, however,
was designed 'from the top down' – to save time. Whenever a problem was
discovered, a lot of redesigning was required in order to fix it."[37] Feynman,
the scientist, continued to explain what he learned:

> [Rocket engine design usually follows a] component system or bottom-
> up design [process]. First it is necessary to thoroughly understand the
> properties and limitations of the materials to be used (e.g., turbine
> blades) and tests are begun in experimental rigs to determine those.
> With this knowledge, larger component parts (e.g., bearings) are
> designed and tested individually. As deficiencies and design errors are
> noted they are corrected and verified with further testing. Since one

tests only parts at a time, these tests and modifications are not overly expensive. There is a good chance that the modifications to get around final difficulties in the engine are not very hard to make, for most of the serious problems have already been discovered and dealt with in the earlier, less expensive stages of the process.

The SSME was handled differently – top-down. The engine was designed and put together all at once with relatively little detailed preliminary study of the materials and components. Now, when troubles are found, it is more expensive and difficult to discover the causes and make changes. For example, cracks have been found in the turbine blades of the high-pressure oxygen turbopump. Are they caused by flaws in the material, the effect of the oxygen atmosphere on the properties of the material, the thermal stresses of steady running, at some resonance at certain speeds, or something else? How long can we run from crack initiation to crack failure, and how does this depend on power level? Using the completed engine as a test bed to resolve such questions is extremely expensive. One does not wish to lose entire engines in order to find out where and how failure occurs. Yet, an accurate knowledge of this information is essential to acquiring a confidence in the engine reliability in use. Without detailed understanding, confidence cannot be attained.[38]

Feynman found another major problem with the top-down approach for the SSME. That is, once an understanding of the fault is obtained, a simple fix – such as a new shape for the turbine housing – may be impossible to implement without redesigning the entire engine. To Feynman the SSME was a very remarkable machine, "built at the edge of – sometimes outside of – previous engineering experience. As expected, many flaws and difficulties have turned up, [but] because it was built top-down, the flaws are difficult to find and fix."[39]

Conclusion

Feynman's observations were originally made by both engineers and managers when the all-up philosophy was adopted in Apollo. Then the problem was an isolated technological one. In the shuttle the reason for adopting all-up testing was endemic to the total project: cut costs. There had been component testing early on in the SSME project, especially with regard to the HPTPs, but willingness to abandon this traditional approach, perhaps prematurely,

Engineering Ethics

was indicative of NASA's overall success-oriented thinking. The success of Apollo had raised the "risk threshold" management was willing to take.

McCurdy's empirical study of NASA's cultural change from 1960 to 1990 also examined NASA's risk-taking behavior. Table 7.1 summarizes McCurdy's results, showing that space flight, in the early years, lent itself to the acknowledgment of high risk. Apollo missions, a contender in an international "space race," were dramatic, short-lived, and spectacular. Thirty years later, the space shuttle, while no less dramatic, was premised (indeed, funded) on the political rhetoric of "cost effectiveness" and "reusability." It was featured as a reliable space vehicle, which, while hardly "safe," had been publicized as such to meet new political goals. Civilian passengers, both senators and teachers, were recruited for its flights as proof of its safety. Within this context, tight budgets, previous successes, and a backlog of unmet deadlines combined over time to promote a less fearful attitude toward adopting a high risk. Psychologists utilize a concept termed "cognitive dissonance" to explain how rationalization to support a difficult decision is adopted once the decision is made – for example, if senators and teachers are flying in the shuttle, maybe it is safe. The adoption of all-up testing for the shuttle project early on did not meet the ethical justification it had in Apollo. The organizational structure of NASA inhibited a corporate understanding of the high risk, prohibited individual engineers from influencing changes, and ultimately invoked irrevocable long-term problems. Neither competency, responsibility, nor Cicero's Creed II can be said to hold up – in hindsight.

Table 7.1. *Attitudes of NASA personnel*

Opinion	Year joined NASA		All
	1951–69	1970–88	

Attitudes toward risk and failure

18. Risk and failure are a normal part of the business of developing new technologies.[a]

Strongly agree			39
Agree			58
No opinion; undecided			1
Disagree			2
Strongly disagree			1
			<1

19. NASA employees are allowed to fail and learn from their mistakes.

Strongly agree			2

Table 7.1. *Attitudes of NASA personnel (Continued)*

Opinion	Year joined NASA		All
	1951–69	1970–88	
Agree			44
No opinion; undecided			23
Disagree			26
Strongly disagree			5

23. NASA rewards people who are technically creative.

Strongly agree			6
Agree			45
No opinion; undecided			19
Disagree			24
Strongly disagree			7

20. NASA has stayed on the cutting edge of new technologies.

Strongly agree			6
Agree			42
No opinion; undecided			17
Disagree			30
Strongly disagree			5

22. Cost constraints have forced us to cut corners in carrying out our programs.

Strongly agree			33
Agree			47
No opinion; undecided			10
Disagree			9
Strongly disagree			1

Attitudes toward management

26. The time we spend on management tends to be time taken away from basic engineering and science.

Strongly agree			20
Agree			47
No opinion; undecided			13
Disagree			18
Strongly disagree			1

27. Engineers make good managers.

Strongly agree			4
Agree			28

Table 7.1. *Attitudes of NASA personnel (Continued)*

Opinion	Year joined NASA		All
	1951–69	1970–88	
No opinion; undecided			34
Disagree			26
Strongly disagree			9

28. Scientists make good managers.

Strongly agree			1
Agree			14
No opinion; undecided			38
Disagree			35
Strongly disagree			12

16. It is fairly easy to change official procedures within NASA once thay are approved.

Strongly agree			<1
Agree			9
No opinion; undecided			24
Disagree			50
Strongly disagree			17

The NASA bureaucracy over time

13. The amount of paperwork has increased substantially since I came to work for NASA.[b]

Strongly agree	45	22	33
Agree	36	35	36
No opinion; undecided	4	25	15
Disagree	14	18	16
Strongly disagree	1	1	1

15. It was much easier to cut through bureaucratic barriers and get things done when I first came to the agency.[b]

Strongly agree	30	3	16
Agree	45	27	36
No opinion; undecided	10	40	25
Disagree	13	26	20
Strongly disagree	3	3	3

Managerial risk taking and communication

24. At the management level, NASA is dominated by people who are cautious and inclined to avoid risks.

Strongly agree			16
Agree			46

Table 7.1. *Attitudes of NASA personnel (Continued)*

Opinion	Year joined NASA		
	1951–69	1970–88	All
No opinion; undecided			22
Disagree			16
Strongly disagree			1

25. *At the management level, the number of people who are cautious and inclined to avoid risks has increased since I joined the agency.*[b]

Strongly agree	21	10	16
Agree	45	32	39
No opinion; undecided	18	44	31
Disagree	15	13	14
Strongly disagree	1	<1	1

17. *People within the agency – including those at different centers – communicate with each other as much as they should.*

Strongly agree			1
Agree			17
No opinion; undecided			18
Disagree			47
Strongly disagree			18

In-house capability

5. *NASA has turned over too much of its basic engineering and science work to contractors.*[a]

Strongly agree			31
Agree			47
No opinion; undecided			11
Disagree			11
Strongly disagree			1

4. *Since I came to work for the agency, NASA has lost much of its in-house technical capability.* (Respondents were told to consider only the technical capability of NASA employees – not the contribution of support contractors.)[b]

Strongly agree	30	11	20
Agree	48	39	43
No opinion; undecided	5	23	14
Disagree	13	24	18
Strongly disagree	4	4	4

Table 7.1. *Attitudes of NASA personnel (Continued)*

Opinion	Year joined NASA		All
	1951–69	1970–88	

NASA testing activities

6. NASA did substantially more testing in the past than it does today.[b]

Strongly agree	26	10	18
Agree	40	29	34
No opinion; undecided	19	49	34
Disagree	14	11	12
Strongly disagree	2	2	2

7. NASA should do more testing than it currently does.[a]

Strongly agree			18
Agree			34
No opinion; undecided			34
Disagree			12
Strongly disagree			2

Interactive failures

21. As the technologies we use have become more complex, the possibility of an interactive failure on a manned or unmanned space flight has increased. (Respondents were told that an interactive failure was "one in which two or more single point failures combine to create a more serious problem.")

Strongly agree			12
Agree			49
No opinion; undecided			23
Disagree			14
Strongly disagree			1

Hands-on work

29. As NASA employees, we have as much opportunity to do "hands-on" work as we want.[b]

Strongly agree	1	4	2
Agree	19	22	21
No opinion; undecided	12	13	13
Disagree	58	47	52
Strongly disagree	10	14	12

30. When I first came to the agency, we did a lot more "hands-on" work than we do today.[b]

Strongly agree	28	6	17
Agree	56	32	44

Table 7.1. *Attitudes of NASA personnel (Continued)*

Opinion	Year joined NASA		All
	1951–69	1970–88	
No opinion; undecided	8	42	25
Disagree	8	19	13
Strongly disagree	0	1	1

The lure of space exploration

31. Working on the space program is still just as exciting as I thought it would be.

Strongly agree	15
Agree	47
No opinion; undecided	12
Disagree	23
Strongly disagree	3

Faith in technical capability

31. NASA has just as much technical capability as its contractors.[a]

Strongly agree	19
Agree	40
No opinion; undecided	11
Disagree	26
Strongly disagree	4

[a]No statistically significant difference between first- and second-generation employees at the .01 level.
[b]Statistically significant at the .01 level.

Source: Howard E. McCurdy, *Inside NASA: High Technology and Organizational Change in the American Space Program* (Baltimore: Johns Hopkins University Press, 1993).

Notes

1. Eugene E. Covert, *Technical Status of the Space Shuttle Main Engine: A Report of the Ad Hoc Committee for Review of the Space Shuttle Main Engine Development Program* (Washington, D C: Assembly of Engineering, National Research Council, National Academy of Sciences, March 1978), p. 2.
2. William P. Rogers, *Report of the Presidential Commission on the Space Shuttle Challenger Accident* (Washington, DC, June 6, 1986), Vol. 2, p. F-2.
3. Charles Murray and Catherine B. Cox, *Apollo: The Race to the Moon* (New York: Simon and Schuster, 1989), p. 158.
4. Ibid., p. 157.
5. Ibid., p. 162.

6. U.S. Congress, Senate, Subcommittee on Science, Technology and Space of the Committee on Commerce, Science and Transportation, *Report of the National Research Council's Ad Hoc Committee for Review of the Space Shuttle Main Engine Development Program*, 95th Congress, 2nd Session, March 31, 1978, serial 95-78, p. 11.
7. Murray and Cox, *Apollo*, pp. 76–81.
8. Ibid., pp. 82–3.
9. Ibid., pp. 152–3.
10. Ibid., p. 161.
11. Ibid., pp. 146–9.
12. Ibid., p. 152.
13. Ibid., pp. 153–4.
14. Ibid., pp. 155–7.
15. Ibid., pp. 158–60.
16. Ibid., pp. 157–8.
17. Ibid., p. 162.
18. Ibid., p. 160.
19. Thomas Nagel, *Mortal Questions* (Cambridge: Cambridge University Press, 1983), pp. 24–38. Nagel presents arguments for three types of moral luck.
20. Covert, *Technical Status of the Space Shuttle Main Engine*, p. 2.
21. Sneed, taped interview in Management Operations Officer, MSFC, *Oral Interviews: Space Shuttle History Project*, Transcript Collection, December 1988, pp. 159–60.
22. Covert, *Technical Status of the Space Shuttle Main Engine*, p. 11.
23. Ibid., p. 27.
24. Subcommittee on Science, Technology and Space, 1978, p. 72.
25. U.S. Congress, Senate, Subcommittee on Science, Technology and Space of the Committee on Commerce, Science and Transportation, United States Senate, *NASA Authorization for Fiscal Year 1980 S.* 357, 96th Congress, 1st Session, February 21, 22, 28, 1979, part 2, serial 96-1, p. 1094.
26. Covert, *Technical Status of the Space Shuttle Main Engine*, p. 12.
27. U.S. Congress, Senate, Subcommittee on Science, Technology and Space, 1978, p. 59.
28. James B. Odem, taped interview in Management Operations Office, MSFC, *Oral Interviews: Space Shuttle History Project*, Transcript Collection, December 1988, pp. 86–8.
29. Subcommittee on Science, Technology and Space, 1979, pp. 1093–4.
30. Ibid., p. 1094.
31. Ibid., p. 1093.
32. M. H. Taniguchi, "Failure Control Techniques for the SSME, NAS836305, Phase I Final Report," Rockwell International, Canoga Park, CA, RI/RD86165 (revised), undated; Harry A. Cikanek III, *MSFC*, "Characteristics of SSME Failures," paper presented at the AIAA/SAE/ASME/ASEE 23d Joint Propulsion Conference, San Diego, June 29–July 2, 1987.
33. Taniguchi, *Failure Control Techniques*, pp. 5–15.
34. Ibid., pp. 5–13.
35. Cikanek, "Characteristics of SSME Failures," pp. 3–4.
36. Sneed, taped interview, p. 176.
37. Richard P. Feynman, *What Do You Care What Other People Think? Further Adventures of a Curious Character* (New York: W. W. Norton, 1988), p. 184.
38. Ibid., pp. 226–7.
39. Ibid., p. 227.

8

A. O. Tischler: An "Ethical" Engineer (Alternatives to Whistle-Blowing)

> *Trustworthiness, with the associated values of honesty, candor, fairness, diligence, loyalty, discretion and competence [are central to engineering]. There is a need for case studies that illustrate what it is like for engineers to embrace these values in the workplace – as well as cases that illustrate obstacles to their fulfillment.*
>
> – Michael Bayless, *Professional Ethics*

Engineers involved in a "Big Science" project such as building the space shuttle take on a variety of roles.[1] The societal, political, and organizational environments in which an engineer works provide opportunities for individual decisions to carry significant ethical import. An engineer, for example, who competes within a federal government hierarchical organization for program funding will be cast amid others who are lobbying for their own priorities. This scenario, described by Thompson as "many hands," characterizes the political arena.[2] If the "engineer" plays the game astutely, he or she may actually win "a piece of the political pie" and then have the responsibility for making decisions to insure that the coveted project or program is completed within a given budget, time frame, and organizational context. Ideally, within these constraints, decisions made will be competent and responsible and will adhere to Cicero's Creed II.

One goal of ethical inquiry into engineering practice is to reflect on the morality of that practice. The building of the main engines of the space shuttle provides an opportunity to inquire into the ethical nature of the decision making at the individual, organizational, and societal level. If reflection leads to the recognition and articulation of an ethical dilemma, a next step would be to propose alternative courses of action to resolve the

171

dilemma; and to pursue a course that does not compromise individual values, the organization's integrity or, the societal "good."[3] As our definition of practical ethics suggests, these goals are not easily discernible in a workday environment. Embedded in the practical concerns of "getting the job done," a project manager may have little patience to stand back and reevaluate the premises on which his or her work is based.

Though not easily accomplished, as this case study demonstrates, this reflective process is nonetheless achievable. Here we critique the work of an individual who serves as a model of an "ethical engineer," Adelbart O. Tischler, noted chemical propulsion engineer and director of NASA's Chemical Propulsion Division within the Office of Advanced Research and Technology from 1964 until 1972. Tischler describes his role in NASA as one of those individuals who bridges the gap between proposed projects and their completion. In spite of his modesty, it is our opinion that his historical track record describes how he maintained moral courage in the face of adversity. The historical setting for the case is the early 1970s, when the decision to build the space shuttle was first being considered. Tischler and his work are evaluated with reference to the principles of competence, and responsibility and his adherence to Cicero's Creed II.

Specifically, the case examines the role Tischler played in evaluating the alternative approaches to developing the space shuttle, in assessing ways to cut the costs in shuttle development, and in carrying out this responsibility as the head of the "cost-cutting" NASA task force. "Some of what the most dedicated advocates of the space shuttle are proposing in schedules, cost and performance is suspect," wrote Tischler to a trusted colleague in a memo in 1969.[4] Given this premise, Tischler painstakingly documented an alternative approach to shuttle development and worked within both the organizational structure of NASA and his professional societies to implement the change he deemed necessary to complete the shuttle project successfully. Personal interoffice memos and letters document the nature of the moral dilemma Tischler found himself in and serve to model ways – other than whistle blowing – to resolve them. Personal communication with Tischler has added incite into how he viewed his involvement. Commenting on the process of decision making in government Tischler wrote:

> Top jobs are often filled by appointing people with better "policy" orientation than engineering acumen. Unfortunately, these people also measure their own success in terms of programs initiated rather than programs completed. Such people seldom appreciate the chasm

between feasibility and practicability, nor what it takes to build a bridge from one to the other. I have spent most of my life working on that bridge I mentioned. Having sponsored the technology programs leading to the SSME I understood what the limitation of such a machine would be.[5]

The historical "retrospectrascope," moreover, points to the political realities that caused Tischler's criticisms ultimately to go unheeded. NASA's current problems prompt one to question the wisdom of these political compromises.

The Principle of Competence

Tischler's educational and professional background clearly rank him a competent engineer – that is, a knowledge expert.[6] He began his professional career as a chemical engineer at the Lewis Laboratory (Cleveland) of the National Advisory Committee for Aeronautics (NACA), the predecessor of NASA. While at Lewis, he completed the Master of Science Degree in Chemical Engineering at the Case Institute of Technology. When NASA was established in 1958, Tischler joined the headquarters staff as chief of the liquid rocket engines, where he started and managed the development of the Saturn-Apollo engines. From November 1961 to January 1963 he served as the assistant director for launch vehicles (propulsion) for the NASA Office of Manned Space Flight (OMSF). Tischler was next appointed director of the Chemical Propulsion Division of the Office of Advanced Research and Technology (OART) in January 1964.

From 1969 through 1972, Tischler served as director of OART's Shuttle Technologies Office. In that position he was in charge of OART's Space Shuttle Steering Group (1969), which was, in effect, commissioned by the Office of Manned Space Flight to develop a base of technological information and experience on which to judge the merits of the various technical concepts for design of the shuttle. In doing that, the group was to tap the specialized expertise residing at NASA's field centers. He received NASA's Exceptional Service Medal, awarded for the job of organizing and directing the team effort to put a firm technological footing under the shuttle development. A group medal was also given to the entire steering team. The cost reduction effort came after the shuttle development had been initiated. Tischler was ultimately selected to head a task force to determine ways of

achieving major cost reductions in the shuttle project through both standardization and management techniques.

Competence, as defined in this text, emphasizes a recognition of both
the limits of one's knowledge base and the acquisition of information sufficient to understand a product's performance characteristics. Tischler
understood both the "state of the art" of building space vehicles and its
limits.

> I have the opinion that even in this era of unprecedented technological
> progress, too large an attempted technological leap can cause us to
> stub our toes by forestalling the availability of the new system, increas-
> ing its cost at the expense of the total space program, and perhaps
> most important, risking obsolescence by first-operational-capability
> time.[7]

Arguing against many of his colleagues, he concluded that "because of the
flow of technological progress . . . the route to low-cost space transportation is evolution." By evolution, Tischler was describing a "pedestrian
approach which entailed detailed studies, fundamental experiments, engineering data determination, model performance evaluation, and full-scale
demonstration, with enough in-house participation so that NASA would
truly develop technical experts among our own staff."[8] Critical of NASA's
claims that reusability would reduce costs and that the shuttle technology
was "off the shelf," Tischler wrote of the shuttle project in 1969:

> What makes high cost? Cost is people. The major part of all program
> funds is spent on man-hours-in-house, at contractors, at subcontrac-
> tors, and at suppliers. Therefore, if we cannot reduce the number of
> people involved in making and operating transportation systems, at
> least on a per-pound-flown basis, then the quest for low-cost transpor-
> tation is hopeless.[9]

Looking at the overall needs that a space program should meet, Tischler
felt that "the idea that a reusable launch vehicle system will serve a single
purpose is preposterous."[10]

Summing up the premise on which his critical observations were based,
Tischler stated in 1970: "We need total visibility of all the things being done
in the name of the shuttle, to make the needed decisions based on scientific
and engineering judgment, independent of political judgment."[11] Though

cautious and skeptical of the proposed shuttle program, Tischler maintained a watchful eye on the aspects of the program for which he was responsible. Working within an organizational framework that did not stress the capability of monitoring the work of its contractors, he was suspicious of non-NASA contractors and carefully reviewed both their work and assumptions. In a memorandum for the record dated, November 14, 1969, he questioned a prime contractor's assumptions in adjusting experimental data.[12] He used this to justify the need for more tests. The specific impulse values used by contractors for the advanced space shuttle engine differed from the four experimentally demonstrated values by more than 2.5 percent at high oxygen to fuel (O/F) ratios. The contractor had corrected upward the experimental data in anticipation of design improvements by an amount which Tischler found to be much greater at the higher O/F ratios than at the lower O/F ratios conventionally used in earlier hydrogen–oxygen engines. He noted that a potential performance deficiency of the order of 1.5 percent from performance projection is possible if the engine is required to operate at a high O/F ratio of 7.0 (since complete mixing and burning of propellant under these circumstances is difficult at ratios above 8). Since such a performance deficiency is equivalent to about 70 percent of the total shuttle payload, Tischler felt that it was worth looking at the consequences and what might intelligently be done to avoid them. The four experimental values showed a serious performance falloff at the high O/F ratio similar to that experienced with other engine systems and propellants which have been operated near the stoichiometric burning regime (O/F = 8).

To Tischler, the sparsity of data on these performance numbers and the doubt that could be raised regarding their projection to shuttle operating characteristics convinced him that a demonstration project sponsored by the Air Force to obtain experimental results on a more nearly prototypical engine should be continued and, in fact, accelerated. Sharply criticizing NASA's anxiety to initiate development of the shuttle, he reasoned, "Proceeding to a multi-billion-dollar vehicle system commitment on the basis of extrapolation of these few experimental data points is not sound."[13]

> High technological risks can not be reduced to manageable proportions by analysis or study. Design analyses leave untouched uncertainties that arise from poor technological information. . . . prototype demonstration serves to identify previously unrecognized or underestimated problems and to resolve uncertainties which could precipitate major changes in cost, availability, or even performance of a new system. A

direct result of experimental engineering work should be a set of preliminary engineering specifications with design criteria and fabrication-
process specifications.[14]

The Principle of Responsibility

If "knowledge is power" and if responsibility accompanies power, then to
satisfy our description of an "ethical engineer" Tischler must have been
responsible, that is, he must have communicated his relevant knowledge to
the appropriate persons at the appropriate times. In fact, numerous examples confirm that he did just that. Tischler used his professional skills as a
writer to communicate concerns regarding virtually every aspect of the
shuttle's development. His articles on the technical status of the shuttle
and of space exploration in general questioned NASA's long-range space
transportation strategy.[15] As was discussed, his design experience led him
to argue for a transitional approach to fully recoverable large space trans-
ports. And, as already demonstrated, he also took issue with the "official"
projections of a low-cost transportation system.

An examination of the events that preceded the publication of the article
"A Commentary on Low-Cost Space Transportation," written by invitation
for the August 1969 issue of *Astronautics and Aeronautics*, demonstrates that
Tischler was cognizant of the importance of his views and of the political
impact they could have. After receiving the galley proofs of his article,
Tischler wrote a brief but eloquent letter to the journal's editor:

> When an editor alters the meaning of a man's introductory remarks he
> presumes, when he arbitrarily interrupts the author's flow of words
> he imposes, when he adds innuendo that the author does not intend he
> offends, but when he changes a sentence to contradict the author's point
> he disqualifies himself as an editor. All these things you have done.[16]

He concluded this letter by demanding that the editor either publish the article as corrected on the galley proof by the August issue, or Tischler would
withdraw it. In a memo to Robert C. Seamans Jr., secretary of the Air Force,
he asked for an honest opinion of the manuscript. He was aware that his
"views . . . were not entirely contradictory but often oblique to . . . reports
prepared more recently by DoD and NASA on the space shuttle. . . . Some
of what the most dedicated advocates of the space shuttle are proposing in

schedules, cost and performance is suspect. My article is not going to straighten that out because it is written to avoid assiduously any quantitative discussion of schedules, cost performance. . . . Still, I don't want to appear to be taking on the U.S. Government in debate, particularly since I work for the U.S. Government."[17] These incidents both show that Tischler took his responsibility to communicate very seriously and thought carefully about what and how he should convey information that he considered vital.

After the shuttle development was initiated by OMSF and because of his knowledge of expenditures and the shuttle, Tischler was appointed to lead a "space vehicle cost improvement" task force established to lower the unit cost of space operations. Reviewing his work as head of this task force illustrates how Tischler worked within an organization that was at times at odds with his own values. It is not surprising to learn that Tischler regarded the issues of responsibility and communications as a critical focal point both for NASA as an organization and for its employees. "Our basic challenge in pulling together the resources and technological information needed for the shuttle, is communication – communication across count-less interfaces: inter-organization, interpersonal and functional."[18] More-over, he understood that engineering decisions tacitly included ethical concerns and believed that communication of NASA's organizational phi-losophies, not just knowledge of technical matters, was crucial. In a memo to George Low, creator of the task force, Tischler hypothesized a relation between NASA's risk-taking attitude and expenditures. He alluded to NASA's "success-oriented" approach to risk taking and concluded that it so pervaded the organization that it escalated costs. He urged a reassess-ment of this approach: As part of that memo, Tischler addressed the com-munication issue: "In particular we need to assess carefully what our attitude toward risk ought to be and then, if it changes at all, express it so that the organization can fully understand and comply with it."[19]

Two months after Tischler sent this memo, he received one from George Low, who was responding to another issue in Tischler's proposed low-cost guidelines:

> I feel that the most important job that NASA must now learn how to do is to design to a lower cost. It is this area where I believe you and your teams should spend most of their time. . . . This area also becomes a rather straightforward engineering problem, and frankly I am somewhat disappointed that we have not made more progress in this area so far.[20]

Low's memo did not show sensitivity to the tacit ethical dilemma involved in cutting costs. That is, for Tischler, lowering costs raised fundamental ethical questions regarding risk. Low, however, characterized this as "a straight-forward engineering problem."

This was not the only area of disagreement between Tischler and Low. Although Low had appointed Tischler to head this task force, it is clear that they did not always agree on ways to cut costs. In November 1972, for example, Low commented on the low-cost guidelines Tischler had constructed and indicated that he had "some qualms about publishing it." While he encouraged Tischler to circulate it within his own committees and task forces, he urged a "second cut" if it were to be distributed more widely. Low's suggestions for revision are indicative of his views: "On management approaches," wrote Low, "the best guideline I could give is to run a 'tight ship,' one that is 'lean and mean.' This means that at each level in the organization – headquarters, centers, contractors, subcontractors – there are only sufficient people so that they are all 'doers' and there are no 'hangers on.'"[21]

Tischler indicated in his paper that he would "limit in-house technical team intercourse with contractors to critical issues." Low, on the other hand, felt that NASA should "only have enough people . . . so that they can only address critical issues." Additional technical teams would be available when needed. "In a 'lean and mean' organization," continued Low, "there is an additional responsibility for each level of management to immediately report problems to the next higher level. If this responsibility is known to all managers and they know that their job depends on this, then one can immediately do away with all sorts of levels for checks and balances, over-the-shoulder onlookers, etc." Low's conception of "cost improvement" comprised two factors: designing to a lower cost and creating a management structure to meet the cost objectives. Seeing the two as separate issues, it was in the design aspect that Low prompted Tischler to concentrate: "taking advantage of reduced weight and volume constraints, standardization, lower parts counts, etc. should have a profound effect."[22]

By February, Tischler had drafted a paper entitled "Low-Cost Space Systems" and sent it to Low to review. "I have read your paper and find that I disagree with so many of your statements that I cannot approve its publication," concluded Low. His frustration at Tischler's critical approach was obvious. "Your basic premise in the first half of the paper is that NASA project managers are bad, that industry is bad, and that everybody's motives are self-serving. I disagree on all counts."[23] Having the opinion that the "multitudes of working groups" that had been supporting Tischler's efforts had results worthy of reporting, Low felt that "instead of the philosophizing which is now in the paper, Tischler should report those results."

As director of the task force and Tischler's immediate supervisor, Low clearly had an impact on what Tischler published. In the May 1973 issue of *Astronautics and Aeronautics*, Tischler's views on "low-cost" space appeared. Incorporating Low's ideas on management with his own previously published views, Tischler described a different "lifestyle" which characterized the low-cost programs. Translating Low's "lean and mean" approach into more "gentle" terms, Tischler explained that "The operating principle seems to be that it is human responsibility, not procedural formality, that underlies success. In these programs, characteristically, small and closely knit and very expert teams exercise substantial authority in exercising work."[24]

Cicero's Creed II

An engineer acts ethically if he or she gathers maximal safety- and failure-related information and communicates this information in an effective manner to the appropriate people. We are now strictly concerned with insuring the safety of the public. As discussed earlier in the "framework," we use the word "harm" to encompass both an individual and an organizational aspect. In addition, the term can be taken literally, identifying immediate and concrete problems. Alternatively, it can refer to more global, abstract, predictive issues. Tischler's work evidences concern for both definitions.

His criticism of the subcontractor is an example of how his work involved a particular subassembly or safety-related monitoring equipment – where the word "harm" needed to be taken literally. Yet, his overall concerns for the design of the shuttle provide us with a less obvious example of an ethical decision in engineering practice. Tischler was concerned with the long-term survival and success of the space transportation system (STS) so that all persons directly and indirectly associated with it would not suffer adverse consequences. The attributes of cost, schedule, and technical performance are present in any project, and trade-offs of one for another are constantly being made. Tischler was of the opinion that it was more important to control the project costs than the schedule. He evidently felt that cost control was the dominating factor in the long-term survivability of the shuttle project and, hence, was protecting the public's enormous investment in the STS. Clearly suspect of the motives of many proponents of the shuttle and displaying a sense of what we have called "stewardship," Tischler wrote in 1969 that "the desire of the aerospace industry, which includes members of government agencies, to build exquisite and innovative equipment does not of itself justify spending the taxpayers' money."[25]

Tischler had the ability and the tenacity to view the space transportation system as a whole. He was an advocate of careful, long-term planning prior to action and spent a considerable amount of time articulating and communicating this. In 1969, while the political decision-making process for "selling the shuttle" to Congress was under way, Tischler argued for "an evolutionary, but pedestrian route to the ultimate low-cost transportation system." He wanted to "do it all within available resources of manpower, facilities and funds." This approach, he insisted, was advantageous because it was flexible and adaptable to "progress and experience." Tischler was evidently dissatisfied with NASA's planning process and felt that the STS was being promoted without regard to providing the practicality of some of its technological approaches. Again, he concerned himself with strategic, long-range issues, which, in terms of not harming, translate into protecting the interests of NASA's employees and, more generally, of the public.

Conclusion

In 1974, a frustrated[26] Tischler "retired" from the space program. He nonetheless continued his memo writing and offered his views to key players in shuttle management. Still the critic, Tischler never gave up his vision of an alternative approach to the development of a manned space program. Although the outspoken views which appeared in the 1969 article and in his personal letters and memos were somewhat modified in his professional writing as the years progressed, he held fast to his ideals. This memo is an example: "The Tug, the principal focus of NASA's upper stage efforts is predicated on the idea that reuse reduces costs. The idea has not been proven for high-orbit payloads by an objective study of either the intrinsic value of payloads as recovered hardware or of the true operational costs of the transport system."[27] On July 25, 1975, Low wrote to Tischler: "Dear Del, it was good to hear from you to know that you are still interested in helping us with our . . . tug concepts. . . . I sent your letter to Manned

Space Flight to get their review and comment. They tell me they have already asked MSFC to evaluate the concept you described. . . . Thank you again for your thoughtful consideration. I hope that you and your family are enjoying your 'retirement.'"[28]

As his colleague Edward Z. Grey indicated, his leaving was considered to be "a real loss" to NASA. Characterizing that loss and the effect that Tischler's alternative to whistle-blowing had on other NASA managers and administrators, Grey wrote that "your enthusiasm, dedication and persistence furnished the spark for the NASA-wide support to do more for less."[29]

Notes

1. Robert W. Smith, *The Space Telescope: A Study of NASA, Science, Technology and Politics* (Cambridge: Cambridge University Press, 1989), pp. 373–93.
2. Dennis F. Thompson, *Political Ethics and Public Office* (Cambridge, MA: Harvard University Press, 1987), pp. 5, 40–1, 44–7, 75–8, 87–9.
3. K. Danner Clouser, "Medical Ethics: Some Uses and Abuses, and Limitations," *New England Journal of Medicine* 293 (8), 1975: 384–7.
4. Adelbart O. Tischler, letter to Robert C. Seamans, Secretary of Air Force, January 2, 1969.
5. Adelbart O. Tischler, Personal communication to Rosa Lynn Pinkus, June 10, 1991.
6. Adelbart O. Tischler, Biosketch, in A. O. Tischler, "Which Way to Shuttle Upper Stages?" *Astronautics and Aeronautics*, July–August 1975: 28.
7. Adelbart O. Tischler, "A Commentary on Low-Cost Space Transportation," *Astronautics and Aeronautics* 7, August 1969: 52.
8. Ibid.
9. Ibid., p. 53.
10. Ibid., p. 52.
11. Adelbart O. Tischler, "Developing the Technological Base for the Space Shuttle," *Aerospace Management* 5 (1), 1970: 59.
12. Adelbart O. Tischler, Memorandum for the Record, November 14, 1969.
13. Ibid.
14. Tischler, "A Commentary on Low-Cost Space Transportation," p. 62.
15. Tischler, 1969, 1970, 1975.
16. Adelbart O. Tischler, letter to John Newbauer, Editor, *AIAA*, June 26, 1969.
17. Adelbart O. Tischler, letter to Robert Seamans, June 2, 1969.
18. Tischler, "Developing the Technology Base for the Space Shuttle," p. 59.
19. Adelbart O. Tischler, memorandum to George Low, September 14, 1972.
20. George Low, memorandum to Adelbart O. Tischler, November 13, 1972.
21. Ibid.
22. Ibid.

23. AD, Deputy Administration Low, memorandum to H, Director, Space Cost Evaluation [Tischler], February 20, 1973.
24. Adelbart O. Tischler, "Low-Cost Space," *Astronautics and Aeronautics*, May 1973: 23.
25. Tischler, "A Commentary on Low-Cost Space Transportation," p. 55.
26. In our original chapter, this word was "disillusioned." When Tischler reviewed it, he changed the adjective to "frustrated." His explanation, included verbatim in Appendix B, indicates his personal account of this "chapter" in NASA history. It again underscored how deeply a moral commitment penetrates one's world view.
27. Tischler, "Which Way to Shuttle Upper Stages?" p. 28.
28. George M. Low, letter to Adelbart O. Tischler, July 25, 1975.
29. Edward Z. Grey, Assistant Administrator for Industry Affairs and Technology Utilization, NASA, letter to Adelbart O. Tischler, July 12, 1974.

9

Cost versus Schedule versus Risk: The Band-Aid Fix: A $3.5 Million Overrun in Test Stand Construction

This chapter examines an incident which occurred at Rocketdyne during the initial research and development phase of the space shuttle main engine (SSME) development.[1] It describes one manager's decision to incur a $3.5 million cost overrun in building a test stand, which was to be used for engine development. This decision guaranteed that the specific project would be completed "on schedule" and in a reliable fashion. However, it had effects within the overall organization that threatened to compromise the long-term safety of the project. Engineers do not work in isolation. Their individual decisions must be examined within the organizational context. The organization, moreover, exists within a societal context. The following case examines the relationship between societal trends in economics, organizational responses to these trends, and individual engineering responses to the organizational priorities. It identifies specific junctures in decision making where ethical decisions were made.

This Pipe Is a Bargain!

As part of the space shuttle's development and testing phase, a large test stand was constructed at Santa Susana for the engine's turbopumps. During the building phase of the test stand, several problems were encountered. They centered on the high-pressure pipes needed to test the turbopumps. The specifications for material for the pipe included an ability to withstand 14,000 pounds per square in atmosphere (psia), a very large diameter, an

ability to withstand cryogenic oxygen and hydrogen, and "great strength."
From a range of materials that would satisfy these conditions, Rocketdyne
chose to use "cast 21-6-9 pipe." Paul Castenholz, SSME manager at Rocketdyne, explained, "It is compatible with hydrogen. . . . it can be used
wrought or cast." Initiating a theme which would become commonplace in
later discussions of building the shuttle, Castenholz emphasized, "Since
cast is half the cost of wrought pipe, we chose cast." In spite of the complex
characteristics of the pipe, Castenholz was pleased that they had obtained it
at "a minimum cost." Additionally, WISCO, the company that made the
pipe, was a "good supplier" and promised a "short delivery time."[2]

Castenholz's references to suppliers and delivery time are worth examining for they provide an understanding of the societal context in which the
initial phase of space shuttle development occurred. In order to satisfy the
various congressional districts supporting the shuttle project, contracts
were distributed "astutely" across the nation. While it is clear that the California aerospace industry was bolstered by the award to Rocketdyne to
build the main engines, the subcontracts to support this unprecedented
technological undertaking were divided piecemeal and involved smaller
amounts of money (see Figure 9.1, which illustrates the distribution of
contracts throughout the country during the early development period).

NASA complained that because the financial rewards to individual companies were not significant, they were not motivated to supply materials on
time.[3] Also, inflation increased costs dramatically within a mere two years
from the date the project started, and the fixed price contracts that NASA
awarded did not enhance suppliers' motivation.[4] Table 9.1 shows the examples
of these cost increases during the three-year period between 1971 and 1974.

In spite of WISCO's impressive reputation, when the pipe was delivered, it had minute cracks on the perimeter and the ends. Normally, when
material is cast, its ends have flaws. WISCO neglected to cut off the ends,
and the pipe had to be sent back. When it was delivered again, the cracks
were absent because WISCO changed its mold technique to get good pipe
throughout. Instead of supplying the lengths of twelve to fourteen feet that
NASA had ordered, WISCO now could only provide seven to ten foot long
segments. Thus the pipes had to be welded and bent twice as much as was
expected. "You normally bend pipe when it is hot," explained Castenholz;
"it turned out that this cast material bends better cold. We didn't know that
until we went through the development program. It cost us a couple of
months in determining that when we did bend it hot, it cracked. We bent it
cold, very well, but that required . . . new tooling."[5] This was not the only
unusual characteristic of cast 21-6-9 pipe. When the welders attempted to

Figure 9.1 Geographical distribution of subcontracts for the SSME.

Table 9.1. *Cost increases*

	1971–2 estimate (based on experience x 1000)	1974 cost (SSME requirements x 1000)	Other factors
Marloy-z liner	$4,500	$20,000	New technology
Duct	650	4,300	Small quantity
Flex joint	2,900	8,400	Small quantity
LPOTP stator	3,100	5,500	Small quantity
HPOTP nozzle casting	900	3,300	Small quantity
HPOTP vane tool	125,000	250,000	Expedite for schedule
Nozzle jacket	25,000	80,000	Weight reduction
Injector post	48	81	

Source: U.S. Congress, House Subcommittee on Manned Space Flight of the Committee on Science and Astronautics, *1975 NASA Authorization Hearings,* 93d Congress, 2d session, 1974, p. 972.

weld the pipe sections together in order to get the required length, there was a high initial rejection rate. In hopes of meeting an approaching deadline, the welders tried to weld too fast and found cracking. It took approximately one month to find a better weld rod. Then, each welder had to be certified. All this took three to eight times longer than originally planned.[6]

The Expense of Meeting the Deadline

The test stand was being built to test the engine's preburner; a test that had been planned for March 1974. Castenholz stressed that both NASA and the Office of Management and Budget (OMB) said that this was a "key date that must be met to assure that the space shuttle will achieve its requirements. . . . Unless overtime and extended effort could be applied, we couldn't achieve that. So, in essence, we proceeded initially to go on overtime. . . .we achieved it, but it did cost us extensively . . . approximately $3,500,000."[7] Examining alternative approaches to the problem, Castenholz reflected, "I have to say in retrospect, if we did not want the schedule in March, we could have certainly put the clamps down and probably have come in relatively close to the original allocated dollars."[8] Describing what the working conditions were on this project he continued,

> We felt all of the pressure, OMB and your needing to make the system.
> . . . We were told this March date was very, very critical, and at the same time, we worked like hell . . . some of these guys were making $1,000 a week, all through the Christmas holidays when the rest of the plant was shut down. And the next two God-damned weeks it rained like hell. We lost all that. And all of a sudden we . . . were at the end of January and we said "we're in yogurt."[9]

Hearing this report, as part of their annual oversight meeting, members of the House Committee on Science and Astronautics asked, "Did you revise your management procedures for budget control so we don't run into a situation like this again?" Castenholz explained the organizational changes that were made:

> We reduced the number of development parts. . . . We take a little higher risk in the program. Instead of having X number of turbopumps, we reduce that one or two and test a little more. . . .
>   There is a risk, for with a major problem we don't have as much hardware as we would have had in an Apollo type program.

> We will do a few less tests because tests are expensive. And we are reducing our manpower to the bare minimum. So we take a little risk in not having development people available if we get into a specialized problem. . . . When we lay off people we lay off designers and engineers . . . so we will not be able to maintain the high level of technical people looking for problems because we don't think we can afford it. We will have to take the higher risk.[10]

Summing up his organization's approach to "paying for the materials," he concluded, "I think that this is the approach that we must make and I think that it is proper to achieve the program within its cost." Obviously troubled by this turn of events, the committee then asked, "What do you foresee on our procurement schedule, and costs problems?" Looking at the grim realities, Castenholz lamented, "we don't see it getting better in any one area. . . . We don't know what the energy shortage impact will be. . . . We are getting NASA support . . . but this isn't able to reduce our costs at this time."[11]

On the bright side of engine development, however, he reported that "the technical aspects of the engine look great. We think the design is good. The engine weight is below what we planned." However, he warned the committee of an additional compromise that was being made in order to keep the cost down: "we shall have to keep the overtime and extra effort down and that will probably jeopardize our schedule." Looking to the overall production of the shuttle, Castenholz, an experienced Rocketdyne manager, confessed, "we are the first out of the box on the space shuttle, and this is probably a sign of the times that will affect all other areas."[12]

Organizational Culture and Individual Decisions: Rephrasing Engineering Decisions in Ethical Terms

The ethical decisions that engineers make take the form of a trade-off between competing values and objectives. These trade-offs can be translated into goals. The goals in this case could be described as:

- Minimize costs.
- Minimize time (stick to the tight schedule).
- Maximize the quality of functioning of the test stand.
- Keep everyone at NASA and Congress as happy as possible.

The ideal set of goals to construct the test stand would have been to build an inexpensive, high-quality test stand within the scheduled time frame. Building the test stand, however, was not achieved in this way. Instead, numerous problems were encountered, contingency funds were depleted, and the organization accepted a higher risk in future endeavors.

Earlier we discussed the issue of how the organizational environment influences an engineer's decisions and conduct. We briefly described Simon's notion of satisficing. One can imagine that when the problems with the pipe were encountered, a number of possible solutions were proposed. One could try to get more money from NASA or Congress; one could try to "slip" the schedule; one could hire more people; one could work around the clock and overspend. These would be "feasible" alternate solutions to the problem, and the last one was thought to "satisfy" the considered outcomes.

Although there are a myriad of considerations both for and against the decision actually made, one can imagine the justification used, by considering both the context in which the decision to overspend was made and the character of the rocket engine development. Recall the political decision to "sell the shuttle," the success-oriented culture of NASA which was validated by the Apollo program, the decision to accept a higher risk by adopting an all-up testing strategy, and Rocketdyne managers' overall professional responsibility to construct the SSME.

It was understood that the state of the engine development program and its schedule would impact profoundly on the final configuration of the shuttle as a whole. Given this premise, one can understand why the schedule for building the engine test stands was chosen to be of the highest priority. By keeping to the schedule and using contingency funds, the managers solved a short-term problem and kept the integrity of their "product" sound. Once the problem was identified, they acted competently. In disclosing the ramifications of their "solution" to Congress they were communicating their concerns about cost and about their acceptance of a higher risk. They were acting responsibly. Those pipes, however, which seemed to be rather innocuous, turned out to bear upon crucial design issues shuttlewide.

Engineers wear many hats and work in various specialized areas. Ethical dilemmas are, accordingly, phrased differently depending on what role one plays in the project. For example, to satisfy the principle of competence, a materials engineer would have to answer if the selection of 21-6-9 pipe was a defensible engineering decision. Was adequate information acquired concerning the performance characteristics of the pipe? The decision to choose cast over wrought pipe was based mainly on cost. The particular materials engineer, as a competent knowledge expert, would have to gather data other than cost in making his or her decision. Given the results of the initial "decision" to bend the pipe hot rather than cold and the impact that the welding speed had on the overall schedule, one could conclude that further characterization was needed. If the cost of the characterization exceeded those allotted to the project (in time and or money), the materials engineer would have to explicitly justify taking the risk of overlooking these steps.

Rocketdyne managers were posed with a different type of ethical dilemma. It involved combining competence and responsibility with respect to safety-risk data and balancing these with the goal of staying initially within the allotted budget. Given the constraints of the shuttle's initial budget, managers invoked a range of cost reduction and budget control strategies. These included the decision to enter into firm, fixed price contracts, and to reduce the number of development parts, tests, and manpower (specifically designers and engineers). After the "pipe crisis" materialized and a significant percent of contingency funds were depleted, the managers sought to remedy the situation by minimizing overtime and extra effort. Each of these decisions involved taking a higher risk, to which the managers admitted. The tacit justification for this risk taking was to save money.

With the investigation of the pipe crisis, both NASA management and the House Subcommittee on Manned Space Flight were faced with a decision that involved Cicero's Creed II. After the fact, Congress heard the detailed account of the events that lead to the $3.5 million cost overrun. The reasons given to explain the loss of contingency funds were inflation, degradation in suppliers, and other factors outside Rocketdyne's and NASA's control. It was predicted that the procurement, schedule, and cost problems were going to continue and that this was "probably a sign of the times."

Future strategies to manage the time and budgetary constraints were posed at the possible expense of compromising the product's performance; a higher risk was the price being paid. But that was a future risk; one that may not occur. Should a reevaluation of the "global" objectives of the space transportation system have been made at this time so as to "recompute" the trade-offs being made between time, funds, and risk? Would it have been appropriate in 1974 to reassess the fate or life-length of the system? The context seemed to encourage a decision making process premised on the assumption that the system would continue "no matter what, right or wrong." This tacit assumption is based on the belief that the engineers would eventually succeed as they had in Apollo. It promoted a context for decision making that encouraged them to consider risk taking as routine.

NASA managers could have been challenged by Rocketdyne engineers and engineering managers to explore if and why that date was so critical. In other words, one could have attempted to find out how the date was determined – that is, what were the true costs of not meeting it? (Sometimes, everything is labeled "critical.") Although cost overruns were problematic and, with hindsight, a portent of future problems, the test stand was completed on time and was a reliable, safe mechanism. In the short term, the schedule and quality of the project took precedent over the cost. The extensive materials characterization that took place after problems developed, in fact, produced a product valuable for future use.

The questioning of competent materials characterization unearthed and documented in this congressional report surfaced many times later in the development of the SSME. Materials for turbopump blades and for pipes in the heat exchanger demanded extensive testing. The reusable spacecraft put particular demands on the selection of materials. Ironically, the test stand, which was thought to be a critical part of the engine development program, was eventually abandoned in favor of using the "engine as the test bed."

Notes

1. The information for this case is contained in the 1975 NASA Authorization Hearings before the Subcommittee on Manned Space Flight of the Committee on Science and Astronautics, U.S. House of Representatives, 93d Congress, 2d Session on H.R. 12689 (superseded by H.R. 13998). Held twice a year, these hearings were Congress's way of monitoring the development of the space shuttle and assuring that the taxpayers' dollars were being spent appropriately. The members of the committee were political appointees, as well as technical engineering experts. When problems occurred during the building of the shuttle, NASA asked this committee for advice as well as for additional funds, if this was needed.
2. U.S. Congress, House, Subcommittee on Manned Space Flight of the Committee on Science and Astronautics, *1975 NASA Authorization Hearings*, 93d Congress, 2d Session, 1974, p. 960.
3. Ibid., pp. 959, 971.
4. Ibid., p. 972.
5. Ibid., p. 961.

6. Ibid., pp. 962–3.
7. Ibid., p. 965.
8. Ibid., p. 966.
9. Ibid.
10. Ibid., pp. 974–84.
11. Ibid., p. 976.
12. Ibid., p. 983.

10

Back to the Drawing Board or Normal Errors: Senate Investigations of the SSME Development Problems

Testifying before the House Subcommittee on Manned Space Flight about why he spent \$3.5 million in contingency funds, Paul Castenholz predicted that the cost overruns in building the space shuttle were not limited to the isolated case of the tests and pipes. Unfortunately, Castenholz was correct. The following extensive case summarizes four congressional reports issued from March 1978 through August 1979.

In December 1977, Senator Adlai E. Stevenson, Chairman of the Sub-committee on Science, Technology and Space, became increasingly concerned about the development problems occurring with the space shuttle main engines (SSME). Stevenson enlisted the aid of Dr. Philip Handler, president of the National Academy of Science. Handler, also a member of the National Research Council, used that organization to carry out Stevenson's charge. In January 1978, the Research Council established the Ad Hoc Committee for the Review of the Space Shuttle Main Engine Development Program. Known as the Covert Committee after its distinguished chair, Dr. Eugene Covert, MIT professor of aeronautics and astronautics, this committee issued two reports – the first on March 31, 1978, and the second approximately a year later in February 1979. Both are used in this chapter. The House of Representatives Subcommittee on Space Science and Applications met twice during this time period to address the main engine problems. That subcommittee was specifically concerned with the shuttle's cost, performance, and ability to meet its scheduled first launch. When paired with the *Covert Committee Reports*, the records of these two

Congressional hearings provide an insightful critique of the progress of the SSME's development and document the ethical dilemmas that NASA engineers and managers experienced.

Primarily descriptive, the case reviews the difficulty that technical experts and the government representatives had in sorting out the cause of the shuttle project's major financial and technical problems evidenced by late 1977. Historian Alex Roland has described this period in the shuttle's development as the time when the "Big Lie" of selling the shuttle to Congress was finally exposed.[1] Even Roland, however, is perplexed. Did the NASA managers really believe they could produce what they promised? Or had they stalled for time, waiting until the project was so far along that it could not be canceled? This case demonstrates how difficult it was to sort out these issues. One must gather and then analyze complete technical facts in order to define the problems and then suggest courses of action. Recall the goals of ethical inquiry in the applied setting. These include identifying moral problems, sorting out alternative solutions, and reflecting on current actions.[2] The investigations of both the Covert and congressional oversight committees can be viewed as "engineering ethics consults." The Covert Committee applied principles of sound engineering practice to examine technical dilemmas that had moral import. The congressional oversight investigations queried NASA's honesty in its testimony and its capability to estimate the costs of the project. Both committees considered going back to "ground zero" if need be to insure the safety of the engines and the long-term success of the shuttle project. By late 1978, cost, which had been such a problem during the initial funding of the shuttle, was no longer the guiding principle in its development. Having invested considerable resources in what was to be the country's sole military and civilian space launch vehicle, Congress was committed to live up to the promises it made to the U.S. public to build a reusable space shuttle.

Categorizing Errors: Are Some Worse Than Others?

Complications were expected to arise in the shuttle project. This case, however, illustrates how complications are ordered in terms of their severity and their "forgivability." Charles Bosk is a sociologist who studied how surgeons train residents to operate, knowing that in the process they will commit errors (some of which could harm or even kill a patient).[3] He provided a framework to categorize the errors that occurred on the academic surgical service he studied. This framework is instructive, for it provides us with terms to interpret the Covert Committee's discussion of problems related to development of the main engines. The technical nature of both engineering and surgery, and the uncertainty inherent in applying techniques as well as the judgment involved in deciding which technique to apply, provide a basis for comparison.

The most forgivable complication on a surgical service is technical in nature, for example, an inappropriate suture closing such as an incision made too far on the abdomen above the appendix. Technical errors, even grave ones, are expected and forgiven if they are reported promptly and if they are not repeated. Repeated mistakes that combine to document a trend raise a red flag of doubt regarding the resident's technical competence. Mistakes that are discovered and not reported cast doubt on the moral character of the offender.[4]

Bosk's book, *Forgive and Remember: Managing Medical Failure,* attests to the maxim that surgeons use in training residents. Surgery, like engineering, is a skill; surgeons, like engineers, are trained in a variety of settings. Several different procedures are appropriate and accepted to treat the same problem. Residents must be attuned to the attending's "preferred operative approach" for a given problem. When a resident uses an approach the surgeon does not favor, Bosk labels this a "quasi-normative error." It alerts the attending surgeon that the resident may not be a loyal, attentive student and is interpreted as showing a disregard for the individual surgeon's expertise. This type of error is again forgivable, but raises the flag of suspicion especially if it becomes a pattern.

Judgmental errors occur when an incorrect strategy of treatment is chosen. "Attending surgeons in charge of devising treatment plans," writes Bosk, "make the most serious judgmental errors."[5] In these cases, judgment is not always incorrect in the absolute sense; the surgeon, given the clinical evidence available at the time, may have chosen an eminently reasonable course of action. An unexpected result – a death or complication – forces the surgeon to consider whether some alternative might have been more profitably employed. Outcome-oriented surgeons balance both clinical results and scientific reasoning to determine how correct judgment is. The two most common judgmental errors are overly heroic surgery and failure to operate when the situation demands. Often understood within the context of uncertainty associated with medical decision making, these errors, once again, are ranked forgivable.

The error not forgiven on a surgical service is one that is moral in nature.[6] Bosk labels these "normative" and categorically states that once a resident is caught lying about results, or fails to report information to an attending in a timely manner, trust, the cornerstone of surgical teamwork, breaks down. The term "normative" refers to the understanding that there are strict ethical principles in surgical practice, namely, truth telling. The principles are constructed to insure that patient care is not compromised. Only by towing the line, being superattentive, and not evidencing a similar mistake will the resident who commits a normative error be reenlisted onto the working team.

Other errors recognized by Bosk were categorized as exogenous (anesthesia did it) or equipment related (the drill broke). While these also provide concepts for comparative evaluations, the categories of error labeled judgment, technical, and normative are most appropriate in this case study.

The Covert Committee

The Covert Committee was formed to specifically determine whether the technical failures occurring in the main engine tests were within the "normal development curve" or a result of "poor judgment." The committee was convened to provide an independent assessment of NASA's progress and to assist congressional oversight. As will be discussed, members of Congress were losing trust in NASA. The fact that the Covert Committee was formed suggests that, as in surgical training, technical errors that had been reported to the Congressional Oversight Committee were forgiven. Only when it became suspect that NASA was "covering up" or committing a "moral error" did the Senate exercise its authority to monitor the program. The following case dramatically demonstrates the difficulty even the experts faced when they tried to sort out "what went wrong" in this complex technological undertaking.

As mentioned, the Covert Committee had two series of meetings, both of which resulted in public hearings before the Senate subcommittee, the first in March 1978 and the follow-up in February 1979. At the time of the Covert Committee's formation, the first manned orbital flight of the space shuttle program was scheduled for March 1979. The February hearings examined, among other issues, if that date could be honored. A second set of hearings discussed long-range issues with the SSME development and operation. In addition to the committee's chairman, Dr. Eugene Covert, other members included Richard C. Mulready, director of Technical Planning, Pratt & Whitney Aircraft Group, who was selected because of his experience with the experimental XLR-129 high-pressure engine; and Dr. Courtland Perkins, chairman of the Assembly of Engineering in the National Research Council.[7] The committee had been carefully selected to include those who were not only experts in their fields, but who also had experience with the type of problem to be addressed – the development of a large, complex technology. Dr. Robert A. Frosch was NASA administrator at the time.

The motivation for the ad hoc review recall stemmed from Senator Stevenson's recognition that numerous unsolved technical problems were delaying the engine development schedule. In his words "delays and technical difficulties prompted the review."[8] (See Appendix C for Covert Committee Technical Assessment.) Given this situation, the ad hoc group sought to determine "whether everything possible (was) being done to assure development of a safe and reliable main engine system" or whether the design of the system would have to be reinitiated from the beginning.[9] These issues are specifically of interest here, for each illustrates how engineers articulated reasons to justify specific decisions regarding risk and testing. The types of factors that carry weight in technical–ethical decision making are defined.

Underlying Assumptions

Underlying the dialogue between the Covert Committee, the Senate subcommittee, and NASA were two opposing philosophies regarding how innovative engineering endeavors should proceed. As previously mentioned, the Ad Hoc Committee first reported to the Senate subcommittee in March 1978 to assess the safety and plausibility of the first manned orbital flight of the shuttle, which was scheduled one year later. Given the severe technical problems NASA was experiencing, the Senate subcommittee suspected that the deadline was not going to be met but pressed both NASA and the Ad Hoc Committee to "name a date."[10] Practically, delays meant that additional funds would have to be allotted. The subcommittee needed a dollar amount to put before Congress. NASA also needed to justify the budget it was requesting. The Ad Hoc Committee stressed that "it [was] too early to predict the exact timing of the first manned orbital flight,"[11] and consistently kept to the adage of "letting the engine be the guide." Considerable attention was given to exploring this approach to scheduling, for it affected how NASA's testing approach was evaluated and also how Congress would approve funds for its fiscal year.

"The Schedule": Post Santa Susana

Placing technical concerns and safety above the political importance of meeting a preselected schedule, the Ad Hoc Committee insisted that "the engine" itself should "tell you what its status is."[12] Stating a rule of thumb in engineering, Dr. Perkins recalled the Minuteman program and severe problems in that missile's second stage. The engineers solved these problems the same day that a report was issued saying that the problems were unsolvable and they would have to start over from the beginning. He described engineers as a "very ingenious lot" and pointed out the possibility that the shuttle's engine problems also may somehow be solved in the next year. Describing what we would term a "tacit ethic," he stressed that his experience told him that pressuring design groups into meeting an absolute schedule forced those persons into a "largely emotional state."[13] The committee was concerned that the schedule was so compressed that a nervous and hurried working atmosphere was being created. This, they said, would inhibit realistic evaluations of problems and encourage Band-Aid fixes instead.[14] It was, they concluded, management's responsibility to maintain a high level of motivation in a group of technical experts while, simultaneously, being cautious and not pushing too far so as to create a group of frantic engineers.[15]

It was the opinion of the Ad Hoc Committee that a 10 to 12 percent schedule extension could be considered reasonable in a project of this magnitude – a month or two of slippage really meant nothing. Moreover, they recognized that since the contract authorization was delayed nine months by the Pratt & Whitney protest, schedule pressures were present from the onset. Their concern was not to pinpoint a date for the first manned launch but rather to assure that the engines fired correctly in the first launch, whenever it happened to take place, thus insuring the safety of the astronauts on board. Clarifying their belief regarding technological development, the committee stressed its concern that a dangerous amount of critical component testing may be omitted in the search to cut corners, save time, and meet a preset launch date.[16]

Component Testing:
The Ad Hoc Committee's View versus NASA

NASA administrator Robert Frosch agreed with the committee that more data were needed before the timing of the first manned orbital flight could be predicted. He also agreed with the recommendation made to shift some hardware from production to development and testing in order to facilitate the problem-solving process. With regard to the concern over decreased component testing to cut corners, Dr. Frosch assured the committee that NASA would not omit any testing for the purpose of meeting the current flight schedule.[17]

Nonetheless, the Ad Hoc Committee was understandably worried about the technical progress being made on the main engines and, especially, about the organizational policies being used to combat the problems. The committee reviewed the traditional process used in new flight vehicle development; progress was accomplished in stages and alternate designs of critical components were developed in parallel until one could be proved, through testing, to be the best. This stepwise method provided the opportunities necessary to test components beyond their usual operational demands. As the committee pointed out, however, NASA deviated from this usual plan and, instead, attempted to reduce testing and redesign efforts by using computer computations.

NASA claimed that the experience to generate these computer models came from the joint Air Force and Pratt & Whitney experimental XLR-129 engine and from the turbomachinery programs of the Marshall Space Flight Center and Pratt & Whitney. For the entire shuttle project, not only for the main engines, NASA had adopted a success-oriented strategy that assumed that components and tests would perform as expected on the first try. Based on this assumption, component testing was shortened and the supply of spare parts was drastically reduced. The objective of this strategy was to save time and money by eliminating parallel development and hardware testing.[18] While the effects of this organizational strategy were felt throughout the entire engine development program, the effects on the high-pressure turbopumps undoubtedly caused the most controversy. As noted, some components – for example, the combustion chamber and the computer controller – were tested as components and were performing relatively well, but the high-pressure turbopumps had severe problems. A component testing facility had been made ready for the turbopumps and testing was initiated. However, the program was later dropped when NASA claimed that progress was not being made fast enough to be useful in the rest of the development program.

Members of the Ad Hoc Committee were not convinced that curtailing the turbopump testing was appropriate. They clarified their insistence on component testing by underscoring that they understood and appreciated the difficulties associated with developing a test rig of the complexity which was necessary to understand fully the turbopump technology, and acknowledged that developing these facilities can possibly take years.[19] In fact, Dr. Frosch claimed that building a suitable test stand was not any easier than getting the engine itself to operate properly and, therefore, a component-development test stand was not necessarily a profitable option. He reported that NASA had decided to reactivate an existing engine test stand and test the turbopumps not as components, but rather as part of the entire engine system, using the engine as the test bed.[20] This plan was carried out during the time between the two hearings held before the Senate Subcommittee, March 1978 to February 1979.

During the second hearing, Dr. Covert strongly criticized NASA's all-up approach and its decision not to accept the committee's recommendation. He again recommended that component test facilities be developed for the turbopumps even though the investment in time and money would be significant. As an example, the committee pointed to the source of a major failure that occurred on the recently reactivated engine test stand. A fire broke out in the main oxidizer valve, resulting in extensive damage to the engine and the test stand. In the committee's opinion, this incident may have been prevented if the proposed component test stand had been available to isolate the problem before exposing the entire engine to the untested components.[21]

The lack of a useful testing rig had made the high-pressure turbopumps the pacing item in the engine development program and forced NASA to compromise, at least temporarily, the initial design and performance specifications. In order to avoid delaying the entire shuttle program, an interim engine, qualified for safe early flights, was now the goal. An ongoing engineering effort would be required to support the shuttle and eventually to reach full power and the longer life requirements. This, the committee claimed, could be more economically and effectively carried out through the use of a component test stand rather than a full engine test stand. Again, for the second time and, although Dr. Covert described the approach as "far from ideal,"[22] NASA defended its philosophy and refused to change it.

Spare Parts

Another consequence of the success-oriented strategy adopted by NASA was the decision to reduce drastically the spare parts inventory. It was reasoned that since failures would not occur, spares were not necessary. Moreover, maintaining inventories of these parts was expensive. Combined with the decision to abandon component testing for the high-pressure turbopumps, the limited number of spare parts and available engine hardware magnified any testing malfunctions and further delayed the engine development program.[23] The engine's rotating machinery and the high-pressure pumps, discussed earlier in Chapter 6, were especially troublesome.

Two test failures that culminated in severe fires in December 1978, one of which was referred to earlier, further highlighted the shortage. The spare parts issue was considered by the Ad Hoc Committee to be even more critical than before. NASA, however, was claiming that spares were not a constraint.[24] In their view, coming up with solutions to technical problems was pacing the engine development schedule. This controversy over spare parts is another issue on which NASA and the Ad Hoc Committee clearly had differing philosophical views concerning not only how to handle the problems encountered, but, more fundamentally, exactly what were the problems and constraints.

Communication: Easier Said Than Done

Communication breakdowns are often discovered to be at the root of many serious ethical dilemmas. In our framework, an "ethical engineer" must be responsible for communicating his or her expert knowledge within the organization. One prerequisite for effective communication involves clarifying and defining the terminology commonly used to discuss professional or technical issues. In Clouser's words, this is the "concept clarification" aspect of applied ethics. Different individuals or organizations may apply their own connotation to the same term. A case where the importance of a word's definition is evident involves the engine preliminary flight certification program and, more specifically, what was meant by a "flight-configured" engine.

While emphasizing the importance of designing safety and reliability into the engine, the Ad Hoc Committee confirmed its definition of a flight-configured engine. To this end, the committee stressed that the total test time accumulated on an engine does not equal the time accumulated on a flight-configured engine since design changes are constantly being implemented.[25] Again, it is management's responsibility to support procedures that insure that the actual flight engines to be tested are equivalent to the test engines to be fired in the preliminary flight certification process.

At the time of the first Senate hearing in 1978, the Ad Hoc Committee was of the opinion that these necessary procedures were in place within NASA.[26] During the time of the follow up investigation, however, the Ad Hoc Committee apparently was not sufficiently satisfied with NASA's definition of "flight-configured." The committee felt that the set of tests planned for the preliminary flight certification was adequate, but that the test engines themselves were not of the same configuration as the flight engines. (Note that "same" does not mean equal in this case, but, rather, it means alike in the essential technical characteristics.) It therefore concluded that NASA was planning to complete engine certification prematurely for the first manned orbital flight and, again, the committee highlighted the necessity for management procedures to deal with conflicts in an organized and prepared manner.[27]

As can be expected in reviews of highly technological systems such as the shuttle main engines, there were instances when the abundance of technical information became confusing to the attending parties. Additional factors contributing to the confusion could be the limited amount of time in which the Ad Hoc Committee was asked to respond to the request, and the nontechnical background or inexperience in a specific area of some of the participants. At any rate, one of the confusing issues involved the material chosen for the construction of the turbine blades. The Ad Hoc Committee felt that the high-strength nickel alloy, MAR-M 246, needed to undergo further material characterization tests.[28]

This is another situation where a clarification of terms is necessary. A "characterized" material implies that it has been studied and its behavior and properties are understood within certain specified temperature and pressure ranges, aging and operating cycles, and the like. It is important to state the upper and lower limits of these conditions under which the material was tested and, hence, is now supposedly understood. Not surprisingly, there was discussion in the Senate hearings concerning the boundaries of the environments to be tested, whether they should extend only to, or perhaps beyond, the operating environment of the engine. NASA reported that further characterization studies were already under way and, when asked by the Senate subcommittee if there were any plans to possibly change materials, it claimed that all of its consultants, including Pratt & Whitney, unanimously and fully supported the decision to use MAR-M-246.[29] Senator Stevenson introduced another confusing point and asked for clarification; he was under the impression that Richard Mulready, of Pratt & Whitney, did not support the material selection. (Later, in a letter written by Mulready, the situation was finally cleared up; he, in fact, stated that the MAR-M 246 alloy would not have been Pratt & Whitney's choice.)[30]

"Component testing" may be defined differently by various organizations. The size of a component to be tested, in terms of complexity, may vary according to the actual systems under development and the responsible organization. Also, organizations may vary in their decisions concerning what relevant data should be generated via the tests. As a corollary, the term "failure" is relative to a particular organization as well. The Federal Aviation Administration considers a turbine blade crack a failure and does not consider the additional time until the crack grows large enough to fracture completely; other organizations, including NASA, include this time in their calculations of safety factors.[31] Some sources claim that component testing, according to their definitions and standards, was not fully carried out for the main engines.[32] The advanced rotating machinery and the high-pressure turbopumps are the most frequently discussed examples.

Different perspectives and agendas can also hinder communication. The Senate subcommittee, for example, was mainly concerned with the congressional budgeting schedule and the status of the engine with respect to when it would be ready for flight. It wanted specific facts, dates, and examples. The Ad Hoc Committee, on the other hand, was primarily concerned with insuring that NASA had assigned the proper priorities with respect to its organizational philosophies and policies, so as to insure the safe and timely development of this complex, innovative technology. The Covert Committee was, in a sense, an "ethics committee." It acted as an ombudsman between NASA and the Senate. It had no authority to enforce its findings and recommendations.

The instances that invoked frustration in the Senate subcommittee involve that committee's attempt to solicit specific examples to accompany the Ad Hoc Committee's testimony. For example, Dr. Covert voiced the committee's concern that the schedule was so compressed that a working atmosphere was being created that inhibited "realistic evaluations" of problems. Instead of providing specific cases in which this was occurring, Dr. Covert, tacitly understanding the situation, replied that his committee indirectly got that impression from the manner in which NASA and Rocketdyne employees answered the Ad Hoc Committee's inquiries. Senator Stevenson also sought the committee's educated estimate of the probability that NASA would be able to maintain the ambitious development and testing schedule. Rather than responding with questions on the odds, Dr. Covert again relied on his knowledge of engineering practice. He explained that with previously untested, highly complex systems, the rate at which successful testing experience is accumulated increases with time. In reality, this was the most accurate statement that could be made. Initially, as was previously discussed, experience is gained at a very slow pace but, as the major flaws are discovered and corrected, the system matures at a much faster rate. Finally, Senator Stevenson questioned the relationship between the high-risk factor, which the Ad Hoc Committee suggested was present, and the tight schedule for the first manned orbital flight. The senator expected, in his words, the "same nonanswer" to his statement. He probably felt he received it! Dr. Covert responded confidentially that the engine itself would provide those answers.[33]

The dialogue just described clarifies both the Ad Hoc Committee's and the Senate subcommittee's objectives. The Covert Committee suggested some general organizational goals for NASA to follow. In our terms, they proposed an ethical framework for the organization. These included management styles that encouraged a productive working environment and strategies for reevaluating development and testing phases. The Ad Hoc Committee was not interested in accusing NASA of wrongdoing. Rather, its short-term goal, as was NASA's, was to insure a safe, first manned mission. Long-term, both economics and the technological success of the overall shuttle project were at stake. Cicero's Creed II was satisfied here, for safety dominated the schedule. Although the Senate subcommittee was also concerned with the shuttle's safe launch and eventual success as well, it had a more immediate and pragmatic task to accomplish. Specifically, it was responsible for reporting to and advising Congress on the status of the engine, primarily as it related to the present and next fiscal year's budget.

Table 10.1 summarizes the Covert Committee recommendations and NASA's response. The charge of the Ad Hoc Committee was advisory, not compulsory. Although NASA depended upon the subcommittee's recommendations for future budgetary coverage, and hence the Ad Hoc Committee's advice, NASA remained the technical expert. The Covert Committee could advise, question, and ask for evidence. It could press NASA to reexamine its assumptions regarding all-up testing and it could monitor the workings of the organization more closely to insure accountability. NASA, in return, could always justifiably rely on its success-oriented philosophy, its past record, and its innovative edge to bolster its position as an expert.

Table 10.1. *Covert Committee recommendations and the NASA responses*

Covert recommendation	NASA response
Delay PFC until more engine maturity	No
Build a component test stand	No
Design, build, and test an improved oxidizer pump	Not now – may be part of "second-generation" engine
Plan more engine tests and hardware	Agreed
Build and test a line-replaceable heat exchanger	No
First manned orbital flight in April–May 1980	Late 1979 has 50–50 probability

Was there sufficient evidence in 1978 for Congress to cancel the shuttle project? The political environment in which the shuttle was funded has been characterized as a coalition of proshuttle advocates consisting of the failing aerospace industry, the Department of Defense, and the scientific community. These interest groups successfully lobbied Congress, counteracting the societal mandates to attend to social reform issues and possibly end manned space exploration. This case indicates how these varied perspectives, in turn, prompted a critical investigation of NASA's progress. Both the Defense Department and the scientific community depended on the launch of the shuttle to carry out their goals. The astute political coalition building that occurred during the initial funding stage was paying off. The tenor of the Covert Committee's investigation was to insure a "safe launch." NASA's philosophy regarding how to balance cost, safety, and schedule had not significantly changed since the Santa Susana test stand was constructed. Then, contingency funds were spent to meet a deadline. Subsequently, Congress was asked for more money. With safety as the overriding value, the Covert Committee viewed both the schedule and the cost as being variable. By allowing "the engine to be the guide," a development strategy akin to the pedestrian, evolutionary approach outlined earlier by A. O. Tischler was suggested. Such an approach, however, would not have been achievable within the original budget and time frame. It took nine years for Congress to grant appropriate development costs and by that time, the decision to use all-up testing had locked engineers into a specific design of the engines – a constraint that would be felt long after the successful launch of the shuttle *Columbia* in 1981.

NASA management maintained its upper hand in terms of justifying its development and testing approach before the scrutiny of the Ad Hoc Committee. Yet, its difficulty in engine development, schedules, and finances continued.

The Yeoman's Task: Mending a Breakdown of Trust

There were major technical concerns and areas where cost overruns were significant. Figure 10.1 illustrates examples that were reported at NASA's monthly General Management Review Sessions, designed to keep top management informed of the status of each placement of the project. The space shuttle main engine project was unanimously described as one of the leading culprits. In 1971, the original cost estimate for the shuttle was reported to be $5.15 billion. In 1979 this estimate was raised by 11.4 percent to $5.737 billion and, specifically, the estimate for the engines was increased by 44.6 percent, which accounted for almost half of the total increase.[34] Even before 1972, the engines had been marked as the pacing component of the shuttle's development. It is not surprising that this component was singled out for a detailed technical and management review.

In June 1979, the House of Representatives Subcommittee on Space Science and Applications held hearings with NASA as it had twice a year since the start of the space program. As usual, the subcommittee's task was to discuss the progress and problems encountered with the development of space technologies. Reflecting the long-time collaborative effort of this committee and NASA, Dr. M. S. Malkin, director of the Space Shuttle Program for NASA, proudly reported in 1975 that NASA has been closer to the House subcommittee than ever before. In that same meeting Representative Don Fuqua, chairman of the House subcommittee, applauded NASA, "You have kept us informed, told us the good and the bad."[35]

The June 1979 meeting, however, was to discuss NASA's recent disclosures regarding significant problems with the shuttle's finances and its flight schedule. This followed the Ad Hoc Committee's investigation that had highlighted problems previously missed or underrated. Representative Larry Winn Jr., who had been involved with the program since its inception, reiterated his support for NASA. "I am thoroughly convinced," Winn stated, "that this expenditure of tax dollars is more than justified." Loyal support, however, did not alleviate his displeasure at what he termed the

General Management Review Sessions
Monthly Project Status

1973

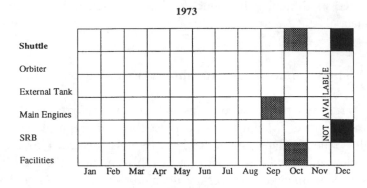

1974

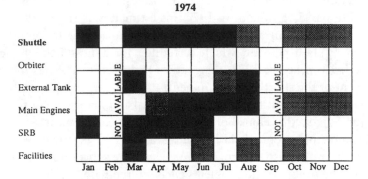

1975

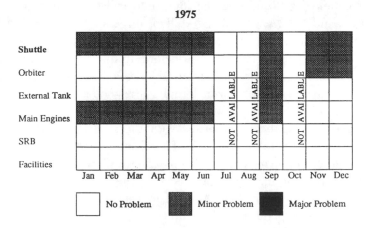

Figure 10.1 Summary of the NASA monthly general management review sessions. These are intended to keep top management informed of the status of each placement of the project.

General Management Review Sessions
Monthly Project Status

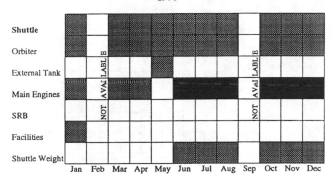

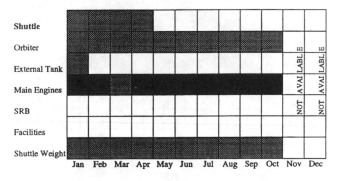

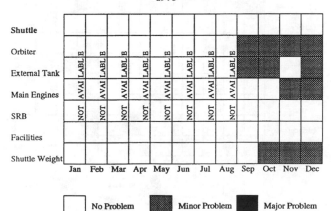

Figure 10.1 (continued)

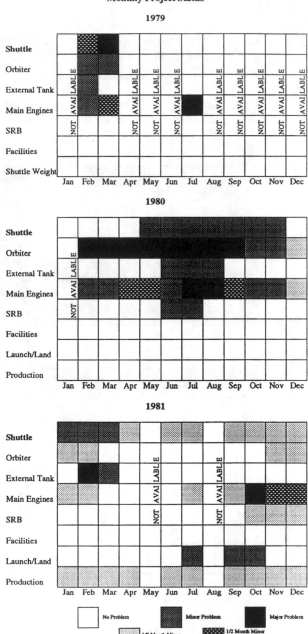

Figure 10.1 (continued)

"yeoman's task" that the congressional subcommittee faced.[36] The recent NASA cost overruns had come as a surprise to Mr. Winn and the rest of the House subcommittee. Only one year earlier, NASA had reported that a 10 percent cost growth and a schedule slip was needed. The dollar amount attached to this change was $185 million. The request, while not popular, was still considered reasonable, given the complexity and size of the shuttle program. However, just a few weeks after these supplements had been fought for and passed in Congress, NASA requested another large overrun and another schedule slip. They asked Congress to reprogram another $80 million for 1979 and for an additional $220 million for 1980. Sensing the gravity of these requests and the misconception of NASA's progress, Congressman Winn clarified his personal anger.

> Time and time again we asked the same question – "are we on the cost plan of $5.22 billion and are we on schedule?" And time and time again we were told that we were on schedule and within the $5.22 billion.
>
> Over a year ago we began to realize that supplemental funding would be required and a slip in schedule would occur. However, this did not seem to be unreasonable. . . . Now . . . we are looking at a substantially larger overrun and a further schedule slip. This makes me feel like a total failure. After spending all these years traveling from one briefing on the Shuttle to the next, I feel like I have totally wasted my time. The visits gave me confidence to go before my colleagues in the House of Representatives and fight for the necessary support to move this program along. I can now see that it was a false sense of confidence.[37]

The subcommittee, in sum, had supported NASA in Congress and now it did not know whether it had been mislead, or whether people were not aware that the program would exceed its budget limits. Of the two, Winn preferred to have been mislead "because it would at least show [that] someone was assessing . . . the program."[38]

The yeoman's task Winn referred to was essentially moral in character. "We must reestablish our own confidence before we can face our colleagues in the House." The series of questions and answers that followed was aimed at that goal. Robert A. Frosch, who had extensively testified before both the February and March subcommittee hearings in response to the Covert Committee's and other outside experts' opinions, provided NASA's defense. Essentially, he highlighted the status of the shuttle program, and justified the fiscal year 1980 budget amendment of $220 million.

The subcommittee's questions were not primarily directed at a technical inquiry of the facts. Rather, subcommittee Chairman Fuqua inquired "when did you first learn about the problem that there might be some accelerated cost increases?"[39] Essentially, Frosch reported that by the time he appreciated that the problem was real, the 1980 budget had been closed and finished. He did not consider renegotiating the original amount requested because "institutionally," he did not have "any valid basis" to request more funds. By "valid," Frosch meant an exact dollar figure. Representative Fuqua responded to this by impressing upon the NASA administrator the importance of communicating in a timely manner. "It seems to me," he advised, "that when you had an informal knowledge . . . that there might be some difficulties . . . [reporting this to either the subcommittee or the OMB] would not have helped the situation . . . but I think the Committee and the Subcommittee would not have been as surprised as we were when you came back to us."[40]

Frosch repeated the adage we have heard often in this case study: "hindsight is a wonderful thing and as I look back, if we had all done something else, then the conclusion would have been different. It is always a question of difficult judgment as to when you cross the line between crying wolf because you think something might happen and informing people because you're pretty sure something might happen."[41] True to the "success-oriented" philosophy that characterized the shuttle management, Frosch had insisted that NASA managers thought that situation could be handled with the funding they already had. He clarified this later as a matter of "institutional pride." In reality, NASA had been routinely using production funds as a source of program reserves for its design, development, test, and evaluation (DDT&E) activities.

Winn was not of the NASA culture and found Frosch's explanation hard to comprehend. Voicing his mistrust, he declared, "after the statements today and the informal meeting yesterday, I am no more convinced that you are answering the questions now than when we met about three months ago." Directing the inquiry to try to understand the motivation for the repeated requests for more money and the way the shuttle project was managed in general, Winn tried to assess the effect that having a reserve in Congress had on NASA's fiscal policy. He painted the following scenario: "The contractor feels, well, we are so far into this contract and we are so far along on the schedule. The Centers feel they are on schedule or could make schedule but they are running short of money. But there's this big fat reserve sitting up there in Washington and we will eat into that for awhile."[42]

Frosch had managed programs both with and without a reserve and felt that requests had been carefully scrutinized and not easy to procure. His concern was rather with the "tightly managed" system NASA operated under that inevitably produced surprises such as the current cost overruns. Specifically Frosch was concerned with the need to revise management and information procedures. While he was aware of the need to make changes, he cautioned the subcommittee that no formal restructuring had taken place, so "we could surprise ourselves again."[43] Displaying his "nervousness" about the $220 million request, he satisfied the subcommittee's queries about sincerity. Mr. Fuqua ended the questioning by stating "we hope that next time you come back that you feel more comfortable with the $220 million."

Conclusion: Technical versus Moral Errors

Although NASA's reputation was salvaged, the additional funds came through Congress with considerable "strings" attached. Trust once broken is not easily restored. Two months after the June hearings, the subcommittee met again and directed its comments to sixteen specific problems. First, it had "serious reservations" about the "adequacy" of the current $220 million request. Given that it had underestimated its budget for four successive years, NASA was now advised to utilize, on an annual basis, a financial assessment team "above the level of the Office of Space Transportation Systems" to assess its calculations. Several other recommendations regarding budget procedures and the estimation of schedules were made. NASA had failed to communicate in a timely manner that its 1980 budget request was not adequate and the subcommittee did not accept NASA's explanation for doing so as "institutional culture." Instead, it concluded that given the chronology of events, "information" presented by NASA during the Subcommittee review of the Space Shuttle Program in the fall of 1978 and in February 1979 "was less than candid." The following obvious recommendation was made: "NASA should keep the Subcommittee fully and currently informed with regard to all events which may significantly affect the Space Shuttle cost and schedule commitments."[44] Furthermore, the urgent and critical national and civilian reliance on the shuttle was cited by the subcommittee as not being adequately recognized. NASA and the Defense Department were to report directly to the subcommittee regarding a comprehensive assessment of both the delays on military and civilian operations.

Each of the recommendations forced NASA to reassess and change the way it had been operating. Prior to the June subcommittee meeting, NASA and its contractors had dealt openly and honestly with the technical problems encountered in the shuttle's development. The follow-up recommendations and strict monitoring imposed on NASA's fiscal authority proposed in August suggest that – as in surgical training – technical errors, reported in a "timely manner," are forgiven. Only when it became suspect that NASA was "covering up" or "being less than candid" did the Congress exercise its authority to closely monitor the project. The "moral error" provoked the breakdown in trust.

Notes

1. Alex Roland, discussion at University of Pittsburgh, June 27, 1989 with authors.
2. K. Danner Clouser, "Medical Ethics: Some Uses, Abuses and Limitations," *New England Journal of Medicine* 293(8), 1975: 384–7.
3. Charles Bosk, *Forgive and Remember: Manging Medical Failure* (Chicago: University of Chicago Press, 1978).
4. Ibid., pp. 37–45.
5. Ibid., pp. 45–51.
6. Ibid., pp. 51–61.
7. Eugene E. Covert, *Technical Status of the Space Shuttle Main Engine: A Report of the Ad Hoc Committee for Review of the Space Shuttle Main Engine Development Program* (Washinton, DC: Assembly of Engineering, National Research Council, National Academy of Sciences, March 1978), p. iii.
8. U.S. Congress, Senate, Subcommittee on Science, Technology and Space of the Committee on Commerce, Science, and Transportation, *Report of the National Research Council's Ad Hoc Committee for Review of the Space Shuttle Main Engine Development Program*, 95th Congress, 2d session, March 31, 1978, serial 95-78, 28-027 O, p. 1.
9. Ibid., p. 3.
10. Ibid., p. 8.
11. Ibid., p. 6.
12. Ibid.
13. Ibid., p. 10.
14. Ibid., p. 7.
15. Covert, p. 16.
16. Ibid.
17. Subcommittee on Science, Technology and Space, March 1978, pp. 58–60.
18. Covert, *Technical Status of the Space Shuttle Main Engine*, pp. 1–2.
19. Subcommittee on Science, Technology and Space, March 1978, p. 1094.
20. Covert, *Technical Status of the Space Shuttle Main Engine*, p. 58.

21. U.S. Congress, Senate, Subcommittee on Science, Technology and Space of the Committee on Commerce, Science, and Transportation, *NASA Authorization for Fiscal Year 1980, S. 357*, 96th Congress, 1st session, February 21, 22, 28, part 2, serial 96-1 43-135 O, 1979, p. 58.
22. Ibid., p. 1093.
23. Ibid., pp. 1093–4.
24. Ibid., p. 1125.
25. Covert, *Technical Status of the Space Shuttle Main Engine*, p. 19.
26. Ibid., p. 22.
27. Subcommittee on Science, Technology and Space, February 1979, p. 1095.
28. Subcommittee on Science, Technology and Space, March 1978, p. 12.
29. Ibid., p. 66.
30. Ibid., pp. 81–2.
31. William P. Rogers, *Report of the Presidential Commission on the Space Shuttle Challenger Accident* (Washington, DC, June 6, 1986), vol. 2, p. F-2.
32. Subcommittee on Science, Technology and Space, February 1979; Rogers, *Reports*, vol. 2, p. F-2; Richard S. Lewis, "Whatever Happened to the Space Shuttle?" *New Scientist* 87, July 31, 1980: 356–9.
33. Subcommittee on Science, Technology and Space, March 1978, pp. 7–9.
34. U.S. Congress, House, Subcommittee on Space Science and Applications of the Committee on Science and Technology, *Oversight: Space Shuttle Cost, Performance, and Schedule Review*, 96th Congress, 1st session, June 28, 1979, serial 31, 50365 O, p. 147.
35. U.S. Congress, House, Subcommittee on Space Science and Applications of the Committee on Science and Technology, 94th Congress, 1st session, *H.R. 2931*, February 19, 20, 1975 [no. 2], vol. II, part 3, serial 49-296 O, p. 1905.
36. U.S. Congress, House, Subcommittee on Space Science and Applications of the Committee on Science and Technology, *Oversight: Space Shuttle Cost, Performance, and Schedule Review*, 96th Congress, 1st session, August 1979, serial U, 49-320 O, p. 1.
37. Ibid., p. 2.
38. Ibid.
39. Ibid., p. 17.
40. Ibid., pp. 19–20.
41. Ibid., p. 21.
42. Ibid., p. 22.
43. Ibid., p. 36.
44. U.S. Congress, House, Subcommittee on Space Science and Applications, August 1979, p. 1.

11

A Judgment Call or Negligence: Maiden Voyage of the Challenger *Delayed Six Months*

Challenger's maiden flight, the sixth space transportation system launch, was originally scheduled for January 20, 1983, but was postponed four times because fuel leaks were found in the orbiter's high-performance liquid-fuel engines. Complications of this magnitude require millions of dollars to remedy; damaged equipment, delays of future flights, and technician's overtime pay are all extremely costly in the space business. Indeed, for this particular case, the *Washington Post* reported that technician overtime alone was up to $3 million a day.[1] Due to the delay, the flight manifest was now in serious jeopardy. Mission 6 along with Mission 8 was to release tracking and data relay satellites whose function was to support *Spacelab*; the high-priority *Spacelab 1* mission was scheduled for September as Mission 9. Mission 10, also scheduled in 1983, was another high-priority flight to carry a Department of Defense payload. Yet, if the hydrogen leaks continued undetected, they could have resulted in a potentially catastrophic failure in flight.

The incident began innocently enough: On December 18, 1982, a short, twenty second flight readiness firing of *Challenger*'s three engines was conducted. Following the firing, a high concentration of hydrogen gas was found in the aft end of the orbiter, creating a potentially hazardous situation in which a fire could break out. After two weeks of leak tests and analyses, NASA still had not determined if the hydrogen had come from a plumbing leak on the inside of the orbiter, was caused by the tanking facility, or was forced into the orbiter by the existing air flow patterns. While concerned, James Kingsbury, Marshall's director of science and engineering, did not expect the incident to affect the launch schedule for the upcoming mission.[2]

In reality, such an incident had been "predicted" by Pratt & Whitney when that company formally protested the original contract award to Rocketdyne. Further, in April 1990, hydrogen leaks still continued to plague the shuttle, resulting in the entire fleet being grounded for five months.

Hydrogen Manifold Leak: Engine No. 1

In order to determine the source of the leaking hydrogen, a second flight readiness test firing with additional instrumentation was conducted on January 25, 1983.[3] That test confirmed that the gas leak was coming from the main engine and resulted in the launch being postponed until at least March.[4] Four days later, and more than a month since the leak was first discovered, a three-quarter inch crack in a hydrogen coolant outlet manifold line on the number 1 main engine was identified as the source of the gas. A pair of Rocketdyne employees, using a sniffer and a mass spectrometer, traced the hydrogen flow until they could feel the gas leak with their hands.[5]

The manifold line carries the cryogenic hydrogen used to cool the main combustion chamber wall to the low-pressure hydrogen turbopump. After the manifold was welded to the main combustion chamber, it was placed on a machine tool for additional work. The operator made a mistake in the machine setup and the tool struck the outlet section, which tore loose from the manifold. The decision was made to hand-weld a new outlet section in place of the damaged one while the unit was still attached to the injector. Under normal circumstances, such welds are made with precise machine welders and then heat-treated to relieve residual stresses in the metal. Since the unit was already attached to the main combustion chamber, heat-treating the small section was not possible because of the effect the process would have had on the combustion chamber.[6]

The welding repair caused the protective copper lining inside the manifold to tear and this, in turn, allowed the flowing hydrogen to contact the outer Inconel 718 layer. The heat used in the welding repair also weakened the Inconel 718 material, further increasing its susceptibility to cracking. Indeed, during the engine test firings, the crack developed in the Inconel layer. A third contributing factor was the residual stresses in the area as a result of having to perform the repair through hand-welding rather than using a heat treatment procedure (which would have damaged the metal on the combustion chamber).[7]

The use of Inconel 718, because it is subject to hydrogen embrittlement, was one of the reasons cited by the Pratt & Whitney Company in protesting the phase C/D contract awards to Rocketdyne (see Chapter 5). NASA had demonstrated to the satisfaction of the investigating Government Accounting Office that the hydrogen embrittlement problem would be controlled by the use of a protective copper coating. Pratt & Whitney had asserted that such a coating method "is, at best, extremely risky and is likely to fail."[8]

In 1975, a NASA memo titled "SSME Hot Gas Manifold Materials" further addressed the decision to use Inconel. The memo stated that, if flawed, crack growth would occur in the material in warm hydrogen at very low stress levels; proof testing capability was, therefore, paramount. A curve was attached to the memo showing the relationship between flaw size that would result in flaw growth and percent of yield strength stress. The curve resembles an exponential decay where a flaw size of approximately 0.05 inches would require 100 percent of yield strength to grow, but a flaw size of about 0.30 inches would grow when subject to only 40 percent of the material's yield strength stress. The crack in the SSME's Inconel grew to 0.75 inches before it was discovered. This particular memo also stated that the curve is applicable to one cycle life only. Obviously, for additional cycles, more serious crack propagation would occur if not discovered.[9]

This was the only known repair of this type made to that point in the shuttle program; no other documented metallurgical or engine repairs were handled in a similar manner. The then associate administrator of NASA, Lieutenant General James A. Abrahamson, reflecting on this repair, stated that:

> Some engineering judgments turned out to be different from what they could or should have been at that point in time, but everybody worked on those decisions. There is a great deal of difference between negligence and a decision that is carefully looked at . . . that did not work out right. [The problems] go back to the fact that we did not have very much hardware available to us, and therefore we were reworking components.

Indeed, having limited spare parts due to budgetary restrictions, shuttle engineers were forced either to repair damaged hardware or to rework older parts.[10]

As a consequence, *Challenger*'s spare engine, although in need of a high-pressure oxygen pump, was chosen to be installed in place of the damaged number 1 engine. A high-pressure oxidizer turbopump was to be taken from the damaged engine. Procedures in place at the time required leak tests to be conducted on the spare engine after it received its "new" pump. During these mandatory tests in February, a leak in the heat exchanger was found. This leak had the potential to cause an explosion and fire in the engine which would have caused a launch abort or even loss of the vehicle and crew.[11] This condition was considered to be much more serious than the initial hydrogen leak in the manifold line.

The heat exchanger had long been recognized as a Criticality 1 component because its single point failure threatened the entire shuttle system. A 1974 NASA memo titled "SSME Critical Items List" recommended that Rocketdyne more thoroughly analyze the effects of a heat exchanger failure. The memo stated: "It appears that directly applicable analytical data are scarce. One reason may be that data derived from experiments adequately representing the configuration and physical parameters of the subsystems involved are not available." The memo supported more testing of this particular critical component at that time.[12] However, just three months before, the SSME project manager at NASA (James R. Thompson Jr.) issued a memo to the engine project manager at Rocketdyne (Paul D. Castenholz) titled "SSME Proposed Cost Reduction Items." Nine of the ten proposed cost reduction items were test reductions and five of these nine were okayed by NASA.[13]

Four years later, but prior to the first shuttle flight (April 1981), the Ad Hoc Committee for Review of the Space Shuttle Main Engine Development Program reported on two fires that were contributing to the delay of the engine development and hence the entire shuttle development. The first fire was in December 1978 and was the result of an explosion in the heat exchanger[14] (see Chapters 6 and 10). Its coil tubing was weakened by repair welding performed nearby while the heat exchanger was still attached to the engine; a leak in the tubing resulted in the explosion.

The heat exchanger is a coil system installed in the engine's hot-gas manifold through which the cryogenic oxygen flows and is heated by the gases in the manifold, converting it to oxygen gas at 3,372 pounds per square inch and 390 degrees Fahrenheit. This heated oxygen is then used for pressurization of the oxygen section of the external tank as well as for the orbiter's oxygen and pogo systems. Although the leaking heat exchanger had performed normally in nine tests, the root cause of the leak was traced back to the component's initial fabrication in 1977. A defective weld was made and eventually broke through during the engine proof tests. As in the hydrogen manifold case, the lack of sufficient spare parts forced the use of old, used hardware. In turn, this was identified as a contributing factor in the heat exchanger leak. At that time, procedures did not mandate proof-testing of the component after any repairs were made. Subsequently, such procedures were established. These mandatory tests detected the potentially fatal heat exchanger leak that delayed Mission 6 in February 1983.[15]

Hydrogen Leaks: Engine Numbers 2 and 3

Less than a month later at the end of February, cracks in the hydrogen lines of *Challenger*'s remaining two engines were found in roughly the same location as those found on engine number 1. Although the crack found in the first engine was attributed to a tooling operator error, the new discoveries led NASA to believe that an underlying generic design defect had been surviving the engine check-out tests in use at that time. In fact, it was believed that the first shuttle, *Columbia*, flew five missions with undetected engine leaks. On all of its flights, higher than expected concentrations of hydrogen gas were detected in the aft section, although the levels were then considered to be tolerable.[16] Preliminary data led managers to suspect that the brazing process used to strengthen the lines was the cause of the problem. It was thought that the temperature of the process could not be adequately controlled and the lines were overheating, resulting in embrittlement of the Inconel 625 metal.[17] That is, as was the case for the manifold leak, the heat treatment may have both damaged protective copper linings and weakened the Inconel material.

Conclusion: Some Dilemmas to Consider

General Abrahamson's comment that "there is a great deal of difference between negligence and a decision that is carefully looked at . . . that did not work out right" identifies the aspect in engineering decision making that attends to the ethical process. His assertion that the problems go back to the fact that "we did not have very much hardware available to us, and therefore we were reworking components" directs attention to the effect that an organizational decision had on individual engineering performance. If, when, or how vigorously individual engineers or associate administrators of NASA should have voiced concern or disagreement with the systemwide decision to reduce costs by cutting down on spare parts poses an ethical dilemma. According to Peter French's concept of responsibility, an individual engineer could have, at a minimum, registered a complaint – communicated to those with decision-making authority the specific assessment of possible harms caused by a shortage of parts. Whether the complaint mobilized a "critical mass" to action, or was attended to, provides information regarding one's power and authority within the organization. It also illustrates the organization's overall decision making regarding risk and its responsiveness to individual concerns.[18]

Caroline Whitbeck points out that engineers need to learn how to register these types of complaints. The educational curriculum she has developed for undergraduate engineers is designed to teach them the skills they need to carry out French's moral action.[19] As A. O. Tischler demonstrated, it takes a combination of technical expertise, authority, and a clear definition of one's goal to communicate concerns (see Chapter 8). The Covert Committee's investigation also emphasizes how an organization's culture and previous experiences affect perspectives (see Chapter 10). Although this cultural influence seems to dominate and permeate decision making, one does not know if or how it can be changed unless one challenges it.

Various technical-ethical engineering issues apparent in this case are listed here to highlight for engineers and managers the types of dilemmas they may encounter and to elicit further consideration regarding how or if they would approach their resolution. Each issue could be researched to provide more information on decision making.

1. The engine leaks and the inadequacies of testing and locating them surfaced in 1983, eight years after the "pipe crisis" was reported to Congress. As a result of the use of contingency funds during that "crisis," NASA made the decision to cut back on overtime and the extra effort being expended on the shuttle's development in order to decrease costs. Manpower was also cut; specifically, engineers and designers were laid off. The long-term ramifications of this decision may have been related to a failure by NASA to develop better leak detection tests, tests that had to be canceled to save costs. Cicero's Creed II addresses the need to assess engineering decisions by considering both immediate and long-term risks and benefits.

2. The hydrogen leak in the manifold was not found after the first test firing but only after the second. Was there any information concerning high concentrations of hydrogen that engineers could or should have effectively communicated to the managers after the first test firing? If

such information was available and communicated, did management act irresponsibly by not following up on the information? In hindsight, it was now thought that there were hydrogen leaks on *Columbia*'s first five missions, but the problem does not appear to have been aggressively pursued. Why was the problem allowed to continue when the consequences of a failure in the region were clearly known? Acting responsibly and competently requires one to communicate what is known and also what is not known; the critical piece of information not known in this case was where the excess hydrogen was coming from.

3. NASA was developing improved leak detection test procedures throughout the shuttle project. Did this development (of test procedures) receive the necessary funds and support to keep up with the design changes of the engines? Or was the development of adequate test procedures stressed only after failures were encountered? In summary, were the leak tests state of the art? Did they satisfy the requirement of competency?

4. Why was the hand-weld repair of the manifold, which was still attached to the engine, the only documented repair of its kind? There were other metallurgical and engine repair decisions made but none was similar to the decision made in this instance. Was this type of repair chosen mainly because of the serious flight scheduling consequences which would result if Mission 6 were significantly delayed? Were the engineers who made the decision to hand-weld fully cognizant of the susceptibility of the hydrogen embrittlement problem with the Inconel 718? Competence and attention to safety by the individual engineer may have been compromised by an organizational decision to accept a high risk.

5. The *Challenger*'s engines were expected to operate at 109 percent of rated power (which was more than *Columbia*'s engines were rated) because *Challenger* was to carry heavier Defense Department payloads. Was the Inconel material chosen for the manifold fully characterized and consistent with the requirements for *Challenger*? The memo discussing crack growth in Inconel (for one cycle life) and reusability conditions indicated that characterization had been carried out. Who had access to the information about how it was used? It was suspected that the manifold outlet line cracks were a generic problem in the main engines, attributable to a design flaw. Did NASA know that the temperature of the brazing (soldering) process could not be sufficiently controlled to eliminate cracking? Was such knowledge available

outside the organization? Was the brazing process state of the art? Each of these issues involves the concept of competence.

6. The lack of hardware, particularly spare parts, was identified in both the manifold and the heat exchanger leak problems as a contributing factor. Early in the shuttle project, when the decisions were being made to minimize spare parts inventory, did the engineers and managers communicate the need for spare equipment to the organization? Were the ramifications of those decisions understood?

7. NASA was understandably interested in keeping expenditures down. However, the cost reduction items, especially as written in the memo discussed earlier, often involved the elimination of tests. What line of authority was required to be followed when tests were eliminated? To what extent were the individual engineers working on the affected components consulted before any tests were eliminated? Were long-term effects on the shuttle's development and flight safety studied?

8. In another memo discussed previously, recommendations were made to conduct tests to assess accurately the behavior of the heat exchanger in its actual operating environment. This was a critical component and hence the data requested could be considered basic safety data. Were these tests ever conducted? This is a particular concern since three months earlier, the memo recommending the deletion of some SSME tests to save money was issued. Was NASA's policy concerning testing flight hardware communicated to all levels within the organization? Since conducting tests and reducing costs (in the short run) are conflicting objectives, were the trade-offs clear to the program managers? To what extent were the risks of not testing considered?

9. The heat exchanger leak was blamed on a defective weld made in the manufacturing process. Was the process state of the art? The decision to use welds as opposed to other types of fasteners was an item protested by Pratt & Whitney. Molds were chosen to reduce the weight. Was this decision well thought out? In the end, the engines weighed less than required by the design criteria. Would it have been advantageous to use mechanical fasteners, although heavier, in certain hard to inspect locations?

10. The hot-gas manifold and the heat exchanger were closely related components. It was well understood that the heat exchanger was a Criticality 1 component whose single point failure could cause the loss of the vehicle and crew. It was reported, after the manifold outlet line cracks were discovered to be a generic problem, that *Columbia*

flew five times with higher than normal concentrations of hydrogen in the aft section.

Notes

1. Thomas O'Toole, "Space Shuttle Friction Is Worsened," *Washington Post*, March 16, 1983, p. A21.
2. "NASA Studies Gas Problem on Challenger," *Aviation Week and Space Technology*, January 3, 1983: 21.
3. James Kukowski, "Second Test Firing of Challenger Engines Set for January 25," NASA News: Release no. 83-4, January 20, 1983.
4. "Costly Slowdown for Space Shuttle," *U.S. News & World Report*, February 7, 1983.
5. "Shuttle 6 Rescheduled for Mid-March," *Aviation Week and Space Technology*, February 7, 1983: 28–9.
6. Craig Covault, "Shuttle Engines Spark Broad Review," *Aviation Week and Space Technology*, February 28, 1983: 18–20.
7. Ibid.
8. Peter Masley, "Space Shuttle Contract: A Techno-Political Snag," *Washington Post*, January 16, 1972, pp. A1, A15.
9. NASA memorandum from EA, Director, Engineering and Development to ES, Chief, Structures and Mechanics Division, "SSME Hot Gas Manifold Materials," June 2, 1975.
10. Covault, "Shuttle Engines Spark Broad Review."
11. Ibid.
12. NASA memorandum from LA2, Manager for Systems Integration, Space Shuttle Program to SA51, Manager, Main Engine Project, MSFC, "SSME Critical Items List," September 9, 1974.
13. James R. Thompson Jr., Manager, SSME Project, MSFC , memorandum to Paul D. Castenholz, Vice-President and SSME Program Manager, Rocketdyne Division of Rockwell International Corp., "SSME Proposed Cost Reduction Items," June 20, 1974.
14. U.S. Congress, Senate, Subcommittee on Science, Technology and Space of the Committee on Commerce, Science, and Transportation, *NASA Authorization for Fiscal Year 1980, S. 357*, 96th Congress, 1st session, February 21, 22, 28, 1979, part 2, serial 96-1, 43-135 O.
15. Covault, "Shuttle Engines Spark Broad Review."
16. Wayne Biddle, "Crippling of 2d Space Shuttle Tied to Design Flaw," *New York Times*, March 2, 1983, pp. A14, A24; "Engine Repairs Delay Launch of Challenger," *Washington Post*, March 2, 1983, p. A5; Craig Covault, "New Challenger Engine Cracks Found," *Aviation Week and Space Technology*, March 7, 1983: 23–5.
17. Covault, "New Challenger Engine Cracks Found."
18. Peter A. French, *Responsibility Matters* (Lawrence: University Press of Kansas, 1992), pp. 71–8.
19. Caroline Whitbeck, "The Engineer's Responsibility for Safety: Integrating Ethics Teaching into Courses in Engineering Design," paper presented at the ASME Winter Annual Meeting, Boston, MA, 1987.

12

Against the Odds:
Ethical Decisions and
Organizational Goals

In Chapter 2 we identified three principles of engineering ethics: the principles of competence, responsibility, and Cicero's Creed II. A five-part scheme, adapted from Simon, to model how the organizational context influences the way engineers make design and performance decisions was outlined in Chapter 4.

1. Decisions are composed of factual and ethical components.
2. The ethical components of a decision are influenced by the decision maker's organizational context.
3. The organization uses influence–authority relations to decompose large tasks into manageable tasks.
4. Because of the complexity of organizational tasks, decision makers satisfice in making decisions.
5. Because organizational tasks are connected through the hierarchy, decisions made in one part of the organization can become constraints for decisions in other parts of the organization.

In this chapter we review the engineering design process surrounding the development of the space shuttle main engines (SSME), and document examples where individual engineers demonstrated very high levels of engineering competence. To meet the design goals, they had to push the limits of the technology far beyond the known state of the art. They employed new materials in the SSME, created new analytic methods (computational fluid dynamics), and proposed bold designs (aerodynamic stability of the delta wing design) for a "production" technology. With only a few

223

exceptions, discussed primarily in the sections on performance decisions and risk analysis, we found little evidence of individual engineers not exercising competence in the shuttle program.

Several management decisions, however, compromised individual engineers' ability to act competently and responsibly and to minimize harm by assessing risk in both carrying out their engineering designs and making performance decisions. This chapter identifies these decisions. Using Simon's model as a template, we first examine the organizational context of engineering design decisions for the SSME.

Engineering Design Decisions

Engineering is a knowledge-based profession which centers on creating, developing, producing, and managing modern technology. This requires both formal training and experience. When technologies like the SSME are designed, a sophisticated organization to coordinate and monitor these tasks is critical for success.

Our cases have identified areas of knowledge where NASA and its contractors were deficient: materials characterization, fracture mechanics, an understanding of quantitative methods of risk assessment, and problems with the mathematical models and accompanying software used to design the shuttle. We briefly discuss these deficiencies. In some instances, engineers recognized their knowledge deficiencies and sought to remedy them. However, without support from management, in the form of outside consultants, new policies, or new employees, these deficiencies could not have been resolved.

What level of technical knowledge did NASA's engineers and the engineers in the contracting organizations need to design the SSME? During the political debates on the initial funding of the shuttle program, NASA leadership, particularly Associate Administrator Dale Meyers, claimed that the technology for the design of the shuttle was largely available. Meyers used the phrase "off the shelf" when discussing the availability of technology required for the SSME design. The SSME would be based on the successful J-2 engine design of the Apollo program.[1]

Although NASA certainly had reason to be confident of its engineering abilities given the success of the Apollo program, Meyers's claims proved less than accurate and were not completely backed by his engineers. At the time the initial engineering work began, the SSME required technology either at or beyond the state of the art in liquid-fuel, chemical rocket engine design. The performance specifications for the SSME exceeded existing technological limits, in some cases by a considerable margin.[2]

James Kingsbury, director for engineering (1975–86) at the MSFC, has commented that:

> The real challenge in shuttle, however, was the SSME. It was an unproven technology. Nobody had ever had a rocket engine that operated at the pressures and temperatures [of] that engine, or the [rotational] speeds of the equipment. That just hadn't been done before, because it demanded some advancements in technology. But the product was a very efficient engine in that it minimized the weight of the engine and optimized the performance.[3]

Not all of NASA's engineering management agreed that the design of the shuttle was an off-the-shelf technology exercise. We have provided the case of A. O. Tischler (at the time director of the Shuttle Technologies Office, Office of Advanced Research and Technology), who, drawing on the experience of the Apollo program, questioned the assumptions of NASA's top management, and argued that the shuttle technology was at the cutting edge, and that the development of such technology does not proceed as expected (see Chapter 8). Tischler argued for an evolutionary approach based on a better understanding of the cargo of the future. He believed that the design of the launch vehicle should be determined primarily by the type and volume of traffic.[4] However, Tischler and his like-minded colleagues were overruled.

It was this "rhetoric" of off-the-shelf technology combined with the technical reality of designing the shuttle that created an environment in which individual design engineers faced a variety of specific dilemmas. That is, their off-the-shelf knowledge base was not adequate to complete the tasks handed down by the organization's management. Typically, this gap would be filled through knowledge gained by a rigorous testing program. However, NASA's decision to save money by adopting an all-up testing philosophy put limits on engineers' prerogative to utilize this approach.

Our cases provide three examples of a knowledge gap concerning inadequate characterization and selection of materials. The first is the hydrogen leaks through cracks in the Inconel 718 metal in the SSME (see Chapter 11). These leaks caused delays in launch schedules and flight cancellations and, finally, the grounding of the shuttle fleet until the leak problems could be resolved. Pratt & Whitney in protesting the contract award to Rocketdyne had predicted such potential problems with that material. Clearly, a better appreciation of the sensitivity of the Inconel 718 material, its protective copper lining, and its performance in the corrosive environment of the SSME would have reduced the very costly technical fixes required once the shuttle reached operational status.

Second, the pipes used in the Rocketdyne test stand required a costly fix, which helped deplete the program's "contingency funds" (see Chapter 9). Management decisions were driven by cost and the cost was driven by the rhetoric of off-the-shelf expectations. The fact that test stand pipe material "bent cold" and needed to be "welded slowly" could have been determined prior to its selection through proper material characterization. The post hoc discovery after problems developed and the use of "contingency funds" suggests that an Apollo mentality still existed at Rocketdyne. If the material had been appropriate, the decision to use it would have been the mark of another all-up testing approach. When it proved to have unexpected characteristics, the contingency fund was tapped to provide a fix that would keep the task on schedule. The shuttle, however, was not Apollo. Once the contingency funds were depleted, there was no guarantee that additional resources would be forthcoming.

A third example is provided by the Covert Committee's investigation into the use of the material MAR-M 246 by Rocketdyne (see Chapter 10). The Covert Committee noted that a more complete characterization of the materials used in the high-pressure turbopump blades was in order. In particular, the engineers needed to determine as a function of temperature such critical information as the number of cycles to failure for both high and low stresses; additional analyses were needed to determine the effects of aging.[5] The controversy over the choice of materials is underscored by committee member R. C. Mulready, director of technical planning for Pratt & Whitney, who stated that the material used by Rocketdyne (MAR-M 246) would not be the material of choice of others because it exhibits instability after fifty hours at operating temperatures of 1,800 to 2,000 degrees. Mulready added that no gas turbine engine manufacturer uses MAR-M 246; rather he recommended MAR-M 247, which was developed to correct that problem or PWA 1422, a derivative of MAR-M 200, which was very completely characterized.

Certain materials used in the design of the SSME had not been used in previous technologies, and therefore their performance characteristics were unknown. Shuttle design engineers faced the dilemma of selecting untried or undercharacterized materials, while at the same time trying to meet severe budget and schedule constraints. The managerial constraints often forced these materials-based design decisions. One result was that less than competent design decisions created even more difficult performance decisions once the shuttle started to fly. Performance decisions concerning the same hydrogen leaks and cracked turbine blades proved far more costly to resolve than the forgone testing during the design stage.

To Robert Ryan, the NASA engineer who reviewed the shuttle's structural design with an emphasis on the engine following the *Challenger* accident, materials characterization is the single most important item in predicting durability. This characterization must match the details and assumptions of the analytical models used in the technical design. The key to understanding system and bracketing limitations of materials is sensitivity analysis.[6] We note that the shuttle's "durability" is far less than called for in the design specifications, and even further than the original concept of a space truck.

New aerospace engineering designs typically seek a 10 percent increase in performance. The increases in performance for the SSME were often far greater than 10 percent. For example, compared to the J-2 rocket, the SSME combustion pressure was four times greater, the rotational speed of the turbo equipment was 40 percent greater, and its thrust was 87.5 percent greater. Thus, not only did the designers of the SSME have to master the state of the art during their design of a production engine, they had to extend the knowledge base by substantial margins without benefit of the intermediate stages of test or prototype experimental engines. In fact, the performance increases were so great that some NASA officials believed the component design approach was surpassed.

James Odom, Shuttle Projects Office, MSFC, reported that:

> In the case of the [SSME], just the sheer velocities and the sheer quantities of hot gas that it takes to drive the turbine, you can't build a facility to drive it. So we had to build an engine and put it all together, because that was the only way you could get high enough temperatures and high pressures and high enough flow rates, was out of an actual engine. What that says is, when you do that, every time you want to test a pump you have to build a whole engine. And when you lose a pump, you lose a whole engine. It's a very difficult way, and a very expensive way, to really push technology as far as we did in that engine program.[7]

It is clear that Odom, and probably many others within NASA, believed that the process of designing the SSME had "outgrown" component testing. This was certainly the case during the early 1970s when critical design decisions were being made. Moreover, the budget and development schedule precluded the design and development of test equipment and methodologies necessary to support the more deliberate approach of component testing. As we shall see, the use of all-up testing had many ramifications for the design of the SSME and the shuttle.

We also observe that for the critical turbomachinery in the SSME, Pratt & Whitney, an unsuccessful bidder and potential contractor, had developed test stands that operated in the performance ranges of the SSME. Thus, while Odom's claim may have been accurate for Rocketdyne, it did not apply to the industry.

Part of the characterization of materials is knowledge of the specialized area of fracture mechanics. The shuttle's design parameters required the use of turbopumps whose components would experience extreme gas temperature and pressure environments. Specifically, these operating conditions subjected the high-pressure-stage turbine nozzles and blades to severe thermal transients that resulted in large inelastic strains and rapid crack initiation. Did the designers fully understand the complexity of the operating environments and the stress and strains under which the equipment would be placed? Did they utilize important knowledge, particularly that from the then developing field of fracture mechanics, which has proved to be critical for understanding and resolving problems? Certain conclusions must be inferred from the redesign work following testing. It appears that the design of these critical SSME components would have benefited from a better understanding of fracture mechanics. According to Harry A. Cikanek, an engineer in the Marshall Space Flight Center, most of the twenty-seven major engine failures during testing were the result of insufficient knowledge of the environment and loads imposed on the components, particularly dynamic loads on the high-pressure turbine blades[8] (see Chapter 6). To again quote James Kingsbury, director for engineering, MSFC:

> Turbine blades were a big problem. The engine has turbine blades and they are very small. An airplane jet engine has turbine blades . . . 12 to 15 inches long and 5 inches wide. Our blades were an inch long and 3/4 of an inch wide. And they ran from liquid hydrogen temperature to . . . +2000 in fractions of a second. So, they saw terrible mechanical and thermal loads."[9]

Kaufman and Manderschied at NASA's Lewis Research Center note that to attain even the short lives required for this system, anisotropic turbine blade alloys must be used. Assessing or improving the durability of these components has been contingent on accurate knowledge of the stress-strain history at the critical location for crack initiation. For example, cracking occurred in the high-pressure fuel turbopump at the airfoil base near the leading edge and in the blade root shank area. These cracks apparently initiated during the first few mission cycles because of the severe thermal transients and were propagated by vibratory excitation.[10]

Review boards attributed much of the blame for a series of early engine test failures in 1977 to inadequate turbine blade design for the high-pressure fuel turbopump. Board recommendations included establishment of allowable sheet metal cracking criteria, instrumentation to characterize the coolant circuit environment, and involvement of sheet metal with the synchronous vibration. In addition, these boards recommended that NASA conduct tests to further quantify the effects of low-cycle and high-cycle fatigue (HCF) for the Inconel alloy used in certain engine applications, conduct a fracture mechanics risk assessment for large aspect ratio subsurface defects, improve x-ray detection techniques, initiate in-service x-ray inspections, and determine the effects of hydrogen on HCF. Another major failure was due to a small leak that developed from a "through" crack in the main combustion chamber coolant discharge duct. After eleven seconds, the crack reached critical size and the duct ruptured, setting off a chain of events that damaged several components and led to the engine severing itself from the test stand. The review board recommended monitoring components to evaluate life degradation against predictions.[11]

Within the high-pressure oxidizer turbopumps, bearings were observed to crack and spall prematurely. Wear analyses indicated that these bearings were subjected to variable, sometimes very high transient axial loads, which were believed to cause the bearings to crack by fatigue. Based on depth of fatigue cracks, the maximum stress on the bearings was estimated to be greater than was desired. As a result, steps were taken both to reduce high transient loads and to improve understanding of bearing failure.[12]

After other test firing, fatigue cracking was also observed on some of the liquid oxygen (LOX) posts that were exposed to hot hydrogen flowing over tubes on its way to the combustion chamber. Design modifications improved the vibration-fatigue problem, but resulted in an increased pressure drop that ultimately shortened the life expectancy of other components. In order to understand the flow-induced vibration problem and develop techniques to avoid detrimental vibration effects, a fundamental study of the post vibration was initiated. Since the detailed flow field (LOX post array) was not known at the time, a thorough evaluation of the LOX post vibration potential was not possible.[13]

These problems and proposed solutions surfaced during the all-up testing of the SSME in the late 1970s, several years after the engineering design phase of the shuttle. Theoretical knowledge and methodology of fracture mechanics were critical in resolving these problems. The question arises: Was this knowledge available to design engineers in the early 1970s?

While certain key papers in the field of fracture mechanics were published in the 1960s, the application of this knowledge was in its early research stages. The complexity of the SSME design and the extreme demands on the materials used in the design make doubtful whether a simple technology transfer was possible. Also, all-up systems management limited testing, and component testing is where knowledge of fracture mechanics would have been most useful.

Mathematical Modeling and Computer Design Software

In order to reduce development time and cost, the SSME engineers relied extensively on analytical and computer models to test alternative designs. Although many of the models embodied in computer software had been used successfully in prior rocket designs, they were not adequate for the shuttle. A primary problem resulted from the assumption that the mathematical relations in these models could be relied on to design to the new performance levels required in the SSME. Instead of the prior practice of using these models to make extrapolations in the accepted 10 percent range, engineers attempted to extrapolate out to the greater performance increases demanded by the SSME specifications. In some cases, this did not work because the underlying engineering phenomena were not a close enough approximation to reality. Engineers had to then develop more sophisticated models and computer software to accommodate the demanding specifications of the SSME.

Robert Ryan has pointed out that:

> The first shuttle flight showed an aerodynamic moment different from prediction, causing the vehicle to loft more than expected. The cause was an aerodynamic distribution shift due to inability to correctly simulate plume effects in the wind tunnel tests. The small shift in aerodynamic distribution had a major effect on the wing load, and hence its

structural capability. . . . In order to fly safely, the trajectory shape had to be changed to fly within revised boundaries. . . . the cost is a several-thousand-pound loss of payload capability. As a result of this effect, all shuttle flights are specifically tailored with stringent day-of-launch commit criteria.[14]

Ryan went on to write that

a part or component responds differently when it is separate than it does when it is a part of the whole. As a result, we must understand interaction and sensitivities to provide the proper design. Verification becomes even more complex in that it must account for these extremes, including failures and their redundant paths, etc.[15]

An example of this is provided by Frank Stewart, SSME deputy manager in the Shuttle Project Office, who commented:

I guess the biggest benefit of the Main Propulsion test was to verify flow dynamics and some of the hydrogen/LOX system, because the ducting is different than you have in a single engine. . . . All these things were all calculated – until you actually fired, you didn't know. One thing that happened down there that we didn't foresee, didn't realize, on the SSME there's a fuel lead. We'd never had a problem firing a single engine, but when we fired a cluster the first engine had enough excess hydrogen gas that wasn't burned when the second engine ignited, we had a mini-explosion.[16]

Kielb and Griffin have pointed out that "before 1975 turbine blade damper designs were based on experience and very simple mathematical models. Failure of dampers to perform as expected showed the need to gain a better understanding of the physical mechanism of friction dampers. Over the last 10 years research on friction dampers for aeronautical propulsion systems has resulted in methods to optimize damper designs."[17]

Kingsbury has observed another type of example: In solving the problems of cracking turbine blades, the engineers "developed a technology . . . for the first time . . . called computational fluid dynamics."[18] It should be remembered that the SSME was supposed to be a production engine, not a research device or even a prototype. These engineers were forced to create, not just extend, new modes of analysis to use in their designs.

The deficiency in analytical methodologies is different from those mentioned earlier. Both fracture mechanics and probabilistic risk assessment were being used in other fields of technology, and their relevance to SSME design could have been foreseen. However, the inadequacy of the computer design models could not be determined until the technology was built and tested. As Stewart said, "until you actually fired, you didn't know."

Ryan has commented that without this type of information, it was not possible to determine analytically – in a preventive fashion – many of the defects in the machinery. The importance of such information led NASA, following the *Challenger*, to develop a comprehensive set of data reduction and data evaluation tools, as well as an automated data bank, which by 1989 contained over 1,000 engine test firings. Hence, by the late 1980s this problem was being remedied. The normal pump characteristics had finally been statistically formatted. Through the use of this database system and accelerometer data from each engine firing and shuttle flight, the pump status had been determined and with an accurate refurbishment schedule, major failures have been recently averted.[19]

A common event in state-of-the-art engineering design is that "you don't discover what you don't know until you try it." The process of making engineering design decisions is intimately connected with testing. To paraphrase Simon, testing transforms engineering judgments into factual components of the engineering design decision. The key to overcoming problems like those discussed previously with the inadequate mathematical models is discovering these limitations as early as possible in the design process. This is usually done through experimentation, including component testing. Herein lies the catch-22 of the building of the shuttle: The SSME design engineers were limited in what components they could test because top-level engineering managers of NASA had decided to adopt the all-up testing approach to design. Problems typically identified early in the design process were not discovered until relatively late. By this time, many of the major design decisions concerning geometries, sizes, and the like were fixed, and correcting design deficiencies became difficult if not impossible. Many aspects of the SSME design could not be changed without considerable expense. Thus the engineers had to resort to technical "Band-Aid" fixes and work-arounds. Even when problems were thoroughly understood and when design solutions were constructed to resolve them, engineers could not implement them.

Another organizational factor contributed to these problems. Starting in 1974, design engineers were being laid off the shuttle program. The reason

for the layoffs was cost. Design engineers are among the most expensive personnel on the program, and under the success-oriented philosophy of all-up
testing, NASA believed these engineers were no longer necessary. Their
designs were completed, and it was time to move to the operational stage and
fly the shuttle. We have talked to experts in the field of aerospace engineering
and design. They agree on the importance of two ingredients for success:
experience and experimentation. The relevant theory is straightforward. The
hard part is working out the mechanical details and determining the performance characteristics of materials and components. It is straightforward to
calculate that turbopumps have to operate at 35,000 rpm. It is another matter
to make a device that actually rotates at 35,000 revolutions per minute, and
will do so reliably for 15,000 seconds. NASA and the contracting organizations certainly had engineering personnel with the requisite experience.
However, the all-up testing methodology adopted by NASA greatly limited
experimentation. Engineers made extrapolations that could not be verified
until late in the development process.

Role of Judgment

Simon's model of decision making stresses the importance of judgment as
part of the factual component of a decision. Engineering judgment was a
prime ingredient in the design of the SSME, perhaps more important than
it normally is in the engineering design process. Virtually all new engineering designs are based, in part, on judgments. The reason is that scientific
knowledge is analytic. It only applies to that part of the design that fits
within the extant knowledge covered by theory and experimentation. For
example, engineers may reduce a design equation to one or two dimensions
and assume the other dimensions are covered by ceteris paribus conditions.
They make calculations based on "standard" or "ambient" conditions. One-
dimensional solutions and controlled conditions do not always fit well with
the real world. By definition, engineering design is a synthetic process
which integrates all pieces of knowledge generated through analysis. Engineers must use judgment to synthesize multidimensional designs that
operate satisfactorily under the conditions found in the real world.

As part of the design process, engineers transform judgments into verified knowledge through testing. In this way, they extend the technological
envelope of knowledge that constrains their designs. The design process is
cumulative. Current design judgments are based on previous judgments

that have been verified. When design engineers transform their judgments into verified decisions through testing, they accumulate knowledge, not judgments. Uncertainty and risk are reduced. If this design-test cycle is limited, the resulting design can be like a house of cards. If a prior judgment turns out to be incorrect, the whole design suffers.

We have consistently mentioned NASA's adoption of the all-up or systems testing approach to engineering design. This design approach requires engineers to design relatively complete systems without thorough component testing, and then perform tests on an entire system or large subassembly. Our mention of it here is to stress the consequence of the management decision to use all-up testing for the individual engineer: It puts far greater reliance on engineering judgment than does a component testing design process.

Engineers had to design large, complex technological systems without the opportunity of transforming their judgments into verified knowledge. We have already cited Odom's discussion of the problems of component testing of the turbines and pumps in the SSME. The SSME experience provides further examples.

In many respects, the SSME design resembles a house of cards. Our interviewees compared it to a "Swiss watch." When the all-up testing phase for the SSME began in 1975, the engines self-destructed, sometimes destroying the test equipment as well. These tests were disastrous and, worse still, little was learned about the reasons for failure. NASA was very worried that it would run out of engines and parts before they were ever installed in an orbiter. The SSME program was almost forced back to the drawing board, except that many key geometry design decisions were already made and could not be changed. The redesign process for the SSME was thus handicapped by earlier engineering judgments that were never tested.

Another example concerns testing the SSMEs at full power of 109 percent, a design specification derived from payload requirements. To quote Orville Driver, deputy manager of the Shuttle Main Propulsion Test:

> In 1984 we were within 10 days [of 109 percent], we had received the engines, we had gone through all the checkouts, everything was checking out real well. This was 1984, this was three years after we did the last test here in January of 1981. NASA Headquarters got to thinking, suppose we do that test down there and we have some problems – we lose an engine or we lose three engines – that would be a terrific impact on the flight program because we are already short on engine hardware.[20]

The engine tests of 1984 were canceled by NASA headquarters to preserve the small inventory of engines. Thus, a decade after the SSME designs had been developed, they were still untested on this key performance specification. In fact, the shuttle had been flown at 104 percent of power, a level that had not been tested on the ground for the full cluster.[21] By the mid-1980s, the shuttle had reached the stage where "operational" concerns were over-ruling key engineering principles of NASA. During Apollo, NASA had always stuck to the principle that testing came before flight. Moreover, during much of the Apollo program, component testing preceded systems tests.

To exercise the principles of responsibility and Cicero's Creed II individually, the engineer must be supported by the organization. Organizational competence, then, is the support of the processes by which individuals can best develop and utilize their knowledge. Much of our discussion of the organizational context of the shuttle's development shows that NASA certainly did not aggressively support engineering design testing, let alone the prior stages of research and development. This, of course, was a direct consequence of the political decisions and severe budget restrictions. Yet NASA leadership believed it could build the shuttle by imposing the discipline of systems design, minimal budgets, and fixed schedules. Engineering managers who questioned and doubted this approach were ignored. This managerial solution to a technological problem did not work. In the end, it proved far more costly, took years longer, and contributed to the disaster of the *Challenger*.

In summary, engineering judgment is critical for engineering design. Design requires a synthesis of facts and theory, and judgment fills the gaps when all the pieces are fit together in design. Judgments are interpolative or extrapolative. The former fall within the empirical database of the materials and technical designs, while the latter extend performance beyond current experience. Testing converts engineering judgments into engineering facts. Our study has found that, because of all-up systems management, the shuttle design relied heavily on extrapolated engineering judgments that proved inaccurate when the SSME finally reached the testing stage.

Organization and Engineering Design

The organizational framework outlined here, and discussed in more detail in Chapter 4, illustrates other important parameters in the engineering design process. We discuss several, including the group process of design, how decision facts and goals can conflict, how organizational conventions constrain design, the role of documentation in design, and the relevance of the distribution of knowledge across an organization.

Technology as complex as the shuttle requires that engineering design be a group effort. The knowledge required to design even relatively simple components exceeds the knowledge of most individual engineers. Therefore, to create a design, a group of engineers must pool their collective experiences and knowledge. Again, it is the responsibility of the organization to ensure that knowledge crucial to the project is available to practicing engineers, in the form of consultants or new people. For example, when engineers encountered problems with the cracking of turbine blades and acknowledged the limits of their knowledge, the organization should have sought outside help. This may have forced a better understanding of the problems and most likely introduced more rigorously the specialty of fracture mechanics much earlier into the design process; it may have reduced the number of engine test failures; it may have led to a better understanding of the limits of the mathematical models. That this did not happen illustrates a failure of organizational competence.

The shuttle's history provides many examples where facts conflicted with NASA's goals. One of the more widely recognized examples of this concerned Tischler's analysis of the launch costs of the shuttle program. Tischler observed that the overhead costs of maintaining the work force and facility at Cape Canaveral added $500 per pound cost of the shuttle's payload, a figure that exceeded the estimated cost of $100 per pound used by NASA's top management to argue that the shuttle's economic goals were sound.[22]

In Simon's terminology, organizations transform their goals into constraints and tasks at lower operational levels. These constraints derive from several factors. Some are passed on from the overall goals of the organization, and the resources available to pursue those goals. Others derive from the nature of the engineering tasks. Still others stem from the organizational climate and culture, or from professional practice. These constraints are often called standard operating procedures (SOP).

The Importance of Organizational Conventions

To coordinate the decision making of even a small design group, engineers adopt conventions about how they interface with each other. The design of the entire space shuttle is described functionally by an extensive set of performance specifications. For example, the shuttle must be able to lift a payload of 65,000 pounds into orbit, and return a payload of 5,000 pounds from orbit. The cargo bay must be sixty by fifteen by fifteen feet; the shuttle must be able to abort a launch if one SSME fails.

These performance specifications have design implications for all systems and components in the shuttle. The launch payload, combined with the weight of the shuttle, determines the required thrust for the main engines plus the solid-fuel rocket boosters (SRB). The total required thrust is allocated to the SRBs and the SSMEs, which in turn specify the performance of an individual SSME. This elaboration of performance specifications becomes a set of interlocking conventions for the design of the space vehicle.

In addition to performance specifications, design engineers must develop a set of interface specifications. The SSME must fit into its receptacle in the orbiter. The size of the receptacle has design implications for both the orbiter and the SSME. In addition to these physical connections between the components and systems of the shuttle, there are data links, instrumentation, and the onboard computer system. All must interface with each other and the other shuttle parts.

Space vehicles like the shuttle are designed with redundancy in as many critical components and systems as possible. For example, there are five computers, some of which are redundant. However, all three SSMEs are needed for lift-off. There is no redundancy in the shuttle's main propulsion systems.

Design engineers adopt conventions about the manufacturing capabilities needed to build the shuttle. High-tech aerospace manufacturing calls for high skill levels in the work force, supported by the latest manufacturing technology. The SSME design obviously required very high levels of manufacturing skill. Some of the welds in the SSME could not be performed by people; instead, special robots made the welds. Serious problems arise when these engineering design expectations are not met. For example, the designers of the ceramic heat tiles for the orbiter body assumed that very skilled persons would install them. Initially, this was not the case, and many of the tiles popped off and had to be replaced.

Quality control is managed through a similar set of conventions. The engineer assumes, based on these conventions, that manufactured parts will be checked with a specified thoroughness. Welds will be tested using specified techniques. However, there is more to quality control than this.

As discussed in the next chapter, NASA managed risk during the design phase of the shuttle program using qualitative methods of risk management. These included hazards analysis and failure modes and effects analysis (FMEA). These methods work best when something breaks during testing. Engineers then attempt to find out why the part broke and design a solution to the problem. This "engineering problem-solving" approach has worked well for nearly two centuries. Many maintain that it is the essence of empirical engineering design. The cycle of design–build–test–break–redesign is "SOP" in so many areas of technology. However, as we have repeatedly observed, the all-up testing doctrine short circuits the cycle and, in a sense, undercuts NASA's existing quality control and safety procedures.

NASA's quality control methods did not require that data about the performance of systems and components systematically be collected. Therefore, during this crucial phase of shuttle development, the statistical data necessary for quantitative risk assessment and quality control were not available. In fact, key NASA engineering managers did not believe in quantitative risk assessment methods. Thus the organizational climate of NASA did not support contemporary approaches to engineering reliability.[23]

The importance of conventions regarding testing and verification are already clear. The two primary approaches, component and all-up testing, create different decision environments for the design engineer. Component testing is a bottom–up, incremental approach. Each phase of the design builds on a successful previous phase. Engineers using component testing identify many problems relatively early in the process. With all-up testing, engineers design from the top down. They first specify the overall system, and then design components and subsystems as part of that system. They test the entire system as a unit, focusing on the interactions of the components and subassemblies, as well as the performance of components. All-up testing requires engineers to rely more heavily on theory, models, and judgment. The design process involves paper analyses or computer simulation rather than working directly with the materials.

Any project as complex as the space shuttle requires extensive documentation. NASA requires its engineers to document their designs in specified ways. Engineers sign off when they complete the design of the component or part. This act generally signifies that they have reviewed all aspects of their design and, to the best of their professional knowledge, the design will perform as expected. If there were uncertainties about how a component would work, NASA design engineers sometimes attached statements of potential problems as part of their sign off. These statements might recommend more testing than had been allowed under the constraints of budget and schedule. They might stress that components had not been evaluated under conditions, such as low temperatures. Obviously, the significance of the sign off depends on other conventions. Specifically, when engineers have had the opportunity to test their designs during component testing, the certification of their designs has greater weight than when they attest to their "paper" or "simulated" designs under all-up testing.

The distribution of knowledge within NASA and across the key contractors is also relevant. Two of the three potential contractors for the SSME, Rocketdyne and Pratt & Whitney, had expertise in different technical areas. Rocketdyne, drawing on its experience with the J-2 rocket, had the lead in the design of large rocket engines, whereas Pratt & Whitney had the lead in the design of the high-performance turbopumps that would be critical for the SSME. A successful design required the expertise of both companies, and NASA had originally planned to let two contracts but was forced for cost reasons to let only one. Left to one contractor and the proprietary nature of key technologies, the winning bidder would not have access to the expertise of the other company. Thus the knowledge base available to NASA and Rocketdyne was limited by the realities of our economic system. When Rocketdyne won the contract, Pratt & Whitney protested because, among other reasons, Rocketdyne had used data from its Saturn design that was not available to Pratt & Whitney, thereby giving Rocketdyne unfair advantage.

Knowledge Base Limited by Hiring Practices

Under normal circumstances, organizations augment or maintain their technical knowledge base by hiring a steady stream of young, recently trained engineers. To a considerable extent NASA and its contractors were not able to do this. Since the middle of the 1960s, employment levels at NASA and its contractors had been in decline as the Apollo program wound down. The aerospace industry was in a severe recession during the crucial early stages of the SSME design. Indeed, we have already noted, this was one of the primary reasons the Nixon administration approved the shuttle program – to put the California-based aerospace industry back to work.

Neither NASA nor its contractors were able to bring large numbers of new engineers into their organizations. Thus the transfusion of new blood that most technological organizations believe is necessary for long-term health had been disrupted prior to the shuttle program. The cohorts of engineers who initiated the design process were from older generations. They were an experienced group of engineers, very familiar with the technology that supported the Apollo program. However, they were not necessarily familiar with the technological breakthroughs that had evolved during the decade of the 1960s. The critical budget limitations also forced contractors to begin laying off design engineers before the all-up testing phase of the program got under way. Thus NASA was deprived of engineering talent to address the problems that became apparent once testing commenced. Once the shuttle program reached an "operational" phase, problems were even more difficult to fix, because the engineering talent was even thinner. Thus shuttle missions with known problems flew without correction, until January 1986. The horror of the *Challenger* accident brought these organizational deficiencies out into the open.

Conclusion

We have argued that the ethical competence of the individual engineer is a function of the organizational context within which the engineer practices. For an engineer to behave in an ethically competent manner, he or she must be part of an ethically competent organization. Such an organization supports the role of the engineer as a knowledge expert. This includes forming design groups with the requisite expertise; separating issues of fact from normative issues in decision making; developing functional conventions that are implemented through the organizational hierarchy; and, most important, supplying the resources so that engineers create empirically grounded designs. Engineering design is an organizational process that depends on clear communication and informed decision making for its success. The monetary and financial setbacks of the space shuttle program, as well as the tragic *Challenger* accident, serve as a reminder that communication and ethics are crucial components to such an organization.

Notes

1. Richard S. Lewis, *The Voyages of Columbia: The First True Spaceship* (New York: Columbia University Press, 1984).
2. Jerry Thomson, "Shuttle: The Approach to Propulsion Technology," *Astronautics and Aeronautics* 9, February 1971: 64–7.

3. James Kingsbury, taped interview in Management Operations Office, MSFC, *Oral Interviews: Space Shuttle History Project*, Transcript Collection, December 1988; p. 439.
4. Adelbart O. Tischler, "A Commentary on Low-Cost Space Transportation," *Astronautics and Aeronautics* 7, August 1969: 50–64.
5. U.S. Congress, Senate, Subcommittee on Science, Technology and Space of the Committee on Commerce, Science, and Transportation, *Report of the National Research Council's Ad Hoc Committee for Review of the Space Shuttle Main Engine Development Program*, 95th Congress, 2nd session, March 31, 1978, serial 95-78.
6. Robert S. Ryan, "Practices in Adequate Structural Design," NASA Technical Paper 2893, George C. Marshall Flight Center, Huntsville, AL, 1989.
7. James B. Odem, taped interview in Management Operations Office, MSFC, *Oral Interviews: Space Shuttle History Project*, Transcript Collection, December 1988, pp. 87–8.
8. Harry A. Cikanek III (MSFC), "Characteristics of SSME Failures," paper presented at the AIAA/SAE/ASME/ASEE 23d Joint Propulsion Conference, San Diego, June 29–July 2, 1987.
9. Kingsbury, taped interview, p. 440.
10. Albert Kaufman and Jane M. Manderscheid, "Cyclic Structural Analyses of SSME Turbine Blades," Lewis Research Center, Cleveland, 1985, N85-27963, and Albert Kaufman and Jane M. Manderscheid, "Simplified Cyclic Structural Analyses of SSME Turbine Blades," in *Advanced Earth-to-Orbit Propulsion Technology 1986*, eds. R. J. Richmond and S. T. Wu, vol.2, NASA Conference Publication 2437 (Huntsville, AL: Marshall Space Flight Center, 1986), pp. 107–24.
11. Cikanek, "Characteristics of SSME Failures."
12. B. N. Bhat, "Fracture Analysis of HPOTP Bearing Balls," MSFC, NASA TM-82428, May 1981.
13. S. S. Chen, J. A. Jendrzejczyk, and M. W. Wambsganss, "Flow-Induced Vibration of SSME Injector Liquid-Oxygen Posts," Argone National Laboratory, NASA Scientific and Technical Facility N85-27951.
14. Ryan, "Practices in Adequate Structural Design," p. 15.
15. Ibid., p. 16.
16. Frank Stewart, taped interview in Management Operations Office, MSFC, *Oral Interviews: Space Shuttle History Project*, Transcript Collection, December 1988, p. 71.
17. Robert E. Kielb and Jerry H. Griffin, "SSME Blade Damper Technology," NASA Technical Report N87-22798, Lewis Research Center, Cleveland, 1987.
18. Kingsbury, taped interview, pp. 440–1.
19. Ryan, "Practices in Adequate Structural Design," pp. 63–4.
20. Orville Driver, taped interview in Management Operations Office, MSFC, *Oral Interviews: Space Shuttle History Project*, Transcript Collection, December 1988, pp. 7–8.
21. Ibid., p. 9.
22. Tischler, "A Commentary on Low-Cost Space Transportation."
23. Trudy E. Bell and Karl Esch, "The Space Shuttle: A Case of Subjective Engineering," *IEEE Spectrum*, June 1989, pp. 42–5.

13

Engineering Ethics and Risk Assessment

"The chance (qualitative) of loss of personnel capability, loss of system, or damage to or loss of equipment or property."

—NASA definition of risk (until 1988)

"This definition clearly implies evaluation of a set of risks based on the chance of occurrence of each of the various consequences described. However, NASA acknowledges, and our reviews have confirmed, that these 'chances' are not formally or specifically estimated; nor are they documented. Rather, STS risks are assessed based on subjective judgments and the approval of qualitative rationales by various board and panel chairmen."

—Committee on Shuttle Criticality Review and Hazard Analysis Audit of the Aeronautics and Space Engineering Board

Are Accidents Normal?

This chapter is concerned with how risk assessment and management methodologies were employed within NASA, particularly during the years between 1959, when NASA was a new "upstart" federal agency, and 1986, when it underwent the grim investigation by the Rogers Commission following the *Challenger* accident. As a consequence of the accident, NASA was asked to review certain aspects of its space transportation system risk management efforts and "identify those items that must be improved prior to flight to ensure mission success and flight safety." The commission also recommended that an audit panel which was to report directly to the administrator of NASA be appointed by the National Research Council (NRC) to verify the adequacy of that effort. In response, the Committee on Shuttle Criticality Review and Hazard Analysis Audit was established.[1]

In Chapter 1 we defined engineering as a heuristic. We also reviewed Petroski's scheme for including failure as an intrinsic part of the engineering design process. Organizational theorist Charles Perrow agrees with Petroski's premise but elaborates upon it by examining high-risk systems in general and explains why accidents (hence failures) are inevitable. No matter how effective conventional safety devices are, Perrow acknowledges that there is a form of accident that is inevitable when high-risk technologies are used. To Perrow, "most high-risk systems have some special characteristics . . . that make accidents in them inevitable, even 'normal.'" Given an interactive, complex system that is also "tightly coupled" (i.e., the failure process occurs very quickly and can't be turned off nor can the failed parts be isolated from other parts), an accident will inevitably occur.[2] Perrow designates these as "normal" or "system" accidents. The shuttle fits Perrow's description. Was an accident predictable?

Robert S. Ryan, whom we have quoted often in this book, proposes that failure and/or problems in the shuttle generally were "not due to undiscovered or missing theory; but to the neglect or oversight of basic principles . . . in project management of a program. These include management criteria, procedures, philosophy, test analysis and communication/documentation." Echoing the theme we have stressed, Ryan contends that all designs, by definition, are designed for failure. As a design evolves and changes, its requirements are forced to change, and this, in turn, causes performance changes. The resulting compromises, Ryan contends, can be thought of as "failures."[3]

Each of these "doomsday" approaches to failure and high-risk technology accept, in sum, that engineering is a heuristic skill. The question we posed earlier regarding the individual engineer's and the organization's responsibility to alert the public to risk factors stressed that a risk-free environment was impossible. How to inform the public about risk and whose decision it was to decide on how high a risk should be accepted were issues we have been exploring. In order to understand this issue fully, some knowledge of current risk assessment techniques and NASA's choice about which ones to use is important.

The Range of Ways to Assess Risk

Risk analysis techniques range from qualitative hazard analysis, failure modes, and effects analysis with critical items list (FMEA/CIL) to probabilistic risk analyses. A complete risk analysis of a complex system would utilize the full range of techniques, with the results from the qualitative stages becoming the input for the more quantitative stages.[4] Figure 13.1, which has been developed by Bell, defines some of the basic terms in risk assessment and analysis.[5]

Risk: the combination of the probability of an abnormal event or failure and the consequences(s) of that event or failure to a system's operators, users, or its environment.

Risk assessment: the process and procedures of identifying, characterizing, quantifying, and evaluating risks and their significance.

Risk management: any techniques used either to minimize the probability of an accident or to mitigate its consequences with, for instance, good operating practice, preventive maintenance, and evaluation plans.

Failure: the inability of a system, subsystem, or component to perform its required function.

Failure modes: the various ways in which failures occur.

Hazard: an intrinsic property or condition that has the potential to cause an accident.

Quality Assurance: the probability that a system, subsystem, or component will perform its intended function when tested.

Reliability: the probability that a system, subsystem, or component will perform its intended function for a specified period of time under normal conditions.

Uncertainty: a measure of the limits of knowledge in a technical area, expressed as a distribution of probabilities around a point estimated. The four principal elements of uncertainty are statistical confidence (a measure of sampling accuracy), tolerance (a measure of relevance of available information to the problem at hand), incompleteness and inaccuracy of the input data, and ambiguity in the modeling of the problem.

Figure 13.1 Risk assessment terminology.

A hazard analysis is a top-down approach in which all potentially unsafe conditions or events posed by the environment, crew-machine interfaces, and mission activities are enumerated. The potential sources of such conditions are identified, and a rationale for the mitigation and/or acceptance of the risk is explicitly provided.[6] That is, identified hazards and their causes are analyzed to find ways to eliminate (remove) or control the hazard (design change, safety or warning devices, procedural change, operating constraint). Any hazard that cannot feasibly be eliminated or controlled is explicitly termed an "accepted risk."[7]

The FMEA employs a bottom-up approach. Starting at the lowest levels of each subsystem, the analyst asks, How can this device/part fail and what are the effects and consequences of such a failure on the component and all other interfacing, interacting components? The consequences of each identified failure mode are then classified according to its severity. For the shuttle, failure modes that could lead to the loss of crew and/or vehicle are classified as "Criticality1" (CRIT-1) or 1R if the item of concern is redundant. CRIT-1 items are then collected on a critical items list (CIL), which serves as a management tool to focus attention on the mitigation or control of the failure mode through redesign, use of redundant components, special inspections, or tests.[8] Each item on the critical items list requires a formal, written rationale for its retention on the shuttle. In this manner, prior to each shuttle launch engineers and managers are required to waive explicitly NASA policy against flying with such items present.[9] While hazard analysis can be used early in the design phase in order to identify potential hazards,[10] the methodology is also recommended as a means of further analyzing the failure modes identified in FMEA process.[11]

In contrast, reliability analysis is used to determine the probability of failure occurrence. Probabilistic risk assessment (PRA) is one format. It is also a top-down technique in which the possible failure mode of the complete system is identified first, and the possible ways that the failure might occur are enumerated. A "fault tree" is developed by tracing out and analyzing the contributory faults, or chains of faults for each fault, until the basic fault (e.g., single component failure or human error) is reached. Probabilities are then assigned to the various basic faults or errors. This enables probabilities for the various failures to be estimated, and their relative contribution to total risk assessed. In theory, the failure modes with the highest probabilities should be addressed first. When used correctly, PRA yields a measure of risk from a chain of events and an estimate of uncertainty.[12]

Proponents of PRA claim that engineering managers make better resource allocation decisions using these quantitative failure estimates. Managers can anticipate potential failures because the probabilities measure the seriousness of potential difficulties. Moreover, these quantitative approaches require the collection of test data, a process that builds as does experience with the technology. Ironically, in the early 1960s the aerospace industry, including researchers at NASA, developed PRA along with FMEA. However, the industry soon abandoned the mathematically based PRA as being too costly.[13] The data collection, testing, and analysis required to support PRA added a whole new set of tasks to the engineering process. During the shuttle program, due in large part to the budget and schedule constraints, NASA managers did not believe that the agency could afford the extra effort required for PRA. We have documented that the shuttle was underfunded during its critical design and development stages. A casualty of this underfunding was that it prohibited the possibility of using contemporary risk assessment and testing methods.

NASA's Approach to Risk Assessment: The Locked-in Syndrome

Throughout its manned space flight history, NASA's efforts to secure adequate safety design, review, and oversight through its organizational structure have revealed a twofold pattern. First, from the beginning, responsibility for safety has been highly decentralized through the "loose federation" that existed among the agency's manned space centers and its contractors. Concomitantly, at the NASA headquarters level, responsibility for safety, reliability, and quality assurance (SRQA) was shifted from an early position of direct accountability to the administrator to a more remote position within the Office of Organization and Management in the early 1960s after James Webb became chief administrator; in 1975, the Office of the Chief Engineer assumed the position of direct accountability again, giving it a more centralized focus. Yet, under both formats – decentralized and centralized – NASA suffered major accidents.[14]

This has led Sylvia Freis to hypothesize that the success of any risk management program within an organization such as NASA depends upon three interrelated aspects. The first is the caliber of responsible individuals; the second, the degree to which middle-level managers and engineers feel constrained about voicing warnings and dissent; and, finally, the ability of SRQA engineers to insulate themselves from internal political pressures to move a program or project forward. Freis is of the opinion that the first two aspects are difficult to ensure in any systematic or organizational way. In a federally funded organization such as NASA, which is subject to political ebbs and tides, the third aspect is crucial. In a very real sense the livelihood of NASA depends upon its ability to prove its programs are or will be successful. "*The pressures on the best-intentioned manager or engineer to neutralize what may be justifiable expressions of unacceptable risk are likely to be considerable, especially in times of fierce competition over programmatic budgets.*"[15]

When NASA began Project Mercury in 1958, a statistical approach was the most widely accepted procedure used to measure the reliability of weapon systems and complex machinery. Although there was no organizational unit expressly assigned the responsibility of reliability and flight safety until 1962, according to Freis, the operational ground rule was: "no manned flight would be undertaken until all parties responsible felt perfectly assured that everything was ready."[16] Redundancy was used to achieve reliability, but controversy soon arose between NASA and the Army Ballistic Missile Agency (responsible for the Redstone missile) over the validity of statistically measured reliability and the degree to which redundancy sufficed to achieve "pilot safety" as distinct from "mission success." The controversy resulted in the dispersement of "pilot safety" and "reliability" guidelines among numerous engineers at NASA's contractors.

In 1962, when the Manned Spacecraft Center was established in Houston, the first office responsible for reliability and flight safety was created. One of five staff offices that reported directly to the associate director, this structure was similar to the one at NASA headquarters, where the Office of Reliability and Quality Assurance reported directly to Associate Administrator Seamans. However, by 1963, when the headquarter's organizational structure expanded under James Webb, the reporting route for that office became more distant, and virtually "disappeared entirely from the NASA organization chart." The responsibility for safety and reliability became decentralized until the tragic *Apollo 204* fire on January 27, 1967. As a consequence of the loss of astronauts' lives, and subsequent unrelenting Congressional pressure, the position of NASA Safety Director was created along with separate Safety and Reliability Offices at each of the manned space flight centers and program offices. In addition, the external Aerospace Safety Advisory Panel (ASAP), which reports directly to the chief NASA administrator and to Congress, was established. In 1975, the Office of NASA Chief Engineer was established with responsibility for SRQA in an effort to exert more centralized control.[17]

The post-*Challenger* NRC audit panel observed that the engineering organizations within each element (e.g., propulsion, orbiter, ground support, and landing operations) project office at the centers were responsible to a project manager and program director for the performance and reliability of the hardware/software systems they developed. In this manner, safety was to be an inherent feature of the system design, development, testing, and production processes. NASA's rationale was, since engineers design the unit or system, test it, certify it for operation, and inspect it after flight, they have the greatest ability to understand and anticipate the ways in which the unit or system might fail. For that reason, NASA engineers had primary responsibility for carrying out the most technical of the safety analyses and for establishing the rationale for retaining critical items identified through the FMEA.[18]

As noted in Chapters 5 and 7, when James Webb brought in George Mueller as associate administrator for manned space flight, NASA underwent more than an organizational change. Mueller brought in a systems management team that had worked on large Air Force programs. Their goals in managing such large programs were efficiency, cost reduction, and decentralization. NASA's culture of in-house work, hands-on activity, and open communication among teams was not regarded as in keeping with these goals. Many specialized NASA engineers were replaced with the subcontractors who were to be trusted to do it right.[19] The traditional culture within NASA did not disappear, however, and the method of verifying contractor work was to conduct extensive flight tests. While Marshall engineers wanted to continue to conduct tests incrementally in spite of Mueller's new approach, the Air Force ballistic missile philosophy placed much less emphasis on flight tests. It relied more on ground tests of increasing complexity to reduce the need for extensive flight testing. McCurdy quotes one of the executives from the Air Force program on his reluctance to adopt the NASA approach: "At the time," the executive from the Air Force program said, "reliability engineers argued that man-rating a vehicle like the *Saturn V* should require some element of statistical confidence. An oft-cited goal was 90 percent confidence of 95 percent reliability." Even that criterion would produce a nearly 50 percent chance of losing a *Saturn V* during the eleven Apollo launches with astronauts on board. Moreover, to verify those numbers, NASA would have to conduct an estimated 45 *Saturn V* test flights.[20] As discussed in Chapter 7, Mueller enlisted Tischler and Disher to estimate the probability of achieving a Moon landing by the end of the decade and their private assessment was a mere 10 percent.

Hence it followed that in late 1963, Mueller introduced his all-up testing approach, announcing that NASA would drop the traditional step-by-step flight tests of rocket and spacecraft components. Instead, its engineers were directed to assemble the three *Saturn V* stages along with the command and service module as if they were ready to fly to the Moon, and conduct just two or three unmanned flight tests of the whole system. In reality, without all-up testing, NASA had only a slim chance of reaching the Moon by the end of the decade. An Apollo executive informed McCurdy that sequential testing "really didn't gain you anything. You didn't decrease the risks by testing sequentially; you only spread the risks out. [NASA might lose one or two rocket stages], but in that case, you were going to be grounded for a while anyhow, so you hadn't lost anything other than hardware. . . . the risk was there in any event."[21]

This rationale was no doubt a reaction also to the published results of a comprehensive probabilistic risk assessment study NASA commissioned the General Electric Company to perform. Its estimate, close to that of Disher and Tischler, of the likelihood of successfully landing a man on the Moon and returning him to Earth, was a less than 5 percent chance of success. NASA's administrator felt that releasing this information would do irreparable harm to the Apollo project. As a result, the use of probabilistic risk assessment was rejected, and NASA turned to the qualitative approaches, primarily relying on FMEA/CIL.[22]

It is not surprising that Will Willoughby, NASA's former head of reliability and safety during the Apollo Moon landing program, declared in 1989 that "Statistics don't count for anything. They have no place in engineering anywhere." To Willoughby (and many of his colleagues at NASA), risk is minimized by "attention taken in design where it belongs." As we have noted, official NASA policy until the *Challenger* accident favored "engineering judgment" over "probability numbers" and resulted in NASA's failure to collect the type of statistical test and flight data useful for quantitative risk assessment.[23]

The success of the Apollo program did not assuage fears about NASA's high-risk testing philosophy. Raymond Wilmotte, a consultant to the advisory panel, initiated a study of the existing and desirable organizational structure of risk technologies in NASA in 1971. He was particularly concerned about the lack of a structure at the lower (component) levels "where many micro-decisions are made which are not transmitted to higher levels." Wilmotte, concentrating on the bias of the individual design engineer, observed a need to develop a more consistent understanding of the roles of design and risk in relation to the several levels of program management. He stressed the importance of "systematic *co-ordination* of design and risk technologies in the decision-making chain." He even suggested tracing the process of design and acceptance of risk through all levels.[24] Wilmotte informed NASA that, "in the field of complex hardware the designer is not likely to be the most effective person for analyzing and pointing out how the design could fail or be prone to cause accidents." Recognizing the human element in this process, Wilmotte reasoned that however conscious the design engineer may be of the importance of reducing the probability of failure or accident, "he will more easily defend the validity of his design than recognize a weakness. Inevitably he is biased. . . . One cannot expect even the experienced designer or operator . . . to do more than take care of the obvious potential deficiencies. . . . The more subtle ones will not come easily to mind."[25]

Wilmotte's warnings were to go unheeded by NASA for the next fifteen years. The all-up philosophy meant that the shuttle would be tested on its first voyage. The shuttle had to fly the first time. There was no sensible way to flight-test individual components of the orbiter.[26] As the shuttle program matured and gained flight experience, NASA's motivation and its ability to eliminate specific problems decreased. As previous cases document, once a component was operational, the use of the all-up testing philosophy discouraged and limited the options engineers had to make major design charges. Moreover, given the overall cost constraints of the shuttle project, Rocketdyne's techniques of cutting costs by eliminating design engineers and "accepting a higher risk" were evidenced throughout the program. Bill Sneed, program planning office director, MSFC, reports that during this "operational" phase of the project: "we were trying to get the engineer off the program so we could get the cost of the program down. . . . And of course if you're making [engineering] changes, that's big money."[27] In addition, NASA's

> decreased tolerance for failure discouraged testing. . . . As political support diminished and the cost of test hardware increased, failure even on a test flight unleashed a barrage of criticism. NASA officials tried to reduce the frequency of failure by conducting fewer flight tests. Rather, they relied upon ground tests, used computer models to simulate flight conditions, and built in self-correcting mechanisms or maintenance capabilities which would allow them to correct failures during the operations phase.[28]

NASA was relying primarily on its subjective methods for assessing risk. Evidence from the NRC report, the SSME Assessment Team Report, the annual reports of the Aerospace Safety Advisory Panel, and details highlighted in our cases define the basic conflict NASA faced. It did not have the resources to both fly and make the engineering changes supposedly mandated by the CIL process. For a variety of reasons, NASA leadership chose to fly, implicitly accepting the risk associated with these problems. On July 4, 1982, after only four test flights of *Columbia*, President Reagan proclaimed that the shuttle fleet was "fully operational, ready to provide economical and routine access to space." To McCurdy, "the belief that any major space flight program could be made "fully operational" or "routine" clashed with assumptions that NASA operated risky systems.[29]

Not all NASA systems accepted this high-risk approach. Richard Feynman has elaborated on how the shuttle software development program differed from the rest of the shuttle design program. In particular, a bottom-up methodology rather than the top-down all-up testing approach was used for the software development. According to Feynman:

> Each new line of code is checked; then sections of code (modules) with special functions are verified. The scope is increased step by step until the new changes are incorporated into a complete system and checked.[30]
>
> Working completely independently is a verification group that takes an adversary attitude to the software development group and tests the software as if it were a customer of the delivered product.[31]
>
> The computer software checking system is of highest quality. There appears to be no process of gradually fooling oneself while degrading standards, the process so characteristic of the solid rocket booster and space shuttle main engine safety system. To be sure, there have been recent suggestions by management to curtail such elaborate and expensive tests as being unnecessary at this late date in shuttle history.[32]

This bottom-up approach proved to be extremely reliable. In over 250,000 lines of code, only six errors have been discovered, and these occurred during *verification* stages, not during an actual flight. An in-depth study of the individuals involved in this program would provide an example of how the organizational stress toward all-up testing was overridden.

Feynman contrasted this approach with that used by the solid rocket booster and space shuttle main engine safety system. Here, he characterized NASA's decision making as "a kind of Russian roulette . . . the shuttle flies . . . and nothing happens. This suggests . . . that the risk is no longer so high for the next flights. We can lower our standards a little bit because we got away with it last time."[33]

Recognize the Risk: Take It or Leave It

Six years after Reagan had called it "operational," the audit panel, in tacitly adopting an "informed consent" model of responsibility, pointedly stated that the shuttle was still a developmental vehicle and, consequently, "risks must be accepted by those who are asked to participate in each flight as well as those responsible for achieving the nation's goals in space." It went on to state that, "such risks should also be recognized by Executive Branch officials and Congress in their review and oversight endeavors." The basis for acceptance of those risks should stem as much as possible from rationally derived criteria. This acceptance also should depend very heavily on the quality of the methodology and the degree of objectivity by which the risks are determined as well as the rigor by which the risks are controlled (managed)."[34]

The panel found that the NASA hazard analyses did not address the relative probabilities of a particular condition arising from failure modes, human error, or external situations. Hazard analyses for some important subsystems had not been updated for years, even though design changes had occurred or dangerous failures were experienced in subsystem hardware. Equally disturbing, there were no formal linkages between the FMEA/CIL and hazard analysis processes and the shuttle engineering change activities. Also data from shuttle inspection, test and repair, and in-flight operations were not always fed back rapidly enough or effectively enough into the risk management process for it to be useful.[35]

In addition, the possibility of human factors or errors seemed to be ignored as a cause of failure modes in the FMEAs. For example, "cannibalization" of parts from one operational shuttle element to fulfill spare requirements in another had become a prevalent feature of shuttle logistics at the Kennedy Space Center, thus introducing a variety of potential failure modes associated with human error.[36]

In total, this systemic lack of formal linkages, information feedback, and ignoring of the potential for human error, which can be grouped under our framework as evidence of a lack of communication, was, according to the NRC panel, attributable to NASA's multilayered system of boards and panels. This organizational structure tended to "lead individuals to defer to the anonymity of the process and not focus closely enough on their individual responsibilities in the decision chain." The sheer number of shuttle-related boards and panels seems to produce a mind set of "collective responsibility."

NASA's Organization: Many, Many, Many Hands

We have generally described NASA's overall organizational structure. To appreciate the complexity, however, a much more detailed look is important. NASA's organizational structure consisted of four levels as illustrated in Figure 13.2. In order to certify that all components of the shuttle were ready, NASA established a formal flight readiness review process, which began at Level IV, the contractor level. At this lowest level in the hierarchy, each of the contractors certified in writing that the elements under its responsibility were flight ready. That certification was made to the appropriate Level III NASA project managers at the centers – Johnson Space Center (orbiter and systems integration), Marshall Space Flight Center (shuttle main engine, solid rocket booster, and external tank) – as part of each center's flight readiness review. At Marshall, the review was followed by a presentation to the center director; at Kennedy the center director chaired the review that verified the readiness of the launch support elements.

The Level II review was the responsibility of the program manager at the Johnson Space Center, who conducted a preflight readiness review. Each space shuttle element program manager endorsed that the manufacture, assembly, test, and check-out of the pertinent element had been satisfactorily completed. The contractors' certification that design and performance were up to standard were included. The Level I flight readiness review completed the process. A mission management team, which supported the NASA associate administrator for space flight and the space shuttle program manager, assumed responsibility forty-eight hours before launch and continued until the completion of the mission. A structured mission management team meeting was held one day prior to each scheduled launch. Its

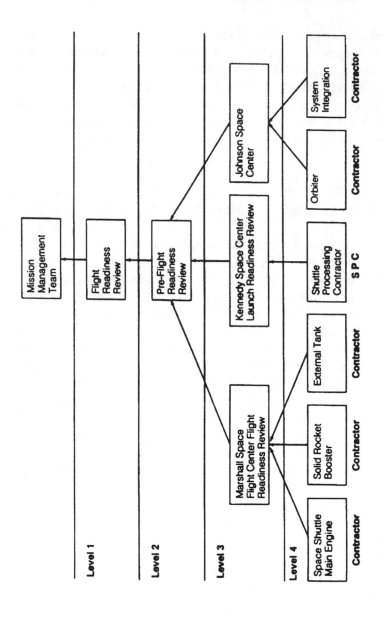

Figure 13.2 NASA's flight readiness review levels. The readiness reviews for both a shuttle launch and mission begin with the contractors and move upward through the centers to the associate administrator for space flight (from Rogers, *Report*, 1:83).

agenda included close-outs of any open work and flight readiness review action items, discussions of new and/or continuing anomalies, and updated weather reports.[37]

The NRC audit panel observed that NASA used six review boards to implement an FMEA/CIL: four Level III review boards (Engineering Review Team, Level-III preboard, Configuration Control Board, shuttle Projects Office Board), a Level II Program Requirements Control Board, and a Level I Headquarters Program Requirements Control Board (see Figure 13.3). NASA also had boards that reviewed design requirements and certification, software, the operations and maintenance requirements and specifications document, the operations and maintenance instructions, the launch commit criteria, and mission rules. There were flight readiness reviews at each stage of preparation with a Launch System Evaluation Advisory Team to assess launch conditions and a Mission Management Team to oversee the actual mission. In summary, the audit panel was dismayed to find that NASA's risk management was embedded in an extensive bureaucracy of committees.

Given NASA's pervasive reliance on teams and boards to consider the key questions affecting safety, the NRC concluded that "group democracy" easily prevailed, with the result that individual responsibility was diluted and obscured. One characteristic of decisions made in this highly participatory democratic process is that the views of experts and specialists can be diluted by consensus.[38]

The National Research Council panel discovered early in its work that the large number of Criticality 1 and 1R items were neither ranked by priority of their importance nor probability of failure, and that NASA did not appear to be making much use of modern analytical techniques in quantitatively assessing probabilities of failures, their effects, and the levels of risk in the program.[39] Although NASA's engineers had primary responsibility for conducting the more technical safety analyses and establishing the rationale for retaining critical items identified through the FMEA, few NASA engineers had any formal grounding in safety engineering techniques and methodologies. As indicated in our postscript case on the *Challenger* accident (Chapter 14), this process was not always followed to the letter.[40]

Consistent with Willoughby's earlier cited dictum, quantitative assessment methods, such as probabilistic risk assessment, were not used directly to support NASA decision making. Furthermore, NASA was not adequately staffed with specialists and engineers trained in statistical methods

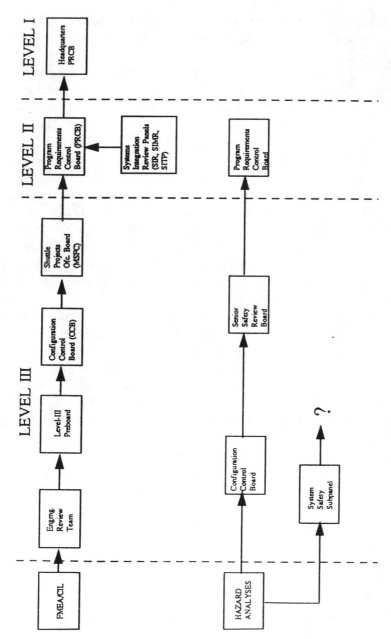

Figure 13.3 Various review boards used by NASA to implement a FMEA/CIL (from Committee on Shuttle Criticality Review and Hazard Analysis Audit of the Aeronautics and Space Engineering Board, January 1988).

to aid in the transformation of complex data into information useful for decision makers, and for use in setting standards and goals.[41] Shuttle risks were assessed based on subjective judgments and the approval of qualitative rationales by various board and panel chairmen and Level II and I authorities, even though quantitative engineering analyses and test data relevant to risk assessment were available and often were used in arriving at what were finally qualitative subjective judgments.[42]

> With such a non-specific (non-value based) risk acceptance process there is little basis for making objective comparisons of the several major risk categories nor for carrying out risk evaluations by independent agencies. Neither can one systematically evaluate the results of efforts to reduce the risk of the various possible losses.[43]

The NRC committee concluded: "Without more objective, quantifiable measures of relative risk it is not clear how NASA can expect to implement a truly effective risk management program."[44] Further, commenting on NASA's avoidance of probabilistic methods for risk management, the NRC panel stated:

> As a result, NASA has not had the benefit of powerful analytical assessment tools that have been developed in recent years, and that are used by other high technology organizations. Without such tools, it would be difficult at best for safety engineers to transform the massive data base which has developed in the shuttle program into specific information regarding what was truly known and what was not known. In addition, the failure to use numerical probability analyses had the unfortunate effect of denying NASA designers the required statistical data base on various types of failures, along with the better understanding of the mechanisms of failures that can be obtained with such data.
>
> While [modern techniques of statistical inference in combination with engineering models of failure modes and system models are] not a panacea, since not everything can be rigorously quantified, they can permit more objective assessment of the varying types and quality of information and data which are available as well as reflect the uncertainties introduced by incomplete data or knowledge.[45]

The panel was equally harsh on NASA's procedures with respect to the critical items list and the mandated retention rationales. It observed that the retention rationale for components on the critical items list sometimes appeared to be simply a collection of judgments that a design should be safe, emphasizing positive evidence at the expense of the negative, and thus did not give a balanced picture of the risk involved. In many cases reviewed by the NRC committee:

- No specific methodology or criteria were established against which these justifications can be measured.
- True margins against the failure modes often were neither defined nor explicitly validated.
- The probability of the failure mode was never established quantitatively.
- Design "fixes" were accepted without being analyzed and compared on the basis of relative risk with the configuration they were replacing.[46]

The NRC panel could find no documented, objective criteria for approving or rejecting proposed waivers. Decision making on critical items list waivers appeared to be subjective, with no consistent, formal basis for either approval or rejection of waivers.

> NASA managers emphasize that Level III engineers and their Level IV contractors are accorded a high level of responsibility and accountability throughout the program and that their opinions and analyses are the real bases for making retention decisions. These engineers bear the burden of proving that the rationale is strong enough to justify retention and waiver of the item. The [NRC] Committee believes that engineering judgment on these matters is not enough. While crucial, it is often too susceptible to vagaries of attention, knowledge, opinion, and the extraneous pressures to be the sole foundation for decision making.[47]

The NRC panel commented that updating the retention rational seemed to many to be considered a routine bookkeeping chore, of secondary importance, even though these rationales were the primary basis for granting waivers. The panel also observed that the retention rationale may sometimes fail to provide data in various important categories such as the effects of environmental parameters:

> The absence of such data, even though it resulted in uncertainty, in the past has sometimes had the effect of bolstering the rationale for retention and providing unwarranted confidence in the readiness reviews. For example, data suggesting that temperature was a factor in O-ring erosion did exist, but the relevant analyses apparently were considered to be inconclusive and these data did not appear in the retention rational. Thus, the rationale implied that there were no data to suggest that temperature was a problem.[48]

When Feynman observed this type of process applied to the SSME criteria, he referred to it as "gradually fooling oneself while degrading standards."[49]

Everything's Critical

Following the *Challenger* accident, NASA and its contractors performed major rework of all shuttle program FMEAs, updating the resulting CILs and reviewing all prior hazard analyses. One immediate result was a large increase in the number of Criticality 1 and 1R items. Prior to the *Challenger* accident, there were 2,369 Criticality 1 and 1R items. However, by November 1987 this number had risen to 4,686. For the SSME turbomachinery, the number of Criticality 1 items rose from 8 to 67. The NRC report noted that NASA had begun to prioritize the most critical items. Before *Challenger* and during the panel's audit, "NASA managers tended to assert that, since all Criticality 1 and 1R items are equally catastrophic in their consequences, all should be treated equally."[50] The panel discovered that prior to *Challenger*, 56 Criticality 1 failures occurred on the orbiter during flight without any of the postulated worst-case effects resulting.

Shortly after the *Challenger* accident, 159 items were selected to undergo mandatory next-flight changes (i.e., redesign, testing or analysis). While all were Criticality 1 or 1R, the NRC audit panel found only a handful had the FMEA/CIL/retention rationale listed as the original source of the change justification (1 of 23 for SSME; 4 of 48 for orbiter). Most of these were long-standing concerns, which had been derived from flight experience or engineering analysis before the accident.[51]

According to Robert Marshall, program development director, MSFC, these post-*Challenger* changes were significant:

> After 51-L [*Challenger*], we changed the design of blades on the tur-
> bine side, changed the damper configuration between blades on the
> LOX pump; we also changed the way we design and treat the blade.
> [We] made some changes in the way we cool the bearings, . . . in the
> sheet metal which guide flow into the pumps, . . . [we made] changes
> to some valve time configurations, how they open and close. [We
> made] some 50 changes to the engine.[52]

Hazard analysis was intended to be a key part of NASA's safety and risk management process. Ideally, to the panel, this top-down approach should have encompassed the FMEA and the other bottom-up analyses, covering the safety gaps that might have occurred. As prescribed by NASA, hazard analysis was to be an important element, complemented by a number of analyses including the FMEA/CIL. It would support the ultimate product of a safety analysis, the mission safety assessment (MSA) that was fed into the deliberations of the various engineering and readiness review boards. In reality, the panel found that:

> Hazard analysis has not played the central role it was designed to play.
> Instead, the main focus has been on the FMEA and its corresponding
> CIL retention rationale. "Hazard analysis as required" is a dead-end
> box, with inputs but no output with respect to waiver approval deci-
> sions. This impression was supported by subsystem project managers,
> engineers and their functional management at the Johnson Space Cen-
> ter. NASA did not use the hazard analyses and mission safety assess-
> ments as the basis for Criticality 1 and 1R waivers.[53]

An examination of NASA's organizational structure indicates why the hazard analysis process had little effect. This procedure was not part of NASA's main decision hierarchy. It was outside the flow of decisions and directives that emanated down through the organization. The NRC report states:

> We found that many engineering personnel, functional managers, and
> some subsystem managers were unaware of what tasks must be done to
> complete the hazard analysis, did not know whether they had actually
> been done, and did not contribute to them. Some, in fact, believed that

HAs were just an exercise done by reliability and/or safety people and that they were redundant to the FMEA/CILs. Their belief appears to be justified, in that these HA activities did not seem to be authoritatively in-line as part of a true hazard control and risk management process. It appears they were carried out in a relatively sterile environment outside the mainstream of engineering.[54]

Another critical problem was the need to provide rapid feedback of information into the flight readiness review (FRR) and commit-to-launch decisions on anomalies detected during inspections, tests and repairs, and in-flight operations. In the past, such information from the previous flight was not available in time to influence the decision to launch the next mission.

Sneed observed that:

Because we were flying the thing at the rates we were, most all of our attention – our management attention, our engineering attention – was on flying the next vehicle. Maybe more so than looking and saying, "well, how did that last one fly?" and "What is wrong with that last one, and what do we do to make it better, to make it more reliable?" . . . it was a very agonizing process if you had a problem with the last flight, saying "O.K. I'm going to fly again on the next flight with that as a known problem." So the element of risk, as you can see, really grew.[55]

The NRC panel concluded (as did the Rogers Commission) that a result of NASA's early decision not to use a specific reliability or risk analysis approach for the shuttle (due in part to the lack of a large statistical database) was that NASA safety organizations were not staffed with professional statisticians or risk analysts, and project engineers were not trained in modern statistical analysis techniques. The NRC did note that, due in part to its interim reports, NASA had taken tentative steps toward the use of probabilistic risk assessment and other techniques. In particular, within a year following the *Challenger* accident, the Jet Propulsion Laboratory was conducting a study of ways to improve the SSME certification process, and trial PRAs were under way for the main propulsion pressurization system.[56]

In short, NASA's organizational practices of collecting and using engineering data did not incorporate the most up-to-date methods. NASA was attempting to manage this complex technological system with the same qualitative decision procedures it had used in Apollo. Yet the complexity of the system overwhelmed the organizational capacity to make critical safety and resource allocation decisions. During the development of the shuttle program, few new and/or young engineers and managers were hired. The organization lacked the experience and training newly educated engineers, who would be skilled in probabilistic techniques, would bring. The old qualitative methods and disdain for quantification of risk dominated NASA's organizational culture, and "statistics . . . [did not] . . . count for anything" until well after the *Challenger* accident.

Post-*Challenger* Approaches to Risk Management

After the *Challenger* accident, NASA established the Office of Associate Administrator for Safety, Reliability, Maintainability and Quality Assurance (SRM&QA) with its director, George Rodney, reporting directly to the administrator. To the Aerospace Safety Advisory Panel, "this is the beginning of an independent 'certification' process within NASA. However, there is recent evidence that budgetary pressures within the shuttle program are causing project directors to propose budget cuts in various SRM&QA activities (e.g., safety documentation associated with the SSME such as FMEA/CILS and Hazard Analyses, and oversight of major STS projects)." The ASAP, upon learning this, recommended that "Across-the-board budget cuts that jeopardize the recently strengthened SRM&QA function must be denied."

The panel went on to recommend that "lists [of recommended changes for reliability and flight and ground safety], and other changes as they are identified, should be prioritized based on attributes of safety enhancement (severity and consequence), cost, schedule and performance and advantage should be taken of risk analysis techniques." Further, it cautioned that "NASA must resist the schedule pressures that can compromise safety during launch operations. This requires strong enforcement by NASA of the directives governing STS operations."

In 1988, NASA issued several policies and directives designed to improve the identification, evaluation, and disposition of safety risks. A new *Risk Management Policy for Manned Flight Programs* (NMI 8070.4) called for a risk management process that included categorization and prioritization of "risks" using qualitative techniques for ratings of the frequency expectation and severity of the potential mishaps. It also proposed quantitative risk analysis to achieve a more definitive *ordering* of risks for risk management purposes. While considering this to be a positive start, the ASAP recommended that "the risk management policies and initial implementing methodologies which have been issued need to be evolved further. Practical quantitative risk assessment and other relative risk-level rating techniques should be actually developed. They should then be applied to help define the risk levels of flight and ground systems."

The Johnson Space Center presented the ASAP with a new approach to hazard rebaselining and rating and a new format for the mission safety assessment report, which used fault trees to identify the potential system mishaps that might result from various hardware or human faults. However, the ASAP felt that "the MSA did not address even the relative risk-levels of the selected potential mishaps. . . . Because all of the items elected for inclusion in the MSA are rated as unlikely to occur and therefore 'safe to fly,' there remain a large number of undifferentiated items designated IHD [improvement is highly desirable]." It recommended that ambiguity regarding risk levels needs to be removed and NASA needs to provide much more objective (quantitative) and data-based risk assessment methodology that will differentiate the "unlikely" events for purposes of assessing the principal contributions to risk.[57]

NASA was now "forced" to adopt probabilistic risk assessment techniques, reimporting them from other agencies, specializing in diverse technological areas. Of particular note was a detailed study of "Practices in Adequate Structural Design" for MSFC conducted by Robert S. Ryan, the Marshall engineer we have quoted often in this book. Ryan noted that "a dynamic data base for the SSME containing information on over 1,000 engine firings was available. Using it with accelerometer data from each engine firing and shuttle flight, the turbopump status is (can be) dynamically determined. This enables refurbishment schedules to be established and thus averts the major problems which plagued the SSME early in its development."[58]

Indeed, in late 1987, Rocketdyne, as SSME contractor, had three important studies under way: a structural audit, which reviewed all of the structural analyses with special emphasis on long-term durability; a weld assessment, in which all critical item welds (over 3,000) were identified and reviewed, and the rationale for retention of each weld reassessed; and a failure trend analyses and reliability model, a result of Rocketdyne's evolving methodologies for analyzing the entire database obtained on the development and flight engines. Failure trend analyses are matched to component failure models using both "failures" and "unsatisfactory condition reports"

where possible adverse "trends" can be quantified as an aid to managing corrective actions. Failure data are also being used to make estimates at selected
confidence levels of the "statistical failure probabilities." The database
(referred to by Ryan) consisted of 49 equivalent engines, 1,000 tests, and
300,000 seconds of hot fire. However, the ASAP noted that for developing a
"risk level," one needs to evolve probabilities for the various consequences
of an engine shutdown during mainstay. This has not been addressed.[59]

As we mentioned, following the *Challenger* accident, the critical items list
for the SSME more than doubled from 73 to 189 items. By the end of 1992,
59 were resolved through design modifications, 27 through software
changes, 15 through inspection, testing, and operations changes, and 20 by
process changes. Rocketdyne and Marshall Space Flight Center indepen-
dently performed system reliability analyses using the engine tests and flight
database. Employing different mathematical methodologies and assumptions, they both achieved similar results. The probability of encountering an
engine shutdown (contained failure) operating at 104 percent power rating
was estimated to 1 in 45 flights; the probability of an uncontained (Criticality 1) failure increased to 1 in 120 flights. Contained failures would result
in the use of the abort mode planned for such incidents, so that crew and
vehicle would be saved. The SSME Assessment Team, created by the Aerospace Safety Advisory Panel at the request of the House Committee on
Science, Space and Technology (*NASA Multiyear Authorization Act of
1992*) to conduct a thorough assessment of the engine, observed: "Because
the analyses cannot and do not take into account the effects of all the special
controls and precautions currently taken with the engines prior to clearing
them for flight, the Team believes that the actual single flight reliability of
the engine is higher. That consideration, coupled with flight experience,
leads the Team to consider that the engine is safe to fly – provided the system of controls is applied vigorously and rigorously."[60]

Does this now mean that all main engine problems were on the way to be
solved? In March 1994, the ASAP reported that the SSME had performed
well in flight but had been the cause of launch delays and on-pad launch
aborts that were primarily attributable to manufacturing control problems.
This caused a shortage of usable turbopumps. Further, "sheet metal"
cracks in the current high-pressure fuel turbopumps have become more
frequent and are larger than previously experienced. This resulted in the
imposition of a 4,250-second operating time limit and a reduction of allowable crack size by a factor of four. To the ASAP, the greatest concern is the
generation of fragments that can, if they strike a turbine blade, cause blade

failure and lead to a catastrophic engine failure, although no such fragment generation has occurred before approximately 5,000 seconds of operation. In addition, engine sensor failures were becoming more frequent and are a source of increased risk of launch delays, on-pad aborts, or potential unwarranted engine shutdown in flight. Finally, the ASAP found that "the SSME health monitoring system comprising the engine controller and its algorithms, software and sensors is old technology. The controller's limited computational capacity precludes incorporation of more state-of-the-art algorithms and decision rules. As a result, the probabilities of either shutting down a healthy engine or failing to detect an engine anomaly are higher than necessary."[61]

The Probability of Failure: Was It a "Sliding Scale"?

Richard Feynman, recall, characterized NASA's decision making with reference to the solid rocket boosters and the SSME's as "a kind of Russian roulette . . . the shuttle flies . . . and nothing happens . . . the risk is no longer so high for the next flights. We can lower our standards a little bit because we got away with it last time."[62] What was the probability of a catastrophic failure "in flight" of the shuttle? Bell and Esch report that, "by the early 1980s, estimates of a catastrophic failure of the shuttle ranged from less than 1 chance in 100 to 1 chance in 100,000."[63] The more pessimistic figures came from working engineers and a series of risk assessment studies that NASA was mandated to conduct by the Interagency Nuclear Safety Review Panel in preparation for the launch of a thermonuclear generator for the Galileo spacecraft. An optimistic NASA had originally planned this mission to be launched to Jupiter by a shuttle with a Centaur upper stage. Three reports were prepared by J. H. Wiggins Company between 1979 and 1982.[64] Wiggins estimated that the risk of losing a shuttle during launch varied between 1 chance in 1,000 and 1 chance in 10,000, and prophetically identified the solid rocket boosters as the element with the greatest chance of failure.

In actuality, Wiggins initially had estimated that a catastrophe during launch would be between 1 in 59 to 1 in 34 based on data from other solid rockets. However, the Space Shuttle Range Safety Ad Hoc Committee, the apparent contract overseer, adjusted these numbers to the higher failure rates "with vague reference to learning curve improvement in solid rocket technology and the improvements commensurate with 'man-rated procurement.'"

Weatherwax and Colglazier (Teledyne Energy Systems), who reviewed the Wiggins study for the Air Force in 1983, further noted that "These estimates have no quantitative justification at all, and this was clearly acknowledged in the [Wiggins] report." They reestimated the solid rocket burnthrough failure rates as being between 1 in 46 to 1 in 71, but noted that these "calculations should be viewed as indicative and not conclusive."[65] Also often overlooked is that most of Wiggins's estimates were given as failure rates per second (or per hour), a factor that would apparently be ignored as NASA continued to rely on the Wiggins estimates, as will be discussed. Thus, in 1985, the Johnson Space Center published its own safety analysis for the Galileo project, and assigned a failure probability of level 2, which meant 1 chance in 100,000 (i.e., 1×10^{-5}).[66]

During his service on the Rogers Commission, Richard Feynman tried to learn more about this "estimate" by conducting several "experiments." In one, he interviewed Louis J. Ullian, a range safety officer with an extensive solid rocket background. Ullian told Feynman that his own experience with unmanned launches would yield an approximately 4 percent failure rate (i.e., 5 out of 127). He estimated that the manned flight failure rate would be 1 percent, assuming that it would be safer. Ullian later adjusted that to a final figure of 1 per 1,000 in light of NASA's claims that the agency was much more careful with manned flights. However, NASA had told Ullian that the probability of failure was approximately 1 per 100,000, a number that neither Ullian nor Feynman could understand. To Feynman, "that means you could fly the shuttle every day for an average of 300 years between accidents – every day, one flight, for 300 years – which is obviously crazy." Ullian told Feynman about his difficulty in trying to talk to James E. Kingsbury, director of science and engineering, Kennedy Space Center, in order to learn the reason behind this failure probability estimate. "He [Ullian] could get appointments with underlings, but he never could get through to Kingsbury and find out how NASA got its figure of 1 in 100,000."[67]

Although Ullian never got to see Kingsbury, Feynman pursued this line of questioning at the concluding Rogers Commission Hearing on May 2, 1986. During that testimony, Kingsbury admitted that the estimate was actually given as a rate per second:

> GENERAL KUTYNA: . . . What was the probability on that [the Galileo mission]?
>
> MR. KINGSBURY: On an order of 10^{-5}. Now that is at any second of time in the flight trajectory, and that was, I believe, misunderstood

by a number of people. If you say what is it over the full spectrum, you have got to divide it by 120. But at any point in time it was on an order of 10^{-5}.

DR. FEYNMAN: 10^{-5} per second, the probability was 10^{-5} per second?

MR. KINGSBURY: No, at any second, in any one second slice, the probability of failure in any one second slice was 10^{-5}.

DR. WHEELON: So the chance of a solid rocket motor failing is about 1 in 10^{-5}?

MR. KINGSBURY: Those were the numbers we had, that is correct.[68]

Although Kingsbury had "clarified" the rate to the commission, two months earlier on March 4, 1986, Milton Silveira, NASA's chief engineer, informed the U.S. House Committee on Science and Technology during a hearing on the Galileo thermonuclear generator that "We think that using a number like 10 to the minus 3 [10^{-3}], as suggested, is probably a little pessimistic. . . . [the actual risk] would be 10 to the minus 5 [10^{-5}], and that is our design objective. . . . We came to those probabilities based on engineering judgment in review of the design rather than taking a statistical database because we didn't feel we had that."[69]

Clearly, NASA management still had a communications problem, and Feynman continued to probe. During a field visit to Marshall Space Flight Center in March 1986 as part of the O-ring investigation, Feynman extended his inquiries to another component – the main engines. He was curious to see if the same slipping of safety criteria and lack of communications would be found elsewhere in NASA. He also wanted to confront whoever would claim that the probability of failure was 1 in 100,000.[70] At Marshall, Feynman met with three engineers and their manager, Dr. Judson A. Lovingood (deputy manager, Marshall Shuttle Projects Office). The engineers were responsible for the main engine (Lovingood was manager of the shuttle main engine project from February 1982 until October 1983, when it was divided into separate development and flight project offices).[71]

After having them explain how the engine functioned, Feynman asked each to write down an estimate of the probability of engine failure in flight. Two of the engineers wrote the equivalent of 1 per 200, and the third wrote 1 per 300. According to Feynman, "Lovingood initially weaseled, and wrote 'cannot quantify.'" When Feynman pressed him, Lovingood then wrote 1 per 100,000, the "official" NASA estimate. Lovingood later sent Feynman a NASA report which attributed the 1 per 100,000 rate to "engineering judgment." To Feynman:

As I can tell, "engineering judgment" means they're just going to make up the numbers! The probability of an engine-blade failure was given as a universal constant, as if all the blades were exactly the same, under the same conditions. The whole paper was quantifying everything. Just about every nut and bolt was in there: "The chance that a HPHTP pipe will burst is 10^{-7}." You can't estimate things like that; a probability of $1/10,000,000$ is almost impossible to estimate. It was clear that the numbers for each part of the engine were chosen so that when you add everything together you get $1/100,000$.[72]

Feynman may have been a little harsh, but not far off base. It is very likely that these "estimates" came from the Johnson Space Center study, which apparently relied on the original estimates provided by Wiggins. For example, Wiggins, using a 1975 Reactor Safety Study[73] and his "engineering judgment," estimated the failure rate per hour of operation for seals and gaskets to be between 10^{-7} to 10^{-4}. As noted, Weatherwax and Colglazier had recommended to the Air Force (in December 1983) that "the Wiggins estimates . . . are substantially low, perhaps by up to a factor of 1,000 in some instances, . . . and the Wiggins results are not suitable for use in the ongoing safety analysis for the Galileo mission."[74] If our supposition is correct, NASA not only used flawed numbers, but often did not realize that these probabilities were given as rates. If anything, it underscores NASA management's lack of faith in probabilistic analyses.

During his "investigation," Feynman was informed of all the engine problems – blades cracking in both the oxidizer and hydrogen pumps, casings blistering and cracking, bearing wear and a subsynchronous whirl (caused by the shaft being bent into a slight parabolic shape at high speed). He counted a dozen very serious problems about half of which were fixed following the *Challenger* incident. Feynman spent a considerable amount of time discussing a high-frequency vibration problem that occurred in some of the engines with the three NASA engineers. They noted that some engines would vibrate so much that they could not be used. To these engineers this problem had not been solved, although NASA had officially stated that 4,000-cycle vibration was within its database. Feynman recalled upon leaving the meeting that "I had the definite impression that I had found the same game as with the seals: management reducing criteria and accepting more and more errors that weren't designed into the device, while the engineers are screaming from below, 'HELP' and 'this is a RED ALERT!'"[75] Noting that thirteen of these major engine problems occurred in its first 125,000 seconds of testing with only three occurring in the second 125,000 seconds, Feynman "guessed" that there may be at least one surprise in the next 250,000 seconds. To him, this would translate into a probability of failure due to the engines of "less than 1/500 per mission."[76]

Risk Assessment and Engineering Ethics

This particular case study emphasizes how the organizational structure and goals of NASA affected the ethical decision-making capability of the individual engineer regarding risk assessment. It also stresses the value-component of how one assesses risk. As a way of summarizing these issues, we will explore them with reference to our ethical framework.

Two complementary approaches to risk assessment and analysis exist: FMEA/CIL process and probabilistic risk analysis. The FMEA/CIL process, adopted by NASA, recognized and attempted to prioritize the "uncertainty" associated with designing, building, and testing its highly complex, innovative technology. Based largely on "engineering judgment," this type of assessment is, de facto, qualitative. Its strengths, indeed, are grounded in an intuitive, creative, judgment-oriented strategy. Precise figures or specific methodologies for computing risks are not part of this strategy. Thus, the individual reasoning and justifications offered for a particular decision – direct lines of responsibility for decision making combined with communication within the organization about why decisions were made – is central to its success. An organizational structure composed of small working groups, with clear lines of communication to each other and to the larger organizational framework would be necessary for its success. Early on, this was characteristic of NASA. By 1972, as A. O. Tischler recognized, NASA's complex management structure did not foster such communication and its policies regarding risk were not well understood.

Given a complex project and an equally complex organizational structure, the strength of the FMEA/CIL approach is also its weakness. As an innovative project becomes more complex and thousands of individual items are considered by their expert teams as Criticality 1, some method for prioritization is needed. Why weren't they prioritized? The application of an all-up philosophy in the building of the SSME masked the need to prioritize. The success of the total project was based on acceptance of a high risk. What seemed risky in isolation was camouflaged when "tested" with all other components. This was what Feynman characterized as a "Russian roulette mentality." If the combined technology worked, then perhaps the individual component was less critical. It was "de facto" prioritizing.

The individual engineer was aware of inherent high risks in his or her assessment of specific projects. The risks were calculated from the perspective of individual responsibility and accountability. What happened when these risks were communicated through the newly created "democratized" organizational structure of boards? NASA management's estimate of 1 per 100,000 reflected a diffusion of responsibility, a bias toward political pressure, and an overall success-oriented perspective. This modified and nullified any calculations provided by individual engineers. The goals of the organization, which were clearly influenced by the political arena, were, in Simon's terms, at odds with the individual's assessment of risk. An individual engineer's risk assessment tends to be biased. It is in part dependent upon the complexity of the task at hand, the consequences of a catastrophic failure, and the organizational structure within which decision making occurs.

At some point in the shuttle project, the FMEA/CIL analysis alone became inadequate to assess risks. Some may argue that it never was adequate, that quantitative risk assessment techniques, which were available, should have always complimented it. NASA management's "can do," "success-oriented" strategy encouraged it to publicize a failure rate of 1 per 100,000 and to ignore the calculations resulting from statistical studies commissioned by NASA, which were too "pessimistic." The statistical measures in and of themselves are not pessimistic, but how individual analysts choose to employ the measures – the assumptions they use – were also based on values. Could individual engineers have gone outside the management structure and adopted quantitative methods to reformulate their assessment? Or was the all-up methodology, risky as it was, the only way to get the shuttle off the launch pad?

Competence may have been violated by both individual engineers and the organization. Although it was a management decision to use qualitative methods, the "competent engineer" should be aware of the range of risk assessment methodologies and, in high-risk circumstances, be able to use them as one check on his or her own bias. Then, a challenge to management carries more persuasive weight. Recall French's plea to understand one's station. An expectation of a competent engineer would be to recognize that there are ways to quantify risk and, either alone or, most likely, with assistance from knowledgeable individuals, estimate that risk, document it, and present it to management. If the organizational management ignores the documentation – the individual assessments of Criticality 1 items – en masse, it sends an overwhelming message to NASA management. The inability to prioritize the Criticality 1 issues, to Feynman, constituted the "red flag" of "too high a risk." NASA's complex and democratic management structure however, masked the individual responsibility issue with intermediate "group decisions." It diluted the appearance of risk while neglecting to negate it by redesign. When challenged in a post hoc review, which included the loss of seven identifiable lives and the visible explosion of the technology, NASA could no longer defend its abstract nature of possible risk of 1 per 100,000 of a catastrophic failure. It had vanished to a harsh but firm 1 per 25. By March 22, 1996, that risk had increased to 1 per 76 with 51 successful post-*Challenger* missions. How would future calculations, whether they were based on probabilities or qualitative assessments, interpret this? In testimony to Congress, seven years after *Challenger*, former astronaut Thomas Stafford used this statistic to document the comparative success rate of the space shuttle, and to strongly urge its continued support.[77]

Notes

1. This chapter draws extensively on two sources – the report of the Committee on Shuttle Criticality Review and Hazard Analysis Audit of the Aeronautics and Space Engineering Board, *Post Challenger Evaluation of Space Shuttle Risk Assessment and Management* (Washington, DC: National Academy Press, January 1988), and the writings of Richard Feynman, the Rogers Commission member and physicist, who wrote an infamous addendum to the Rogers Commission report criticizing how NASA assessed risk.
2. Charles Perrow, *Normal Accidents: Living with High-Risk Technologies* (New York: Basic Books, 1984).
3. R. S. Ryan, "The Role of Failure/Problems in Engineering: A Commentary on Failures Experienced - Lessons Learned," NASA Technical Paper 3213, George C. Marshall Space Flight Center, Huntsville, AL, March 1992, p. 1.
4. Ibid., p. 24.
5. Trudy E. Bell, "Managing Murphy's Law: Engineering a Minimum-Risk System," *IEEE Spectrum*, June 1989: 23–5.
6. Walter C. Williams, *Report of the SSME Assessment Team* (Washington, DC: NASA, January 1993), p. 7.
7. Committee on Shuttle Criticality Review and Hazard Analysis Audit of the Aeronautics and Space Engineering Board, p. 56.
8. Williams, *Report of the SSME Assessment Team*, p. 8.
9. Committee on Shuttle Criticality Review and Hazard Analysis Audit of the Aeronautics and Space Engineering Board.

10. Bell, "Managing Murphy's Law," pp. 26–7.
11. Committee on Shuttle Criticality Review and Hazard Analysis Audit of the Aeronautics and Space Engineering Board, p. 56.
12. Eric J. Lerner, "An Alternative to 'Launch on Hunch,'" *Aerospace America 25*, May 1987: 40–4.
13. Ibid., pp. 40–1.
14. Sylvia Freis, draft report, "NASA: Safety Organization and Procedures in Manned Space Flight Programs," NASA History Office, March 24, 1986, unpublished manuscript.
15. Ibid.
16. Ibid.
17. Ibid.
18. Committee on Shuttle Criticality Review and Hazard Analysis Audit of the Aeronautics and Space Engineering Board, p. 17.
19. Howard E. McCurdy, *Inside NASA: High Technology and Orgainzational Change in the American Space Program* (Baltimore: Johns Hopkins University Press, 1993), p. 92.
20. Ibid., p. 94.
21. Ibid.
22. Trudy E. Bell and Karl Esch, "The Space Shuttle: A Case of Subjective Engineering," *IEEE Spectrum 26*, June 1989: 42–5.
23. Ibid., p. 42.
24. Raymond M. Wilmotte, "The Structure of Risk Technologies in the Decision Making Process," memorandum to the Aerospace Safety Advisory Panel, July 1971.
25. Raymond M. Wilmotte, "An Introduction to the Basic Problem of Risk Management," paper presented to the Aerospace Safety Advisory Panel, July 6, 1971.
26. McCurdy, *Inside NASA*, p. 148.
27. William Sneed, taped interview in Management Operations Office, MSFC, *Oral Interviews: Space Shuttle History Project*, Transcript Collection, December 1988, p. 168.
28. Howard E. McCurdy, *Inside NASA*, p. 149.
29. Ibid., p. 144.
30. Richard P. Feynman, *What Do You Care What Other People Think? Further Adventures of a Curious Character*, as told to Ralph Leighton (New York: W. W. Norton, 1988), p. 234.
31. Ibid.
32. Ibid., p. 235.
33. Michael Collins, *Liftoff* (New York: Grove Press, 1988), p. 228.
34. Committee on Shuttle Criticality Review and Hazard Analysis Audit of the Aeronautics and Space Engineering Board.
35. Ibid., p. 56.
36. Ibid., pp. 64–5.
37. William P. Rogers, *Report of the Presidential Commission on the Space Shuttle Challenger Accident* (Washington, DC, June 6, 1986), 1:82–3.
38. Committee on Shuttle Criticality Review and Hazard Analysis Audit of the Aeronautics and Space Engineering Board, pp. 68–70.
39. Ibid., p. 11.
40. Ibid., p. 17.

41. Ibid., p. 56.
42. Ibid., p. 36.
43. Ibid., pp. 36–7.
44. Ibid., p. 37.
45. Ibid., p. 55.
46. Ibid., p. 42.
47. Ibid., p. 44.
48. Ibid., p. 43.
49. Feynman, *What Do You Care What Other People Think?*, p. 223.
50. Committee on Shuttle Criticality Review and Hazard Analysis Audit of the Aeronautics and Space Engineering Board, p. 45.
51. Ibid., p. 51.
52. Robert Marshall, taped interview in Management Operations Office, MSFC, *Oral Interviews: Space Shuttle History Project*, Transcript Collection, December 1988; pp. 122–3.
53. Ibid., p. 47.
54. Ibid., p. 50.
55. Sneed, taped interview, p. 166.
56. Committee on Shuttle Criticality Review and Hazard Analysis Audit of the Aeronautics and Space Engineering Board, p. 56.
57. Aerospace Safety Advisory Panel, *Annual Report* (Washington, DC: NASA, March 1989).
58. Robert S. Ryan, "Practices in Adequate Structural Design," NASA Technical Paper 2893, George C. Marshall Space Flight Center, Huntsville, AL, 1989, p. 64.
59. Aerospace Safety Advisory Panel, *Annual Report* (1989), pp. 2–21.
60. Williams, *Report of the SSME Assessment Team*.
61. Aerospace Safety Advisory Panel; *Annual Report* (Washington DC: NASA, March 1994), pp. 9–10; 26–7.
62. Collins, *Liftoff*, p. 228.
63. Bell and Esch, "The Space Shuttle," p. 44.
64. J. H. Wiggins Co., *Space Shuttle Range Safety Hazards Analysis*, Technical Report 81-1329, 1981; *Development of STS Failure Probabilities MECO to Payload Separation*, Technical Report 79-1359, 1979; *Development of STS/Centaur Failure Probabilities Liftoff to Centaur Separation*, Technical Report 82-1404, 1982. As cited in R. K. Weatherwax and C. W. Colglazier, *Review of Shuttle/Centaur Failure Probability Estimates for Space Nuclear Mission Applications*, Teledyne Energy Systems, Timonium, MD, December 1983, AFWL-TR-83-61 p. 5.
65. Weatherwax and Colglazier, *Review of Shuttle/Centaur Failure Probability Estimates*.
66. *Space Shuttle Data for Planetary Mission RTG Safety Analysis*, Johnson Space Center, Houston, TX, February 15, 1985, pp. 3–1, 3–2; Bell and Esch, "The Space Shuttle," p. 44.
67. Feynman, *What Do You Care What Other People Think?*, pp. 179–80.
68. Rogers, *Report*, 5:1635.
69. Bell and Esch, "The Space Shuttle," p. 46.
70. Feynman, *What Do You Care What Other People Think?*, p. 181.
71. Rogers, *Report*, 5:911.
72. Feynman, *What Do You Care What Other People Think?*, pp. 182–3.

73. Reactor Safety Study, *An Assessment of Accident Risks in U.S. Commercial Nuclear Power Plants*, WASH-1400 (NUREG-75/014), Nuclear Regulatory Commission, October 1975.
74. Weatherwax and Colglazier, *Review of Shuttle/Centaur Failure Probability Estimates*, pp. 75, 77.
75. Feynman, *What Do You Care What Other People Think?*, pp. 184–5.
76. Ibid., pp. 228–9.
77. Thomas P. Stafford, Lt. General USAF (Ret.), Statement to the House Science, Space and Technology Committee; Space Subcommittee on the Future of the U.S. Space Launch Capability, February 17, 1993.

PART III

Postscript

14

The Challenger *Accident: Engineering Design and Performance Decisions*

*I can't recall a launch that I have had where there was 100 percent cer-
tainty that everything was perfect, and everyone around the table would
agree to that. It is the job of the launch director to listen to everyone, and
it's our job around the table to listen and say there is this element of risk,
and you characterize this as 90 percent, or 95, and then you get a consensus that that risk is an acceptable risk, and then you launch.*

– Major General Donald J. Kutyna, USAF, member, Presidential
Commission on the Space Shuttle *Challenger* Accident

*Since the earliest days of the manned space flight program that I've been
associated with and Mr. Armstrong has been associated with, our basic
philosophy is: Prove to me we're ready to fly. And somehow it seems in this
particular instance we have switched around to: Prove to me we are not
able to fly.*

– Robert L. Crippen, astronaut, to Presidential Commission
on the Space Shuttle *Challenger* Accident

In our original discussions regarding the choice of a space shuttle component for the study of engineering ethics, we explicitly chose not to comment on the *Challenger* disaster and, in particular, the O-ring "case."
Ethics, as we have maintained throughout this report, is critical reflection
into the nature and grounds of morality. An ethicist working in a practical
setting is neither a policeman nor a reformer. Rather, he or she is an acade-
mician traversing complex ground with the aim of aiding practitioners – be
they physicians, engineers, nurses, or administrators – to identify their own

value stances, recognize alternative values, identify moral issues inherent in their practice, and attempt to resolve moral dilemmas with the aid of principles grounded in their practice and better understood by referencing philosophical theory, law, and other pertinent bodies of knowledge. Tackling issues surrounding the *Challenger* accident, we reasoned, could give the appearance of moralizing or, even worse, "witch-hunting." This might jeopardize our larger concern. As our ethical framework was constructed and discussed with reference to the four cases that focused on the development and testing of the main engines, we reconsidered examining the launch of the *Challenger*. It became apparent that, when our framework was applied to the Rogers Commission report and to the events which led up to the decision to launch, new and instructive insights could be gained. This book, and this final case, document a chapter in history which will be ultimately useful, if it serves to heighten our sensitivities to the ways in which the opportunity for moral harm and ethical judgment are intertwined in decisions to foster technological progress.

The *Challenger:* Two Ethical Dilemmas

The tragic *Challenger* accident and the subsequent public attention to its causes provide detailed examples of ethical dilemmas in both engineering design and performance. While the cause of the accident was systematically traced to a faulty design, however, the resultant investigation, particularly that of the Presidential Commission on the Space Shuttle *Challenger* Accident, chaired by former secretary of state and attorney general William Rogers, found that this inadequacy was known since 1977. The five-volume report of the Rogers Commission, including the two volumes of hearings, provide a rich database for research and study.[1] In particular, the Rogers investigation provides an example of engineers struggling with two types of ethical dilemmas involving design and performance decisions.

The design dilemma involves the design and testing of the solid rocket booster joint. The performance dilemma concerns the decision to launch the *Challenger* on January 28, 1986, in light of the extremely low temperatures predicted for that morning. During a twenty-four-hour period, having designed, built, and tested a complex system for a variety of conditions, engineers now faced a decision as to whether it would function under conditions that were neither fully anticipated nor completely included in the design criteria. In order to analyze these dilemmas, it is important to understand the organizations and technology involved.

As an "undercard" to this main examination, we have taken one sidelight of the Rogers Commission investigation, the testimony of Richard Cook, a NASA budget analyst. By working closely with NASA's propulsion engineers, Cook learned that the O-ring mechanism in the solid rocket booster (SRB) threatened the safety of the shuttle and that, if it failed to seal, the results would be catastrophic. He wrote two memos that captured and summarized the engineering problems with the SRB – one six months prior to the *Challenger* explosion, the other just six days after it occurred. Although Cook was not an engineer by training, he provides an example of adherence to our three principles of competence, responsibility, and Cicero's Creed II. It is perhaps the vantage point he held "outside the main stream" of engineering decisions that enabled him to understand and articulate both technical and ethical problems. We illustrate this in the following two sections, taken from two early Rogers Commission hearings held on February 10 and 11, 1986, less than two weeks after the accident and before its cause had been determined.

Undercard I: The Case of Mr. Cook – What the Budget Analyst Found: First Memo

Richard C. Cook joined the government in 1970, first at the Civil Service Commission, followed by the Food and Drug Administration, and then the White House Consumer Affairs Council (during both the Carter and Reagan administrations). After a stint in industry where he worked on a defense intelligence hardware project, Cook joined NASA in the summer of 1985 as a resource analyst in the Comptroller's Office. He was assigned the external tank, SRB, and Centaur upper stage. One of his first tasks was to see if the SRB had any outstanding engineering problems which might require either additional funding or a change in the funding profile. In order to do this, Cook explained:

> We have to keep pretty much in touch with the project people in the Office of Space Flight, and we also go on field trips down to Marshall or Kennedy or other places. . . . And then when issues arise that look like they might be budget threats, we have got to report back on it and

try to come up with some kind of estimate with the program office of what it's going to cost to repair this type of thing.[2]

In meetings with engineers in the Office of Space Flight, it became apparent to Cook "that there were some real concerns with the O-ring problem at that time, concerns from an engineering standpoint which . . . had flight safety implications and potentially major budgetary concerns, because . . . if you fix something like this you've got quite a range of cost implications." This was particularly true for the solid rocket booster where development was almost closed. "If something came along [on the SRB] we would have to think real hard, work with the Office of Space Flight on figuring out how to cover something like that. And we felt that the O-ring problem – and I think it was our impression from the Office of Space Flight – was a potentially major budget hit. . . . Every month the O-ring problem was on the list of budget threats from the summer [1985] on into the fall."[3]

On July 23, 1985, Cook wrote a memorandum on "Problems with SRB Seals" to his supervisor, Michael B. Mann, chief, STS Resources Analysis Branch, Office of Comptroller.[4] That memo stated in part:

> Earlier this week you asked me to investigate reported problems with the charring of seals between SRB motor segments during flight operations. Discussions with program engineers show this to be a potentially major problem affecting both flight safety and program costs. . . .
>
> Engineers have not yet determined the cause of the problem. *There is little question, however, that flight safety has been and is still being compromised by potential failure of the seals, and it is acknowledged that failure during launch would certainly be catastrophic.* There is also indication that staff personnel knew of this problem sometime in advance of management's becoming apprised of what was going on. . . .
>
> It should be pointed out that Code M [Office of Space Flight] management is viewing the situation with the utmost seriousness. From a budgetary standpoint, I would think that any NASA budget submitted this year for FY 1987 and beyond should certainly be based on a reliable judgment as to the cause of the SRB seal problem and a corresponding decision as to budgetary action needed to provide for its solution. (Emphasis added)

As Cook explained it, "there was a lot of concern about how [to] get redundancy back in that joint without having to throw away half a million dollar SRB segments" through redesigning and recasting. A thirteen-month lead time was needed if a new segment was ordered from the manufacturer, an obvious problem in terms of the launch schedule.[5] Softening his prediction of disaster, he recalled that "there was a flight in the fall of 1985 where there was no erosion at all, and I reported that back to my management. In fact, I was reporting at that time that it looked to be as though the fix for the O-rings might be less serious than was earlier indicated."[6]

The Rogers Commission first learned about Cook from a *New York Times* article based on the July 1985 memo. It's headline was: "NASA Had Warning of a Disaster by Booster" (February 9, 1986).[7] The article provoked a number of questions at the next day's commission hearings and led to Cook's testimony at a subsequent session.

What follows is a select transcript from the February 10, 1986, hearing.

CHAIRMAN ROGERS: . . . You will notice at the end of that he [Cook] ties safety to budgetary considerations. He said "I would think that any NASA budget submitted this year for fiscal year 1987 and beyond should certainly be based on a reliable judgment as to the cause of the SRB seal problem and a corresponding decision as to budgetary action needed to provide for its solution." Do you know whether any such action was taken or consideration was given to his memorandum on that point?[8]

MR. [L. MICHAEL] WEEKS (deputy associate administrator for space flight – technical; NASA): I can state authoritatively that no action – I think this is true of Mr. [Jesse W.] Moore [associate administrator for space flight] as well, because I didn't see this memorandum until yesterday [in the *New York Times*].

CHAIRMAN ROGERS: Do you know whether anybody else took it seriously then?

MR. WEEKS: We certainly were alert, as you will see as we go through this whole chronology, you will see that we were alert to a problem, but we had not identified a precise amount of money that we thought would be required to fix it. . . .

CHAIRMAN ROGERS: . . . but we want to be sure that we face the facts. The fact is you have a memorandum, and Cook says certain things he thinks should be done. All I want to do is find out what was done. If it wasn't done, tell us why and we will understand and the record will be clear. That's all.[9]

DR. [SALLY K.] RIDE [astronaut and physicist]: . . . What sort of threat to the budget was it being considered as? In other words, were people thinking of it as a threat because they needed lots more O-rings, or were they thinking of it as a threat because there was a potential redesign of the solid rocket?[10] . . . I guess what I'm concerned about is, you're saying you might want a potential redesign because you were concerned at some level about erosion of the seals, and if there's any concern if the O-rings go you've lost the solids, and if you've lost the solids you've lost the flight. So that seems like a fairly serious consideration.[11]

CHAIRMAN ROGERS: I'm really less interested in that [documentation of decision to fly] than whether there were two schools of thought, whether some people were saying we should stop and others thought it was such a serious safety consideration that we should stop and correct it, no matter what the budgetary considerations are, and other people say, no, it costs too much, or we're not worried about the safety aspect, or it has some safety features but we're not very aware of them? Do we have that kind of a discussion? Because just these charts don't really help us too much.

MR. MOORE: To my knowledge – and anybody else in the room can address the question that you asked – to my knowledge, there was no concern on the part of anybody here who said we should stop flying because of the budget threat potential and so on.

CHAIRMAN ROGERS: Was there anybody who said we ought to stop for a little while and slow down and take the following corrective steps before we fly?

MR. MOORE: No, sir.

VICE-CHAIRMAN [NEIL] ARMSTRONG [astronaut]: But what I'm trying to understand here, this charring item on the chart is on there, and that says that there was a concern of some sort and Mr. Moore is telling us that it wasn't a safety-of-flight-concern. And what I'm trying to understand is, what might have made it a safety-in-flight concern?[12]

MR. [DAVID] WINTERHALTER (acting director, Shuttle Propulsion Group, NASA headquarters): Firstly, if I thought at the time that it was a real safety-of-flight issue, it wouldn't have been a budget threat. It wouldn't have appeared on this list. It would have appeared as a mandatory change, a make-work change, that we would say we don't do any more flying, we don't do any more testing, until we make some changes. . . .

DR. RIDE: What amount of erosion would have given you a problem to call it a safety-in-flight issue?

MR. WINTERHALTER: Well, we had test results on this and, even with the erosion on the secondary ring, which was the only instance we saw, we had a safety factor sizewise of over two to one in our tests.

DR. RIDE: What does that mean in terms of the amount of time?

MR. WEEKS: Sally, I don't think that you should get the idea that we weren't deeply concerned about that first instance of the secondary O-ring having erosion.

VICE-CHAIRMAN ARMSTRONG. I find myself not really understanding the feeling of the people that were involved in this.

Undercard II: The Case of Mr. Cook – What the Budget Analyst Found: The Second Memo

On February 3, 1986, six days after the *Challenger* accident, Cook prepared a second memorandum to Mann, who had requested that he pull together all the possible budget implications. Admittedly written in "the heat of the moment," and prior to the O-rings having been determined to be the cause of the accident, Cook noted that:

> There is a growing consensus that the cause of the Challenger explosion was a burnthrough in a Solid Rocket Booster at or near a field joint. It is also the consensus of engineers in the Propulsion Division, Office of Space Flight, that if such a burnthrough occurred, it was probably preventable and that for well over a year the Solid Rocket Boosters have been flying in an unsafe condition. This has been due to the problem of O-ring erosion and loss of redundancy caused by unseating of the secondary O-ring in flight. . . .
>
> Even if it cannot be ascertained with absolute certainty that a burnthrough precipitated the explosion, it is clear that the O-ring problem must be repaired before the Shuttle can fly again. . . .
>
> Given these facts, it is my considered opinion that NASA is facing a suspension of the Shuttle flights due to SRB problems of a minimum of nine months and possibly as long as two years or more. This assumes that the agency makes a rapid decision to proceed with the required SRB improvement program, along with improvements in Thiokol's safety management.

Eight days after preparing this memo, Cook was called as a Rogers Commission witness. Also appearing at that session was Cook's immediate supervisor, Michael Mann. It was at this session that Cook revealed the existence of the second memo.

> CHAIRMAN ROGERS: . . . You say at one point [February 3, 1986, memo] when you are referring to the engineers . . . "It is also my opinion that the Marshall Space Flight Center has not been adequately responsive to headquarters' concerns about flight safety, that the Office of Space Flight has not given enough time and attention to the assessment of problems with the SRB safety raised by senior engineers in the Propulsion Division. . . . And that these engineers have been improperly excluded from investigation of the *Challenger* disaster."[13]
>
> MR. COOK: . . . I was amazed that when this incident occurred the engineers in Washington were over there in their offices getting the data on the investigations from the newspaper and the media, and now and then phone calls from guys down at Kennedy about what was being found. . . . These were the top propulsion engineers who prepared reports for the Office of Space Flight and for the Administrator and for us. I just couldn't understand why that group wasn't down there going through the data and looking at the photos and everything else. . . .
>
>   The only thing that I would urge [the commission] would be that as much as you can to get just the ordinary working guys, such as me and the engineers and the guys from the Marshall S&E Lab, and if you can get them in from Thiokol, just the ordinary engineers who break these things down, who look at them, who call each other on the phone and say hey, look what I found here. You've got to take a look at this."[14]
>
> MR. MANN: . . . I discussed it [Cook's July memo] with my immediate supervisors, and I had it sent back to the Office of Space Flight, and then I also went to discuss with the engineers if the memorandum really reflected the situation as they saw it. And in those discussions . . . I got the feeling that maybe the memo overstated the concerns, that there were quite a few actions being taken within the program office to resolve the issue, that there were extensive reviews going on, as there are in almost any technical type of review.[15] . . .
>
> CHAIRMAN ROGERS: Were you surprised at the contents of the February 3 memorandum?

MR. MANN: Frankly, I was. I had specifically asked for a cost estimate, and I specifically needed a number, this is a $20 million problem, this is a $100 million problem, to integrate that with other estimates that we were doing on the rest of the program, and frankly, the memorandum didn't provide that kind of information. It had some useful insights into the particulars of the program, such as how many motors we had in inventory, but it didn't really come down to what I was trying to accomplish.

CHAIRMAN ROGERS: Any other questions?

[No response.]

CHAIRMAN ROGERS: Thank you very much, Mr. Mann.

Here we see the reverse of the situation illustrated in the Applegate memo in the "DC-10 Case" (Chapter 3). Applegate, recall, used the language of liability to describe his concerns with the faulty design of the DC-10. Cook actually uses moral language in his description of the consequences of the O-ring problem. In contrast, Mann, his supervisor, wanted only a dollars and cents accounting of what this problem meant; after all, that is what Cook's assignment had been.

Morton Thiokol, Inc.: The Company and the Contract

In November 1973, the then Thiokol Chemical Corporation was awarded the sole source contract to produce the first thirty-seven sets of solid rocket boosters. In NASA's proposal evaluation process, Thiokol had been rated fourth under the design, development, and verification factor; second under the manufacturing, refurbishment, and product support factor; and first under the management factor. Its overall rating was second, but due to its substantial cost advantage, Thiokol was awarded this sole-source contract over its higher rated competitor. As further justification, its SRB joint design was considered a plus: "The Thiokol motor case joints utilized dual O-rings and test ports between seals, enabling a simple leak check without pressurizing the entire motor. This innovative design feature increased reliability and decreased operations at the launch site, indicating good attention to low cost and production."[16]

In the period following the booster contract award, Thiokol became the largest private employer in Utah with 6,400 workers at its Wasatch plant.

Further, the potential of securing additional guaranteed long-term government contracts was very high. When the shuttle was declared operational in 1982, the projected number of flights translated into a sharp increase of boosters procurement.[17]

The actual flight frequency indicated that the first series of boosters would be consumed by the end of 1986. Therefore, in late 1985 (prior to the *Challenger* launch) Thiokol was actively negotiating for the $1 billion second-phase contract which would cover the next sixty pairs of boosters, well aware that for at least two years other aerospace companies had been lobbying Congress to break Thiokol's booster monopoly. Although Thiokol was assured of a short-term extension of its contract, NASA had proposed on December 26, 1985, to develop procedures for a second source for later procurement. This initiated a three-week period of intense negotiations between Thiokol and NASA over the terms and duration of the extension, with a major negotiation session scheduled for Tuesday afternoon, January 28, 1986.[18]

The Solid Rocket Booster and Its Field Joints: An Overview

The shuttle's solid rocket booster resembles a high-tech, giant firecracker – it is a metal cylinder, 116 feet long, 12 feet in diameter, filled with solid propellant, which is a viscoelastic rubbery material (see Figure 14.1). Following ignition, the fuel burns until depleted, thus providing rocket thrust as the exhaust gases pass through the end nozzle. The twin boosters were designed to provide 80 percent of the total thrust at lift-off, the remaining thrust coming from the three shuttle main engines. As noted, this design was the result of compromises over the more costly original concept of a manned, reusable first stage using the same liquid-fuel engines as the orbiter.

The solid rocket booster contains four segments. Since there was no reasonable way to ship the fully assembled rocket from Utah to Florida, rather than having been assembled during manufacturing, these must be stacked together at the launch site in order to form the solid rocket motor. In order to assure a complete seal of the motor casing during this field assembly, a "field joint" was designed as shown in Figure 14.2. The rim of the upper segment has a tang – an extension that fits into the clevis, a groove in the lower section. A total of 180 pins hold the two sections in place, and sealing

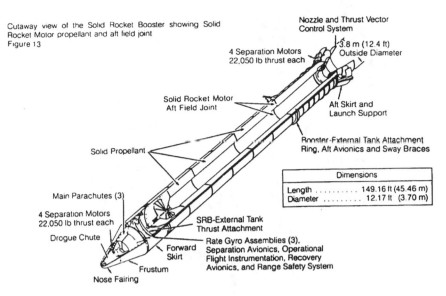

Figure 14.1 NASA drawing of the solid rocket booster showing main sections (from Rogers, *Report*, 1:56).

is provided by two flexible rubber O–rings (0.280 inch in diameter; +0.005, –0.003). However, an extremely small gap of 0.005 ± 0.004 inch (design tolerance) will remain between the tang and the inside leg of the clevis. The width of this gap dictates the amount of O–ring static compression during and after assembly. However, the gap at any location on the circumference of the two sections is influenced by the size and shape (concentricity) of the segments and the loads applied. Consequently, a maximum gap of 0.033 inch is possible prior to ignition. Following ignition the size of the gap is primarily determined by the motor pressure as well as external loads and other joint dynamics.

Zinc chromate putty is applied to the composition rubber insulation face prior to assembly. The putty serves as a thermal barrier, preventing direct contact of the hot combustion gas with the O–rings. As the combustion gas pressure displaces the putty in the space between the motor segments, a mechanism is created that forces the O–ring to seal the casing. That is, the displacement of the putty acts somewhat like a piston and compresses the air in front of the primary O–ring, forcing the O–ring into the gap between the tang and the clevis. This process is known as pressure actuation of the

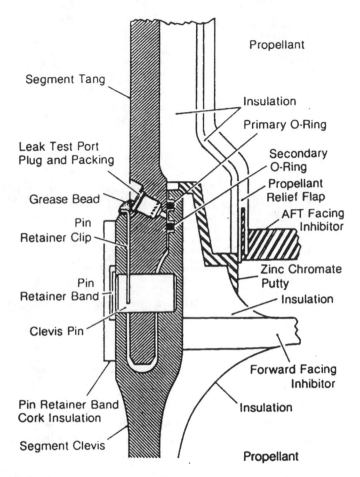

Figure 14.2 Cross-section view of the solid rocket motor joint showing position of tang, clevis, two O-rings, leak check port, and putty (from Rogers, *Report,* 1:57).

O-ring seal.[19] If the hot gases are able to "blow by" the putty and primary O-ring, the secondary O-ring was designed to provide a redundant sealing function.[20] The seal is tested statically by a "leak check" in which it is actuated at a specified test pressure by forcing air through the leak test port. However, as can be seen from Figure 14.2, the leak check forces the primary O-ring to seal in the wrong direction. Thus, following ignition it must be forced to travel rapidly the very small distance across its groove and then seal in the opposite direction, as the motor goes to full pressurization (900 pounds per square inch [psi]) in 600 milliseconds (0.6 second).

The pressure actuated sealing must occur very early during the solid rocket motor (SRM) ignition transient. As was evident as soon as testing began, the gap between the tang and clevis increases as pressure loads are applied to the joint during ignition due to a phenomenon denoted as "joint rotation" (described later). This unanticipated situation was not fully understood nor was its significance appreciated by all concerned parties until after the *Challenger* accident. Simply put, it meant that if pressure actuation is delayed and the gap opens substantially, the possibility then exists that the rocket's combustion gases will blow by the O-rings and damage or destroy the seals.

What now seems clear is that sealing is best accomplished when the actuating pressure is behind the entire O-ring surface to force it across its groove. If either the O-ring groove is too narrow, or the initial "squeeze" placed on the O-ring by the tang edge (prior to ignition) compresses it sufficiently so that it contacts both groove surfaces, pressure actuation may be inhibited. This situation is greatly complicated when the gap begins to open following ignition. In this case, the compressed O-ring must be sufficiently resilient to track across the groove as well as expand toward its original, uncompressed shape. This must occur at the same rate as the "joint rotation" opens the gap, so that the ring extrudes into the opening between the tang and clevis surfaces. This sealing is assisted by the solid rocket motor's increasing pressure buildup. However, the O-ring's resiliency is *affected by temperature*. Thus, at low temperatures, a compressed O-ring may not be able to return to its original shape fast enough. Further, if initially the clevis and tang are extremely close or have metal-to-metal contact, the O-ring may be fully compressed in its groove, contacting all three groove surfaces. In that case, the motor pressure cannot actuate the seal, but instead holds it in the groove as the gap continues to open. These situations are illustrated in Figure 14.3. Joint rotation can cause the gap openings to increase approximately 0.029 and 0.017 inch for the primary and secondary O-rings respectively.[21] Another complication is the effect of the grease that is heavily applied to the O-rings to prevent damage, but introduces more uncertainty to the actuation process, especially at low temperatures.

The Design Decision

In September 1977, when the SRB first underwent static "hydroburst" tests of the motor casing under simulated firing, the "joint rotation"

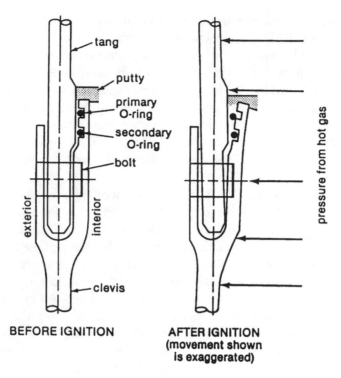

BEFORE IGNITION   **AFTER IGNITION**
(movement shown
is exaggerated)

Figure 14.3 "Joint rotation," showing inside of clevis bending away from tang (from Martin and Schinzinger, *Ethics in Engineering,* p. 88; reproduced with permission of McGraw-Hill.

phenomenon was first observed. Contrary to the design concept, the joint tang and inside clevis bent away from instead of toward each other, as shown in Figure 14.4. It occurred because, as Ryan explains:

> At ignition, the SRM becomes a pressure vessel due to the expanding burning propellants, reaching over 900 psi in 600 milliseconds. This large pressure has two effects: (1) the SRB case expands longitudinally, and (2) the case expands radially from the large pressure induced hoop stress. Due to the tang clevis configuration and the fasteners (pins) which lock the two segments together, these two effects cause a rotation between the tang and clevis at the O-ring location opening up the joint O-ring area.[22]

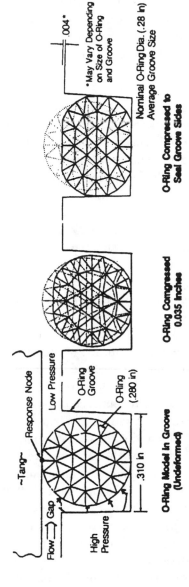

Figure 14.4 O-ring under three different conditions. In undeformed position, gas under high pressure should force O-ring to move to the right and extrude into the groove on right. If O-ring is compressed (sqeezed) too much because of reduction of gap between tang and clevis, this will inhibit and eventually block high-pressure gas flow from getting behind the O-ring and activating the seal. (From Rogers, *Report*, 1:61.)

Hence, joint rotation increased the size of the gap and reduced, rather than increased, the pressure on the O-ring within milliseconds after ignition. Thiokol engineers, who reported the test findings to Marshall, did not believe that the results indicated the existence of a potentially serious problem. Consequently they did not schedule further tests, nor did they contemplate any design changes.[23]

However, initial reaction at Marshall was rapid and totally opposite. Glenn Eudy, chief engineer of the Solid Rocket Division, felt that a serious problem did exist and recommended a full redesign rather than a quick fix to Alex McCool, director of the Structures and Propulsion Laboratory (September 2, 1977). Leon Ray, another Marshall engineer, prepared the report "SRM Joint Leakage Study" (October 21, 1977), which characterized the Thiokol position of recommending no design change as "unacceptable." Ray believed that joint rotation could result in seal leakage, and recommended that the best option for a long-term fix would be to redesign the tang and reduce the clevis tolerance. To Ray, "joint rotation" meant that the secondary O-ring was not redundant.[24]

Ray authored a memorandum (January 9, 1978) for John Q. Miller, chief of Solid Rocket Motor branch (MSFC) to Eudy, which described the problems with the joint seal. It noted that "all situations which could create tang distortion are not known, nor is the magnitude of movement known."[25] A year later (January 19, 1979), having not received a response, Miller sent a second memorandum: "We find the Thiokol position regarding design adequacy of the clevis joint to be completely unacceptable." Miller noted that joint rotation caused the primary O-ring to extrude into the gap, "forcing the seal to function in a way which violates industry and government O-ring application practices."[26] Thiokol did not receive copies of either memo; nor was a reply from Eudy to Miller documented.[27]

Additional static motor tests in July 1978 and April 1980 further demonstrated that inner tang and clevis relative movement was greater than originally predicted. However, Thiokol continued to question the validity of these joint rotation measurements and their effect on the ability of the secondary O-ring to perform as designed.

In 1980, due to serious concerns with the shuttle's overall progress, NASA created a Space Shuttle Verification and Certification Committee to conduct a complete flight worthiness study. Among its findings were:

• The booster's leak test pressurized the primary O-ring in the wrong direction; ignition would have to move the ring across the groove before it sealed.

- The effect of the insulation putty was not certain.
- The redundancy of the O–rings is of concern and required verification.

The report noted that "the Committee understands from a telecon that the primary purpose of the second O–ring is to test the primary and that redundancy is not a requirement." The panel recommended that NASA conduct full-scale tests to verify the field joint integrity. NASA's response was that these concerns either had been or were being satisfied by its testing and upgrade program.[28]

The solid rocket motor certification was deemed satisfactory by the panel on September 15, 1980. Two months later, the solid rocket booster field joint was classified as Criticality 1R (redundant). This implied that NASA believed that the secondary O–ring would pressurize and seal even if the primary O–ring did not. However, the critical items list carried the caveat: "The redundancy of the secondary field joint seal cannot be verified after motor case pressure reaches approximately 40 percent of maximum expected operating pressure."[29] That is, the redundancy was only true for a very short period at ignition but would not hold thereafter.

It did not take long for the feared problem to appear. On the shuttle's second flight, STS-2 in November 1981, indications of in-flight erosion of a primary O–ring were found. Although the damage would be the worst ever found in a field joint, it was not reported in Level I flight readiness review for STS-3; nor was it entered on the Marshall problem assessment system as were other anomalies. Thiokol engineers believed that the erosion's cause was most likely attributable to problems with the putty. That is, blow holes in the putty allowed jets of hot gas to focus on the primary O–ring, "impinging" and destroying a portion of it. Thiokol engineers calculated that the maximum possible impingement erosion was 0.090 inches, and that a lab test proved that an O–ring would seal at 3,000 psi when erosion of 0.095 inches was simulated. This "safety margin" was used to approve shuttle flights while explicitly accepting the possibility of O–ring erosion.[30] Since the tested pressure was three times the actual maximum motor pressure (1,004 psi), this margin was somewhat greater than first might be apparent.

Following additional testing of a new lightweight motor case, Marshall finally accepted the conclusion that the secondary O–ring would not be functional after motor pressurization due to joint rotation. Consequently, the joint was reclassified as Criticality 1, rather than 1R, by Marshall on December 17, 1982. That change was approved by the Marshall Configuration Control Board (chaired by Lawrence Mulloy) on January 21, 1983, and forwarded to Level II. Thus, in effect, Marshall now *officially* considered the joint to consist of only a single functional O-ring. The retention rationale needed to fly with this design was provided by Howard McIntosh, a Thiokol engineer: It noted that the

> joint concept is basically the same as single O-ring joint successfully employed on the Titan III solid rocket motor. . . . Full redundancy exists at the moment of initial pressurization. However, test data show that a phenomenon called joint rotation occurs as the pressure rises, opening up the O-ring extrusion gap and permitting the energized ring to protrude into the gap. This condition has been shown by test to be well within that required for safe primary O-ring sealing. This gap may, however, in some cases, increase sufficiently to cause the unenergized secondary O-ring to lose compression, raising questions about its ability to energize and seal if called upon to do so by primary seal failure. Since under this latter condition only the single O-ring is sealing, a rationale for retention is provided for the simplex mode where only one O-ring is acting.

The Rogers Commission observed that the McIntosh document from which the rationale was taken had included the conflicting statement: "This [initial testing] . . . indicates that the tang-to-clevis movement will not unseat the secondary O-ring at operating pressures."[31]

Based on this rationale, the required criticality waiver (to fly) for the joint was signed off and sent to Level I where it was approved on March 28, 1983. Glenn R. Lunney, the Manager of the National Space Transportation Program Office (Level II) who approved change from 1R to 1, testified that:

> As a result of those tests, it was concluded that it was possible under certain circumstances of extreme dimensional tolerance not to have the secondary O-ring seal. In that case we would be left with just one seal, that is the primary O-ring acting as the seal for the SRB case-to-case joint. In that case, then, we would have been dealing with not two

seals, as we originally thought, which is why the R was on the nomen-
clature, but rather one seal and therefore, it was changed from 1R
redundant to 1. . . . At the time I was involved in that [approval of the
waiver], I was operating on the assumption that there really would be
redundancy most of the time except when the secondary O–ring had a
set of dimensional tolerances add up, and in that extreme case there
would not be a secondary seal.[32]

Thus, it appears that although Marshall *formally* acknowledged the joint to
no longer be operationally redundant, *informally* this may have not been
the case. The changed designation never became widely recognized, and, as
described by Lunney, some believed that the seal was redundant for all but
exceptional cases. Further, Roger Boisjoly, then a Thiokol design engineer,
later stated that Thiokol management did not make this change in redundancy known to its engineers.[33] As is illustrated under our discussion of the
performance decision, testimony to the Rogers Commission demonstrates
the degree of confusion that remained and supports the premise that both
NASA management and Thiokol still considered the seal to be redundant.
In addition, a number of Thiokol and Marshall documents continued to
refer to the seal as 1R.[34]

During the tenth shuttle flight – STS–41–B on February 3, 1984 – damage to two primary O–rings was found. One of these was on a nozzle joint,
which also had two O–rings, but was a slightly different design from the
field joint. Thiokol reacted by informing Marshall that its recent tests indicated that the secondary O–ring would not unseat, providing confidence
that the secondary was an adequate backup. At least some Marshall engineers, including Keith Coates and Ray, were not convinced. In a memorandum to his superior, Coates noted that Thiokol had not examined the
observed .065 inch erosion's effect on the ring's sealing capability.[35]

At the Level III flight readiness review for the next mission (STS 41–C)
held at Marshall, Thiokol reiterated that its earlier test results predicted
that the maximum possible erosion would be 0.090 inch, and sealing integrity had been demonstrated with a simulated erosion depth of 0.095 inch at
3,000 psi. Marshall's problem assessment system report incorporated this
statement and concluded "therefore, this is not a constraint to further
launches."[36]

The next mission, 41–C, was approved by Level I "accepting the possibility of some O–ring erosion due to the hot gas impingement." According to

Thiokol, the most likely cause would be blow holes in the putty, which could develop during the joint leak test. This was believed to be due to the higher stabilization pressure (200 psi versus 50 or 100 psi) now used in the leak check procedure, and not the fault of the putty. However, a concerned NASA deputy associate administrator, Hans Mark, directed Marshall to conduct a formal review of the joint sealing procedures. That directive had been preceded by a letter from General James Abrahamson, then associate administrator for space flight, to William Lucas, Marshall director, requesting that the center develop a plan of action to improve NASA's ability to design, manufacture, and fly solid rocket motors. Abrahamson noted that NASA was flying solid rocket motors whose basic design and test results were not well understood. It was now clear that NASA in general and Marshall in particular were very concerned about the putty erosion/blow hole problem.[37]

By now two types of O-ring erosion were of concern:

- blow-by in which the O-ring didn't seal, enabling hot gases to "blow-by" the O-ring; and
- impingement where the surface of the sealed ring is struck by hot gases, which would erode a portion of it, possibly destroying its ability to continue sealing.

Blow-by and impingement erosion occurred together for the second time during flight 51-C, January 24, 1985. This was also the first flight in which a secondary O-ring had eroded on the nozzle joint. The launch temperature for that flight was fifty-three degrees Fahrenheit, the coldest to date. Marshall's assessment of the damage was: "O-ring burns were as bad or worse than previously experienced . . . design changes are pending test results."[38]

On January 31, 1985, Mulloy, the Marshall solid rocket booster project manager, sent an urgent message to Larry Wear, Marshall solid rocket motor manager, which directed that the flight readiness review for the next flight:

> Recap all incidents of O-ring erosion, whether nozzle or case joint, and all incidents where there is evidence of flow past the primary O-ring. Also, the rationale used for accepting the condition on the nozzle O-ring. Also, the most probable scenario and limiting mechanism for flow past the primary on the 51-C case joints. If MTI [Thiokol] does not

have all this for today, I would like to see the logic on a chart with blanks TBD [to be determined].

A copy of this request was sent to Thiokol as a directive to prepare a detailed briefing for the next flight readiness review.[39]

Eight days later on February 8, 1985, Thiokol engineers presented Marshall with their most detailed analysis to date of the O-ring erosion problem. The resolution for the next flight (51 D)   accept the risk even though the seal could be lost. This was based on the concept of "maximum expected erosion" and "maximum erosion experienced" as the result of their tests and modeling. During the Rogers Commission hearing, Feynman and Walker (professor of applied physics, Stanford University) pointed out that "these models really were not precise models. There is a considerable amount of variation, because all of the parameters were beyond your ability to measure or know"[40]

Nevertheless, Thiokol's engineers justified accepting damage to the primary based, in part, on the assumption that the secondary O-ring would seal, even with erosion. As noted, this basic assumption was at odds with the critical items list where the joint had been rated Criticality 1 for over two years, since officially the primary O-ring was classified as a single point failure. Thiokol also stated for the first time that the low temperature enhanced the probability of blow-by, noting that "51 C experienced worst case temperature change in Florida history."[41]

Two weeks later at the Level I flight readiness review for 51-D (February 21, 1985), the Thiokol report had been reduced to a one-page chart, which stated that the erosion and blow-by experienced were an "acceptable risk because of limited exposure and redundancy." No mention was made of the effect of temperature on 51-C, or the detailed analysis of the O-rings.[42] As the shuttle continued to fly, some form of erosion problem was found on each of the next four flights, 51-D, B, G, and F.

In particular, erosion on Flight 51-B (April 29, 1985) provided an ominous counterexample to Thiokol's "maximum expected erosion" analysis. A primary O-ring (nozzle) had eroded 0.171 inch and had not sealed. This was in contrast to the Thiokol model justification and Marshall acceptance that the maximum primary O-ring erosion would be 0.090 inch for the field joint and only 0.070 inch for the nozzle.[43]

Although the secondary ring had sealed, it had eroded by 0.032 inch. Marshall management realized that "this erosion of a secondary O-ring was a new and significant event," according to Mulloy. Consequently, Mulloy

and the Marshall Problem Assessment Committee placed a "launch constraint" on the shuttle, so that a launch could not proceed until sufficient
rationale was provided that the problem would not occur during prelaunch,
launch, or flight. Having put on this constraint, Mulloy then proceeded to
waive it for each succeeding flight. Mulloy explained the reason for the
waiver was based on an assumption that the O-ring in question had not
sealed during the pressure leak check which was performed at 100 psi, and
was therefore not capable of sealing. For all leak checks performed at the
higher pressure of 200 psi, "test data indicated [that the compressed air]
would always blow through the putty, and in always blowing through the
putty, we were guaranteed that we had a primary O-ring seal that was capable of sealing . . . and we already had that on the field joints." However,
although the blow holes in the putty were an indication that the ring would
seal and tended to prevent delays in pressurization after ignition, they also
created a pathway for the hot ignition gases to reach the O-ring and erode
it. Indeed, O-ring damage data indicated that it was much more prevalent
when the higher pressure was used for leak checks.)[44]

The last mention of the erosion problem at a Level I flight readiness
review was for flight 51-F (July 2, 1985). The erosion was attributed to the
above mentioned leak check at 100 psi. Again, the "organizational filtering
process" resulted in no mention that the 0.171 inch erosion exceeded the
maximum erosion prediction of 0.070 inch, nor was the launch constraint,
which had been placed on this flight and then waived, mentioned at this or
any succeeding Level I flight readiness review.[45]

The extent of the erosion on 51-B caused Thiokol engineers to also
become increasingly concerned with the problem. On July 22, 1985, staff
engineer Roger Boisjoly wrote his first memorandum predicting that
Thiokol might either lose its NASA contract or that there might be a flight
failure unless the company came up with a timely solution to the problem.
This was quickly followed by a second memorandum: "O-ring Erosion/
Potential Failure Criticality" on July 31, 1985, to Robert Lund, vice president of engineering, which reiterated Boisjoly's concern that a flight would
be lost unless the field joint was redesigned; Boisjoly recommended this be
given number one priority with a team established to solve the problem.[46]

In response to Marshall's heightened concerns, Thiokol's Brian Russell
(Special Projects Engineer – Solid Rocket Motors) wrote to James Thomas
Jr. at Marshall, responding to two specific questions which had been raised:

> Per your request, this letter contains the answers to the two questions
> you asked at the July Problem Review Board telecon.

1. *Question:* If the field joint secondary seal lifts off the metal
mating surfaces during motor pressurization, how soon will it
return to a position where contact is re-established?
Answer: Bench test data indicate that the O-ring resiliency (its
capability to follow the metal) is a function of temperature
and rate of case expansion. MTI measured the force of the O-
ring against Instron platens, which simulated the nominal
squeeze on the O-ring and approximated the case expansion
distance and rate.

At 100°F: The O-ring maintained contact. At 75 °F: The
O-ring lost contact for 2.4 seconds. At 50 °F: The O-ring did
not re-establish contact in ten minutes at which time the test
was terminated.

The conclusion is that secondary sealing capability in the
SRM field joint cannot be guaranteed.

2. *Question:* If the primary O-ring does not seal, will the secondary seal seat in sufficient time to prevent joint leakage?
Answer: MTI has no reason to suspect that the primary seal
would ever fail after pressure equilibrium is reached, i.e., after
the ignition transient. If the primary O-ring were to fail from
0 to 170 milliseconds, there is a very high probability that the
secondary O-ring would hold pressure since the case has not
expanded appreciably at this point. If the primary seal were to
fail from 170 to 330 milliseconds, the probability of the secondary seal holding is reduced. From 330 to 600 milliseconds
the chance of the secondary seal holding is small. This is a
direct result of the O-ring's slow response compared to the
metal case segments as the joint rotates.

Please call me or Mr. Roger Boisjoly if you have any additional questions concerning this issue.[47]

However limited these tests were, Thiokol's data clearly indicated that
the colder the temperature, the less the resiliency. Thus Russell, on behalf
of Thiokol, informed Marshall that "secondary sealing . . . cannot be guaranteed," although "there was no reason to suspect that the primary O-ring
would fail after motor ignition."

The three "probability" estimates had been provided by Boisjoly in
order "to rework that information into a more concise form and to try to

give probabilities to those zones for the August 19 presentation at NASA Headquarters.[48] These estimates attracted the attention of commission member Walker, who questioned both Boisjoly and Mulloy on their meaning:

DR. WALKER: . . . [To Boisjoly] By high probability did you mean 95 percent or 75 percent?

MR. BOISJOLY: In the beginning on the basis of the limited test we had, I felt there was 100 percent if it happened within the first 150, 170 milliseconds because there would be an impingement problem and then the margin would truly be a margin of impingement and not blow-by.

DR. WALKER: But you didn't use the word likely. You used high probability.

MR. BOISJOLY: Yes.

DR. WALKER: Which implies that there is some chance.

MR. BOISJOLY: Well, there always is.[49]

DR. WALKER: . . . [To Mulloy] What does high probability mean [in the Russell letter]? Does it mean 75 percent, 80 percent, 82 percent?

MR. MULLOY: I don't know. I can't quantify that.

DR. WALKER: But surely you are basing your decision to proceed on this assertion that the secondary seal has high probability of working.

MR. MULLOY: Yes, sir, and the reason was that the secondary seal would be energized in the zero to 170 millisecond or 330 millisecond time frame.

DR. WALKER: Does that mean 90 percent of the time or 70 percent of the time?

MR. MULLOY: I don't know, sir.

DR. WALKER: But you are basing your decision on that.[50]

As Boisjoly noted, Thiokol (and Marshall) engineers briefed NASA headquarters personnel on August 19, 1985. The result of this comprehensive meeting was recommendations, which included:

- The lack of a good secondary seal in the field joint is most critical and ways to reduce joint rotation should be incorporated as soon as possible to reduce criticality.
- Analysis of existing data indicates that it is safe to continue flying existing design as long as all joints are leak checked with a 200 psi

stabilization pressure, are free of contamination in the seal areas, and
meet O–ring squeeze requirements.
* Efforts need to continue at an accelerated pace to eliminate SRM seal
  erosion.[51]

The next day, Lund established an O–ring task force at Thiokol, requesting
that it recommend both short- and long-term solutions. Two days later,
Arnold R. Thompson, Thiokol's supervisor of structures design recommended a short-term solution of increasing both the thickness of shims
used at the tang and clevis and the diameter of the O–ring, given that prom-
ising long-term solutions would require several years to solve this "acute"
problem. By the end of August, Thiokol and Marshall were still not in
agreement concerning the magnitude of the joint rotation problem. Thus,
it was proposed to design a mutually acceptable "referee test" a standard
procedure in such cases to determine the joint operating characteristics.

Marshall's science and engineering director James Kingsbury informed
Mulloy by letter that the O–ring seal problem required priority attention
from both Thiokol and Marshall.[52] One month later on October 4, 1985,
Boisjoly again warned Thiokol management in a third memorandum; this
one was concerned with the lack of support for O–ring team effort. Concurrently, on October 1, 1985, Robert V. Eberling, manager of Thiokol's
solid rocket motor ignition system, complained in a memorandum to Allan
McDonald, manager of the Space Booster Project: "HELP. The Seal task
force is constantly being delayed by every possible means. . . . This is a red
flag." Following Flight 61-A (October 30, 1985), which experienced nozzle
O–ring erosion and field joint blow-by, Eberling told the task force members that shipments of boosters should stop until the problem was fixed.
Eberling later admitted to the Rogers Commission that he had not voiced
these concerns to the right people.[53]

On December 12, 1985, Thiokol's Russell wrote McDonald requesting
closure on the "O–ring erosion critical problems." His justification included
test results, future test plans, and the work of the task force. McDonald
responded by a written request to NASA for closure, noting that the problem would not be fully resolved for some time.[54] Thus, even though at least
two Thiokol engineers, Boisjoly and Eberling, had recognized the seriousness of the problem, and no corrective action had been taken, Thiokol
requested closure. Russell later informed the Rogers Commission that he
understood that the director of engineering [Kingsbury] at Marshall had

wanted the issue closed. Boisjoly explained the particular way this action was viewed at Thiokol:

> Basically, closeout may be a poor term because nobody was going to close it out. Instead of tracking the same problem through two or three different channels, we were going to track the problem now in the highlight of the task team, plus the flight readiness reviews, so there was no need to track it in several different locations. It was the same problem. . . . So it was a simple matter of bookkeeping to take it out of one area because we were already doing it in another area.[55]

However, in response to the closeout request, an entry was placed in the Marshall problem reports noting that "contractor closure received" on December 18, 1985. Five weeks later, on January 23, 1986, a second entry was incorrectly placed in the problem report indicating that the O-ring problem was considered to be closed. Mulloy testified to the Rogers Commission that "the people who run this problem assessment system erroneously entered a closure for the problem on the basis of this submittal from Thiokol." Mulloy's reaction to the McDonald request was that "we are not going to drop this from the problem assessment system because the problem is not resolved and it has to be dealt with on flight-by-flight basis." He indicated that the people who ran the problem assessment system erroneously entered a closure for the problem on the basis of the submittal from Thiokol. The result was that the O-rings were no longer an open launch constraint. Consequently, it didn't come up in flight readiness review for the next launch, 51-L (January 28, 1986) as an open constraint.[56]

The Performance Decision

At approximately 2:30 P.M. eastern standard time, on January 27, 1986, Eberling, Boisjoly, and other Thiokol engineers at Wasatch grew concerned about the low temperatures predicted for the next morning's *Challenger* launch. Eberling noted that "we were way below our data base and we were way below what we qualified for." Following an hour meeting, McDonald, serving as Thiokol's liaison at the Kennedy Space Flight Center, was informed. McDonald asked Eberling to involve Lund, vice-president of engineering, and to prepare charts for what was to be a teleconference with NASA management.[57]

That "well known" teleconference actually occurred in two phases. The first phase began at 5:45 P.M. and involved engineers and managers from Kennedy, Marshall, and Thiokol. Thiokol's engineers initially proposed that the launch be delayed until noon or later due to the uncertain effect of the low temperature on the O-rings. Due to the seriousness of this recommendation, it was agreed to reconvene later that evening with all involved parties present. Thiokol was to assemble and transmit the relevant data to NASA. Judson A. Lovingood (deputy manager, Marshall Shuttle Projects Office) recommended that William Lucas (director of Marshall Space Flight Center) and Kingsbury (director of science and engineering at Marshall) be included, and that NASA plan to go to Level II if Thiokol recommended not launching.[58] Lund summarized the first phase as follows:

> The roles kind of switched, and so after making, or listening to the verbal presentation in the afternoon, they asked what Thiokol's position was, and I looked around the room, and I was the senior person, and I said I don't want to fly. It looks to me like the story says 53 degrees is about it. And of course, we were requested then to go back and do something more and prepare detailed charts to show that in more detail. And so we got busy then, and I gave assignments to a dozen or so people to go out and generate data that would in a workmanlike manner show the rationale and show the data that we had so that everyone would understand all the data and where it came from.
>
> And we spent the next couple of hours beating the bushes trying to put together that data. And so we did that and began transmitting charts even late then, and then went through that rationale. And all of this time we were preparing the data, the data were coming in, and I was trying to put together what I was concluding out of all this because there was some additional data . . . being generated, and trying to understand and to absorb all of the data . . . there, to again see what my thought processes were.[59]

The second or main phase of the teleconference began at 8:45 P.M. Lund, as stated, had prepared several charts, including a hand-lettered one:

RECOMMENDATIONS:
- O-ring temp must be ≥ 53°F at launch development motors at 47° to 52 °F with putty packing had no blow-by

- SRM 15 (The best simulation) worked at 53°
- Project ambient conditions (temp & wind) to determine launch time[60]

In essence, Thiokol's engineers were recommending not to fly since the temperature was outside of their database. Their concern related to the uncertain effect of temperature on resiliency, specifically the timing function for sealing: Would the O-ring be able to seal as the gap opened? Boisjoly indicated that "I was asked to quantify my concerns and I said I couldn't. I couldn't quantify it. I had no data to quantify it, but I did say I knew that it was away from goodness in the current database."[61]

Mulloy testified that Joe Kilminster, Thiokol booster program manager, when questioned during this phase could not recommend launch based on the engineers' recommendation. Mulloy's response indicates the confusion between the actual and perceived criticality classification. Mulloy stated:

> I asked Joe Kilminster [Thiokol program manager for boosters] . . . what his recommendation was, because he is [who] I get my recommendations from in the program office. He stated that, based on that engineering recommendation, that he could not recommend launch. At that point I restated . . . the rationale that was essentially documented in the 1982 critical items list, that stated that . . . we were flying with a simplex joint seal. And you will see in the Thiokol presentation that the context of their presentation is that the primary ring, with the reduced temperatures and reduced resiliency, may not function as a primary seal and we would be relying on secondary. . . . My assessment at that time was, that we would have an effective simplex seal, based upon the engineering data that Thiokol had presented, and that none of those engineering data seemed to change that basic rationale.
>
> Stan Reinartz then asked George Hardy, the Deputy Director of Science and Engineering at Marshall, what his opinion was. George stated that he agreed that the engineering data did not seem to change this basic rationale, but also stated on the telecon that he certainly would not recommend launching if Thiokol did not."[62]

It was then that Kilminster requested a five-minute caucus.

The five-minute caucus, which began at approximately 10:30 P.M., lasted a half hour. Thiokol's engineering managers had challenged their engineers to produce the data on which their concerns were based. Collectively, they examined the records from the twenty-three previous shuttle launches in which the boosters had been recovered and the field joints examined, as well as a series of test data of the joints. Figure 14.5 presents one of the datasets the Thiokol engineers examined in an effort to determine if a relationship between temperature and resultant O–ring anomalies (damage) existed. Note that these data only include the seven cases in which an anomaly had been found; it does not contain the other sixteen cases in which no O-ring damage occurred (see Figure 14.6.) Note that damage was observed at both the high end (75° Fahrenheit) and the low end (53° Fahrenheit).

Following this apparently inconclusive debate, Thiokol management decided to reverse the initial recommendation not to launch, and, overruling

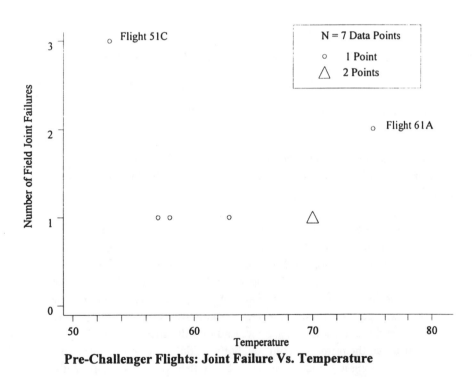

Pre-Challenger Flights: Joint Failure Vs. Temperature

Figure 14.5 Data presented during *Challenger* "Telecon" which includes only flights in which damage occurred.

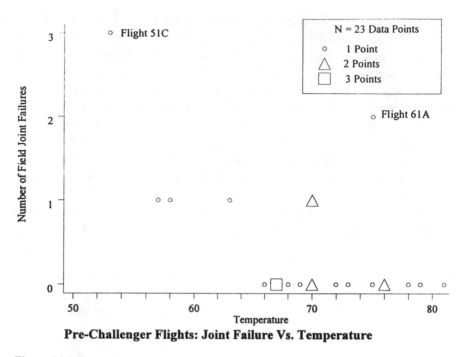

Pre-Challenger Flights: Joint Failure Vs. Temperature

Figure 14.6 Data of O-ring damage versus temperature including those flights in which no damage occurred.

most of their involved engineers, agreed to proceed. At Mulloy's request, Kilminster signed the launch readiness report for Marshall Space Flight Center.[63] Jerald Mason, senior vice-president of the Wasatch Operations, testified:

> So we had our caucus, in which we revisited all of the things we had talked about before. And we recognized two primary things . . . the worst experience we had in erosion was 0.038 [field joint] and we know from tests that the O-ring would seal with over 0.120 [inch] of erosion. So our first thought was that if we had more erosion on the primary because it took longer, it would still seal, even if it were eroded three times as much. So we said we still had a reasonable expectation that the primary would seal, but we didn't have absolute data that said how long it would take to move. So we then said, what happens if it doesn't? And we took the second point, which was that the secondary

was in position and did not have to move. So we felt that the primary probably would seal, but if it didn't the secondary would because it was already in position."[64]

To Mulloy, the field joint was considered a nonredundant, simplex seal and Thiokol's data had indicated that the primary O-ring would seal. In contrast, Thiokol seemed to be concerned about the reduced resiliency of the primary, and thus a reliance on the secondary O-ring. Boisjoly testified that:

> I believed and I still believe and I believed that night that there isn't anybody on the face of this earth that can tell you exactly the mechanism that happened in that joint. And even before the fact, you don't understand if it's going to rotate and walk up and delay or either slide because of its stiffness and delay. But the timing function that I spoke of that night had to do with the fact that I was afraid that that timing function could throw us in from an ignition transient at the start to somewhere after that start time, and that is what my major concern was about. . . . It is eroding at the same time it is trying to seal, and it is a race between, Will it erode more than the time allowed to have it seal?[65]

Lund told the Rogers Commission:

> We have dealt with Marshall for a long time and have always been in the position of defending our position to make sure that we were ready to fly. . . . But that evening I guess I had never had those kind of things come from the people at Marshall. We had to prove to them that we weren't ready, and so we got ourselves in the thought process that we were trying to find some way to prove to them it wouldn't work, and we were unable to do that. We couldn't prove absolutely that that motor wouldn't work.[66]

As noted, there was considerable confusion concerning the degree of redundancy of the seal. The following indicates not only this confusion among Thiokol and NASA personnel, but also the difficulty that the commission had in trying to clear it up.

DR. COVERT: Larry [Mulloy], I'm confused, and could I go back about four meatballs here. What is a simplex seal?

MR. MULLOY: No redundancy.

DR. COVERT: That's a single O-ring?

MR. MULLOY: Yes, sir.

DR. COVERT: Is this common notation, or is it named after Charlie Simplex or something?

MR. MULLOY: Perhaps it is an unfortunate phrase. I guess I consider "simplex" singular and "duplex" dual, and I was using it in that context.

DR. WALKER: So you're talking about the secondary seal?

MR. MULLOY: Yes.

DR. WALKER: In other words, if the first one didn't work the second one would?

MR. MULLOY: Yes, that is correct. And the engineering rationale you will see for accepting the situation was counting on the secondary seal, and that will be developed. Yes, sir.[67]

DR. RIDE: Okay. I guess I was just looking at the CIL, and it says that . . . the primary O-ring is, I heard now is a single point failure, because you can't count on the secondary O-ring. Is that a fair assessment of the CIL?

MR. MULLOY: That is correct.

DR. RIDE: And then I guess my question was, in your discussion the day before launch and the evaluation of the effects of the cold temperatures on the O-ring, if you were going to base your decision on the CIL, it seems that you would have to assume that the cold temperature affected the secondary O-ring, but not the primary O-ring, since the primary O-ring is the criticality one.

MR. MULLOY: . . . I did not base my decision on the CIL. The CIL states that we have a simplex – the rationale for a simplex seal. We do not have a redundant seal. . . .

DR. RIDE: A simplex seal where the one O-ring was the primary O-ring?

MR. MULLOY: No. The rationale, you will see, says we were counting on the secondary O-ring to be the sealing O-ring under worst case conditions and the worst case analysis that is presented here.

DR. RIDE: But doesn't the CIL say you can't count on the secondary O-ring?

MR. MULLOY: Yes, and you have to see the engineering development for the rationale that states that if the cold effect on the primary O-ring . . . is it reduces the diameter of the O-ring. It also reduces the resiliency of the O-ring. . . . we went a step further and said, suppose under worst case conditions the cold effect caused the primary O-ring to be totally ineffective. If the primary O-ring is ineffective, the secondary O-ring, which is in a position to seal, will be pressure actuated in the time before the joint rotates.[68]

Later in that Rogers Commission session:

> MR. MASON: The reason that [CIL] was changed from the 1R to 1 was this very rationale here, which said that after early on, after the 170 or 330 milliseconds, you didn't have a redundant seal, and so it was changed because it was not redundant all the time. But it didn't really remove the redundancy at ignition.[69]

Still later in that session:

> MR. LUND: . . . Our concern was if the actuation time increased, the threshold of secondary seal pressurization capability is approached; and if the threshold is reached, then the secondary seal may not be capable of being pressurized.
> MR. WAITE: But in terms of the discussion we've been going through this morning, if you're talking about a primary seal, it is the secondary seal that is the primary seal.
> MR. LUND: There is a scenario that would draw that conclusion.[70]

While Thiokol personnel at Wasatch, Utah, were caucusing, McDonald, who was at Kennedy, had a discussion with Mulloy and the other two NASA participants. McDonald argued that if the motor was supposedly qualified for 40 to 90 degrees, then it was his impression that the launch commit criteria were based upon "whatever the lowest temperature, or whatever loads, or whatever environment was imposed on any element or subsystem of the shuttle. And if you are operating outside of those, no matter which one it was, you had violated some launch commit criteria." He told the Rogers Commission that "I still didn't understand how NASA could accept a recommendation to fly below 40 degrees."[71]

When Kilminster signed the launch recommendation, he was unaware that McDonald, who was with Mulloy during the teleconference, had informed the NASA official that "I wouldn't sign that, it would have to come from the plant because normally I am responsible for telling whether the flight goes or not."[72] Neither did Thiokol management inform NASA that, at best, "the engineers were reasonably evenly split on whether to launch or not launch."[73] (When questioned on the extent that Thiokol's engineers disagreed with the final decision that evening, Mason told the Rogers Commission that they were split. Feynman then pressed the issue, and Lund admitted that Thiokol's two best seal experts, Boisjoly and Thompson, were opposed, a third was in favor, and Lund was uncertain about the fourth engineer most knowledgeable about the seal.)[74]

Data presented during the teleconference did not include any statistical analysis of the past history of O-ring damage due to temperature. A focus of the conference was on those flights where damage had resulted. However, by comparing *only* the flights that exhibited O-ring erosion as a function of temperature, no relationship is evident (as can be seen from Figure 14.5). If all previous data had been examined (Figure 14.6), those present would have observed that damage occurred for only three of the nineteen flights in which the launch temperature was above 66 degrees Fahrenheit compared with damage for all four flights with a temperature below 64 degrees Fahrenheit. The Rogers Commission noted that "consideration of the entire launch temperature history indicates that the problem of O-ring distress is increased to almost a certainty if temperature of joint is less than or equal to 65 [degrees Fahrenheit]."[75]

Apparently no one, neither Thiokol nor NASA engineers and managers, examined the data as presented in Figure 14.6. Clearly the data indicate a trend. Noted Wiley Bunn, director of reliability and quality assurance at Marshall, after the fact: "we had that data. It was a matter of assembling that data and looking at it in the proper fashion. Had we done that, the data just jump off the page at you."[76]

The commission pointedly concluded that: "careful analysis of flight history of O-ring performance would have revealed the correlation of O-ring damage and low temperature. Neither NASA nor Thiokol carried out such an analysis. Consequently they were unprepared to properly evaluate the risks of launching 51-L mission in conditions more extreme than they had encountered before."[77]

In summary, a majority of those Thiokol engineers most knowledgeable about the field joint recommended not to extrapolate beyond their database. Even given this recommendation, the company's engineering managers felt that the data did support extrapolation with an acceptable risk and recommended launching.

As *Challenger* lifted off, puffs of smoke emanated from one of the field joints on the right booster rocket. The color and dense composition of the smoke suggested that grease, joint insulation, and rubber O-rings were being burned and eroded by the hot propellant gases. At 58.788 seconds into the flight a flame appeared on the right booster in the area of the aft field joint. As the flame plume grew in size, it was deflected onto the surface of the external fuel tank and a strut holding the booster rocket. The hydrogen in the tank caught fire, the booster rocket broke loose, smashed into *Challenger*'s wing and then into the external fuel tank. At 76 seconds into the flight, *Challenger* was totally engulfed in a fireball.[78]

Postscript to the *Challenger*

Much has been made of this series of events and their context. As noted, Thiokol was renegotiating its contract with NASA at the time of the *Challenger* flight. Obviously, the successful renegotiation of this "sole-source" contract was of great importance to the company's management. Thiokol, recall, was awarded the original contract, in part, because its innovative O-ring mechanism brought the cost of the SRB into a low-bid category. The inherently weak hardware design had been documented since 1977, yet the shuttle had been successfully launched at each mission. When Marshall's engineering management requested data on the field joint on January 27, 1986, did Thiokol managers view the situation with alarm? If the hardware could not stand up to the rigor of the operational schedule, including winter flights at low launch temperatures, how would the company be viewed during contract negotiations?[79]

The Marshall Space Flight Center (and all of NASA) was under pressure from within the Reagan administration to justify its existence. While the Reagan White House was adamant that pressure was not exerted on NASA to launch in time for the president's State of the Union address that evening, it is not unreasonable to assume that NASA's leadership perceived such pressure. No doubt, NASA would have wanted to provide the president with a high publicity event – the "Teacher in Space" project.

The popular press created heroes and villains out of the Marshall and Thiokol engineers who participated in the launch decision. The Thiokol engineers who have publicly admitted that they strongly opposed the launch have been "canonized," while their administrators and the NASA project managers have been severely criticized for ignoring these courageous engineers. This approach to analyzing the performance decision makes good news copy and good drama. Indeed, one TV "docu-drama" that featured the debate was produced.

In actuality, none of the engineers view themselves as heroes. Boisjoly left Thiokol, sued his former employer, and, after a difficult period of being out of work, went on the lecture circuit. In contrast, McDonald chose to stay at Thiokol, and was promoted to vice-president of engineering. According to McDonald: "When we were at the Congressional hearings, they attacked both him [Boisjoly] and me for not doing more than we did. You can let the things you could have done keep playing on your mind, but it's totally impractical. I can't do anything about what's happened, but by God I can help make sure it never happens again."[80]

In the aftermath of *Challenger* and the Rogers Commission, McDonald expressed, what for him, was a tacit ethic involving his responsibility as an engineer: "It's everyone's personal and professional responsibility to voice a disagreement as well as an agreement to management. People who are silent and passive are hurting the engineering profession. If anything came out of this, it may be that people will feel more comfortable about voicing their concerns – without worrying about being out on the street."[81]

Whether or not one feels "more comfortable" disagreeing with management will be a personal preference. Engaging in real-life discussions that involve ethical dilemmas (i.e., self-interest versus public safety) is inherently *uncomfortable*. Having a professional maxim or guideline to follow in spite of the discomfort is one lesson that can be learned from McDonald: specifically, knowing how to articulate and document concerns both technically and ethically; understanding the power or limits one has in the organization; and communicating one's concerns comprise an ethical response.

Was the *Challenger* Disaster a "Normal Accident"?

Hans Mark, a former deputy director of NASA who left his position six months before the *Challenger* accident, has observed that he participated directly in twelve shuttle launch decisions. In each instance, there was always one group of engineers who, like those at Thiokol, advised against the launch, claiming that the shuttle would experience a catastrophic failure. "Sometimes we took their advice and postponed the launch, and at other times we went ahead and flew in spite of the advice we were given. The mere fact that a group of engineers opposed the launch because they were afraid one of their subsystems would not work was not enough to cancel the launch." Mark observed that the shuttle is such a complex and advanced technology that top management at NASA grew accustomed to dealing with these claims of disaster.[82]

It is interesting to note that for the *Challenger* launch NASA's top decision makers (Level I) were never informed of Thiokol's concerns. Marshall officials chose not to pass this information on to their superiors. Thus Mark's experience may actually be magnified many times. How many cases were there like the O-rings that never became an issue at the highest levels of NASA? No doubt the organizational hierarchy of NASA was a considerable impediment for negative information reaching the top of the organization. The normative concerns of decision makers in the hierarchy overrode the factual concerns and judgments of those below them in the chain of command. Similar questions can be addressed to Thiokol. Did the company downplay the early signs of a flaw because the O-ring design had been central in securing the NASA contract? Did the "low-cost," "high-risk" mentality reflect the general success-oriented perspective common in NASA and its contractors? Why did its key engineers believe that the O-rings were redundant two years after the criticality had been changed, and nine months after a primary O-ring had eroded substantially and not sealed? Why were they unaware that a launch constraint was placed on the field joint, nor what a launch constraint meant?

Larry Mulloy, in *Oral Interviews: Space Shuttle History Project,* documents how a pervasive organizational culture can affect the decision making perspective. Interviewed on April 25, 1988, two years after the accident and the Rogers Commission, Mulloy commented on flight 51-B (April 29, 1985), in which an O-ring was completely burned through and a secondary O-ring eroded (case to nozzle joint). He noted that the analysis and testing on the field joints was expanded in order to determine if the conditions were tolerable. "We concluded that it was," Mulloy reported.

> It's in the record and I truly believe that if there was a fatal error made . . . in engineering judgment, it was accepting that kind of a condition where you've completely destroyed a primary O-ring and accepted damage to the second and concluded that that was an acceptable thing to continue to fly with.
>
> In retrospect, after you have something like the Challenger accident and you look at your judgments . . . you throw away . . . all that good engineering and analysis and tests and said "there's something

drastically wrong when something that you think isn't supposed to get any damage at all sustains that kind of damage, and you conclude it's O.K.

Recall Feynman's "Russian Roulette" description of how test results were interpreted. Mulloy describes the same phenomenon.

> I think that was . . . where we continued, we were starting down the road in accepting that situation anyway. And then when we got that little warning and decided that, that's O.K., too, we started pressing further on down that road to the point where, inevitably, you're going to have an accident; if, if, if the thing that we were doing in parallel to eliminate that problem didn't get incorporated before the odds caught up with you. And that's exactly what happened; that's exactly what happened.[83]

Disregarding the engineering test data is not an easy act. In spite of McDonald's "don't look back" philosophy, Mulloy and others still struggle with the decision. Again, they struggle because it was, at base, a moral decision that ultimately cost human lives. In Mulloy's words:

> There was an attempt made by Roger Boisjoly at Thiokol to make a case that the situation could be worse at lower temperatures than what we had experienced before. Unfortunately none of the data, when you looked at all data of all the temperatures we had flown at, correlated with that. And also the three data points that he had, that he was basing his case on, were very simple tests – one of them at 100 degrees, one of them at 75 degrees, and one of them at 50 degrees. And based on that simple test, it said that the O-ring wouldn't function at 75 degrees, and we knew it would. We also knew it would function at lower temperatures than that because it had functioned at lower temperatures than that. So there just wasn't anything beyond, it was just rolling the dice once too often. I don't mean to say that in terms of every time we flew it was a roll of the dice – it wasn't and that wasn't the way the decisions were made. It was a calculated risk each time. And it was taking that calculated risk one time too many.[84]

Recounting the grim events that were revealed by the investigation, Mulloy summarizes:

So the investigation showed that the accident was caused by a failure of the joint in the Solid Rocket Motor, and it took an incredible series of events in order for that to happen. There are six joints on the shuttle. There's three in each motor. One failed – five didn't. The one that did had some peculiarities about it. And those are documented in the *Presidential Commission Report.* There were things found out in the investigation that nobody knew before, and therefore could not have had any information to make any decision other than what was made. The Commission concluded that at the point where the joint failed is where you had the maximum, or minimum gap between the mating parts, and therefore the greatest compression on an O-ring. Up to the point of the accident, what all of my engineers at Marshall were concerned with, and what Thiokol engineers were concerned with, was to make sure you had the O-ring squeezed when you seated it – you made a seal by squeezing it. What happened here is, the point it failed was where it had the maximum squeeze on it, you had to have something else happen, then, to have that failure. Before the pressure that's built up in the motor gets to the O-ring, in order for it to fail, the joint has to open up and cause the metal to move away from the O-ring. The only way that could happen is if the putty, which is in the joint upstream, or on the motor side where the pressure is, if the putty held the pressure longer than anybody knew that it would. So the gap opened up, the O-ring didn't return, the putty held the pressure up to a point. Then when you got in this configuration the pressure broke through the putty and blew right by both the O-rings. And, from the film that you see from the liftoff of *Challenger,* then it sealed. It puffed by and then it sealed for a while. The last incredible event (not incredible), but the last thing that has to happen to you in order to have that joint fail, after having these two things happen – maximum squeeze, low temperatures (which made this even slower to return, but it wouldn't have mattered if it was 75 degrees, it wouldn't return either according to test data) – the next thing that had to happen was we hit the highest windshears that we'd ever experienced at 60 seconds into flight, where the max-Q (maximum dynamic pressure) is encountered. We had the maximum.

We hit the highest windshears that we had ever encountered. And that load puts a load down on the joint which causes it to rotate and then open up again, which then let the gases go by and cause the burn through. And that's what happened to *Challenger.*[85]

Conclusion

Our role in presenting this rather extensive, but important and instructive case, is not to take sides. Rather, our concern focuses on the methods by which Thiokol and NASA engineers faced two major ethical dilemmas. One involved continuing to fly with a critical component which might require redesign; the other, agreeing to launch under conditions that were beyond the scope of both field experience and the original design criteria. Clearly differences in perception of the seriousness of the problem existed among the involved parties. Misunderstanding of technical data, miscommunications based on specific meanings of common terms, and an overall organizational tolerance of high risk contributed to how decisions were made. These went far beyond the politics and the immediate constraints of weather and "glory" which accompanied the *Challenger* launch decision.

We have intentionally not reduced this case to a simple scenario which focuses on the launch decision. That would not do justice to the involved engineers, or enable the reader to appreciate fully the complexity of this problem and the degree of uncertainty these engineers struggled with month by month as the evidence of a serious problem mounted. As we illustrated with the DC-10 case in Chapter 3, much is lost when the material is greatly condensed, or the case writer uses the scenario in order to make a point. The result is too often a case in which there are "white hats" and "black hats." As we hope we have demonstrated with the *Challenger* case, the lines of demarcation between the "good" and "bad" guys are often extremely blurred. We would encourage the interested reader who would like to gain additional insight into the people and events surrounding the *Challenger* accident to examine the wealth of material contained in the Rogers Commission report.

In presenting the *Challenger* material we have chosen to set the stage by introducing Richard Cook, a budget analyst who, during a short span of time through his meetings with the involved NASA engineers, was able to grasp the full seriousness of the O-ring problem and the impending accident if the problem were not addressed and resolved. Then, having seen his worst fears realized, six days later while the cause of the accident was far from being settled, Cook was able to again synthesize knowledge from these same engineers and reiterate the concern that "the Solid Rocket Boosters have been flying in an unsafe condition . . . due to the problem of O-ring erosion and loss of redundancy caused by unseating of the secondary O-ring in flight." Cook predicted that this condition would have to be corrected before the shuttle flew again, and he urged the commission to "get to the ordinary engineers" from Marshall and Thiokol in order to determine the cause of the accident. Indeed, it is these engineers who provided much of the insight into this problem, and much of this case is based on their writings and testimony. Yet, even to Cook, the question must be asked, Could you have done more? Should you have done more? Would anyone have listened?

Throughout this book we have examined three principles of engineering ethics and shown how they apply to each case. The first principle is *competency*: Engineers are obligated to know as much as is reasonably possible about the technology of concern. How well did the Thiokol and NASA engineers fare by this criterion? Three areas of concern with respect to the *Challenger* include: knowledge of the joint, its design and mechanism of operation; knowledge of the materials, particularly the O-ring and how it was affected by temperature, pressure, wind, and ice; and knowledge of data analysis.

For example, Thiokol engineers were concerned about how the rubber compound (Viton) used in the field joint O-rings would function under low-temperature conditions. The predicted launch temperature that morning would be below the specifications the engineers had followed when designing the field joints. Their very limited test data showed a loss of resiliency in the rubber compound as temperature dropped. The question remained: As the rubber O-ring stiffened, would either the primary or the secondary ring still seal the joint under stresses of launching the shuttle?

During the Rogers Commission hearings, Richard Feynman demonstrated the well-known glass transmission temperature phenomenon – that is, the rubber compound used in the O-rings becomes brittle to the point of cracking when immersed in ice water, which approximated the predicted launch temperature. Were the materials used in the field joint fully characterized over the range of conditions of actual use of the *Challenger*? Were the properties of this material, particularly its response to temperature, sufficiently understood by those who designed the field joint to enable them to make a decision about operating outside the design temperature range? Did engineers fully understand the joint's functioning within the normal temperature range?

Others have also criticized the joint design. For example, Kamm, a design and consulting engineer, has panned both the design of the groove and the leak check port. As noted, the port's placement may enable any air under high pressure that escapes past the primary O-ring to tunnel through the heat-insulating putty and establish a path for the hot motor gases to reach the O-ring following ignition.[86] The Rogers Commission observed that when the leak check pressure was 50 or 100 psi, only one field joint O-ring anomaly resulted in nine flights compared with over half of the flights that had anomalies when the pressure was raised to 200 psi.[87]

Kamm has also critiqued the redesigned joint that has three O-rings in series.[88] That design and the process NASA went through to assure its acceptability are described in Appendix D. In spite of Kamm's concern, the new design is the result of a process that would be followed by an "ethical organization," adhering to our three principles. James Kingsbury, director of science and engineering at Marshall (1975–86), commented in *Oral Interviews: Space Shuttle History Project* that:

> One of the things that we were trying to do was recover the inventory [in redesigning the joint]. We had $200 million worth of hardware that we really didn't want to throw away. And so we were trying to come up with a design that would permit us to use that inventory. It wasn't like starting over. If we were starting over we would have done it differently than that has been done now. But what's been done is perfectly all right. It's in no way hazardous, as far as I'm concerned. And then there were really three prime candidates after all that good shaking out. Two of the three were taken essentially through qualification, and then one of the two was selected to fly. So, there were two redesigns, basically, requalified.[89]

Our third point is that the Thiokol engineers had trouble organizing and understanding the engineering data at hand, specifically whether there was a relationship between temperature and O-ring damage. Their "analysis" of the O-ring performance and temperature data, which focused on examining the individual cases, found no empirically supported evidence that a relationship existed between these two factors. The Rogers Commission hearings, and examinations of other investigative reports, suggest that these engineers were unable to statistically analyze the O-ring versus temperature data. A simple plot of these data like that in Figure 14.6 is suggestive: All four launches below sixty-five degrees Fahrenheit exhibited O-ring thermal distresses; in contrast only three of the nineteen cases above sixty-five exhibited thermal distresses. What does this say about a temperature effect?

In presenting these data to management on the evening of January 27, 1986, Thiokol engineers focused on the seven cases in which problems had occurred and seemed to ignore the cases for which there were no problems. In doing this they missed an opportunity to see the true pattern in the data. Specifically, if the zero failure cases are ignored as in Figure 14.5, a "U" shaped relation occurs for the resultant seven cases. Any analysis of these six cases will not be conclusive. The engineers uniformly noted that they could not prove that a relationship existed between O-ring "failure" and temperature, even though they did not statistically analyze the data in order to ascertain if a relationship did exist.

If the data had been presented as in Figure 14.6, a simple calculation would produce a statistically significant correlation between temperature and O-ring anomalies that occurred. For example, Figure 14.7 presents the same data points as Figure 14.6 with two curves fitted to the pre-*Challenger* data points. One curve is the Median Polish curve, an empirically generated curve that connects medians of subgroups of data points. This curve shows a pronounced relation between the number of incidents of erosion and temperature, and essentially classifies the data point of Flight 61-A (seventy-five degrees Fahrenheit) as a statistical outlier. The second curve is the linear regression line through these data. Again, there is a pronounced relation, with a correlation of -0.56, which is statistically significant at the .05 probability level. Thus, using relatively simple techniques, it would have been possible to establish the statistical existence of the O-ring erosion–temperature relation.

Using a more sophisticated logistic regression technique, we constructed another probability estimation model shown in Figure 14.8. Fitting this probability model to the pre-*Challenger* data points, and projecting to thirty-one degrees Fahrenheit (the temperature at launch), we estimate that at least one field joint would suffer blow-by or impingement with a probability close to 1.0. The estimated damage probability at fifty-three degrees Fahrenheit is about 0.8, a high probability given the consequences of a joint failure.

From the record, it is evident that both the involved NASA and Thiokol engineers lacked the statistical knowledge that might have been critical in making the O-ring performance decision. When the joint rotation problem first occurred, NASA engineers were the first to recognize its seriousness. Then for a significant time interval, Thiokol engineers did not appreciate that the joint had been reclassified from Criticality 1R to 1, and was no

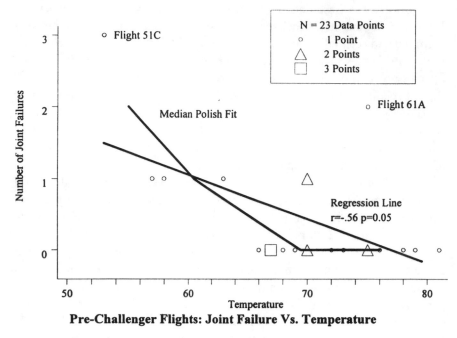

Figure 14.7 Two curves fit to O-ring damage data. Median Polish Fit indicates that flight 61A is an outlier; linear regression indicates that significant correlation with temperature is found.

longer considered to be redundant. Further, the Thiokol engineers did not completely understand the performance of their material relative to the mechanism of the joint, particularly as both related to temperature effects. In addition, there was Kamm's concern about the design of the groove in which the O-ring is inserted.

The resultant manner in which the performance data were used to make the decision is clearly of concern. Thus we conclude that not all the engineers who participated in the design and decisions regarding the field joint performance provide good examples of upholding the ethical principle of competency.

The principle of individual competency also extends to acknowledging, indeed declaring, knowledge areas of incompetence. No one is expected to know everything required to design or make performance decisions about a product. However, as knowledge experts, engineers must understand the

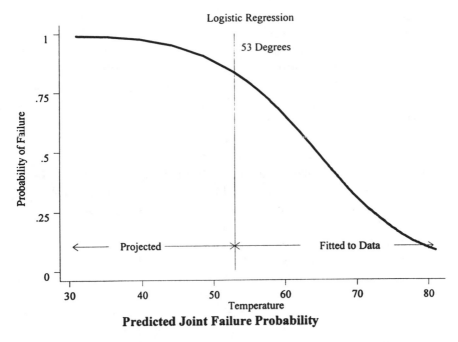

Figure 14.8 Logistic regression model fit to O–ring damage data. This model indicates that the probability of some O-ring damage is 0.80 or greater as temperature drops below 53 degrees, approaching 1.00 at freezing (32 degrees).

limits of their competency. When technical problems extend beyond one's competence, at a minimum, the engineer should acknowledge the need for help and, more directly, seek it.

Several Thiokol engineers have acknowledged that they did not understand what was needed to fix the field joints, or how to predict what would happen under cold conditions. Over a two–year period, they had sought help, within their own company, and from NASA. In this sense, they had met the principle of competence – that is, they communicated their concerns to those in the organization who had authority to demand more resources for tests, or an extension of the schedule. Another example is provided by NASA's Ray and Eudy, who visited two O-ring manufacturers in an effort to "seek opinions regarding potential risks involved."[90]

Parallel to the individual principle of competence is an organizational principle of competence. Although it is unreasonable to expect that individuals or even design teams will possess all required knowledge to solve particular technological problems, it seems reasonable to expect that such knowledge exists within the organization or set of organizations responsible for the product. One of the strengths of a formal organization is the ability to provide technical knowledge across a broad range of subjects. When an individual facing an ethical dilemma seeks help from the larger organization, the ethically competent organization should attempt to fill the knowledge gap. In those cases where the organization does not possess the needed knowledge, it must go outside to obtain it.

The responses of Thiokol and NASA to the designers of the SRB field joints were less than exemplary. In reality, the first indications that the field joint design was problematic occurred in 1977. This triggered a sequence of requests by concerned Thiokol and Marshall engineers to redesign the joints. Somehow, these requests for a redesign fell through bureaucratic cracks in both organizations. At times when Marshall engineers were calling for a redesign, Thiokol attempted to provide evidence that it was not needed. On the night of the *Challenger* launch teleconference, NASA managers requested that Thiokol engineers do what they had already stated they didn't know how to do, and document why it would be unsafe to launch. Although the Thiokol engineers apparently lacked statistical skills for analyzing their field joint performance data, the record indicates that the involved NASA engineers were no more adept at solving such problems. Since the responsibilities of NASA engineers concerned reviewing technology, and in making performance decisions, they should have been expected to be more skilled in making these types of decisions. In summary, we have identified problems of organizational competency in this case.

Our second principle of engineering ethics is *responsibility*. When engineers identify an ethical dilemma, they have an obligation to voice their concern. The case of the O-rings has identified several responsible engineers, within the ranks of both Thiokol and NASA. One of these engineers, Allan McDonald, has made the important observation, that even within the group of design engineers at Thiokol, the *majority* did not participate in the difficult decisions regarding the performance of the SRB field joints. Instead, they elected to avoid the controversy, and said nothing. Mr. McDonald has criticized this silent majority for just that.[91] In choosing to remain silent, these engineers did not meet the principle of responsibility.

There is also an organizational counterpart to the principle of responsibility. Even the most responsible engineers who strongly express their concerns when facing an ethical dilemma will have no impact if they are part of a non-responsive organization. Joseph Trento, a journalist who has followed the space program, claims that Marshall management did not tolerate individuals who publicly expressed concerns about safety or performance. To Trento, the Thiokol engineering managers chose to override the profound concerns of their engineers after perceiving pressure from NASA management.[92] Organizational responsibility and responsiveness were clearly lacking.

Our third principle of engineering ethics is *Cicero's Creed II*. This dictum implies that engineers should understand and characterize the risks associated with technology. In that respect the organizational system of NASA and its contractors was unable to satisfy this principle. We could not find evidence that either NASA or its contractors fully understood how to treat engineering data stochastically. As discussed in Chapter 13, the high-risk, success-oriented culture of NASA went against a probabilistic approach to decision analysis.

To NASA, risk was a qualitative label, not a parameter to be estimated and used to design and make decisions. In taking such a perspective, the organization seemed to ignore the existence of probabilistic phenomena. Did the engineers of NASA and its contractors really live in a deterministic world where "statistics don't count for anything"?[93]

Statistical quality control was developed during the 1930s, and adopted by several industries, notably outside the United States. However, some of the core U.S. engineering disciplines are just starting to include statistics as part of required curriculum in their program, although it remains less a part of everyday practice. NASA engineers may be competent, even responsible, but it will be difficult for them to meet the principle of Cicero's Creed II until they can realistically assess the risks of the technologies under development.

The solid rocket booster problem began with the faulty design of its joint and increased as both NASA and contractor management first failed to recognize joint rotation as a critical problem, then failed to fix it, and finally accepted it as a necessary flight risk. Thiokol didn't accept the implication of early tests that the design had a serious and unanticipated flaw – the inner part of the clevis bent away from the tang, thus reducing the pressure on the O-ring in the milliseconds after ignition. NASA didn't accept the judgment of its engineers that the design was unacceptable and, as the joint problems grew in number and severity, NASA minimized them in its high-level management briefings and reports. Thiokol's position was that the condition was not desirable but was acceptable. As tests and then flights confirmed damage to the sealing rings, the reaction by both NASA and Thiokol was to increase the amount of damage considered "acceptable." When management finally suggested a redesign of the joint, no clear-cut recommendation was made, nor was the shuttle grounded until the problem was solved.

In summary, looking for heroes and villains in the O-ring story misses the real lesson to be learned from the *Challenger* accident. NASA and its contractors were deficient in their abilities to manage the risks. We have identified several areas of technological knowledge that were missing from NASA and its supporting organizations, including risk management and assessment skills. Moreover, the political pressures and underfunding that surrounded the birth of the shuttle program caused NASA to manage the entire shuttle program in ways that exacerbated these knowledge gaps. Specifically, the all-up testing methodology curtailed the experience engineers were able to gain about their components and systems. It reduced the amount of test data available to measure reliability and risk, even in the qualitative manner favored by NASA management. In short, it tended to create a technology that was beyond control of rational decision making.

Yet, even on that night, the launch decision was not the classical "no brainer. We have gone to considerable lengths to describe this case, the data that were available, and the thinking behind the decisions. Yes, we have documented a number of the major problems that hampered the decision makers. However, the reader should recognize that much of the technical discussion could be reduced to the question of whether a piece of rubber, 12 feet in diameter and slightly more than a quarter inch in cross-sectional diameter would move approximately 0.02 inch across a groove and extrude into a gap of approximately the same length in less than 0.2 second when the temperature was at least 20 degrees Fahrenheit colder than previously experienced. Engineers who were very familiar with the problem said not to extrapolate or launch – we don't understand enough about the mechanism. In contrast, managers said that our understanding of the mechanism, the data, and how they were presented justifies extrapolation and the launch.

Notes

1. William P. Rogers, *Report of the Presidential Commission on the Space Shuttle Challenger Accident*, vols. 1–5 (Washington, DC: Government Printing Office, June 6, 1986).

2. Ibid., 4:377.
3. Ibid., p. 378.
4. Richard C. Cook, memorandum to Michael B. Mann, July 23, 1985, in Rogers, *Report*, 4:391.
5. Ibid., p. 379.
6. Ibid., p. 382.
7. Phillip M. Boffey,"NASA Had Warning of a Disaster by Booster," *New York Times*, February 9, 1986.
8. Rogers, *Report*, 4:256.
9. Ibid., p. 257.
10. Ibid., p. 261.
11. Ibid., p. 262.
12. Ibid., p. 263.
13. Ibid., p. 388.
14. Ibid., p. 389.
15. Ibid., p. 397.
16. Rogers, *Report*, 1:120.
17. Malcolm McConnell, *Challenger: A Major Malfunction* (Garden City, NY. Doubleday, 1987), p. 177.
18. Ibid., p. 181.
19. Rogers, *Report*, 1:57.
20. Richard S. Lewis, *The Voyages of Columbia: The First True Spaceship* (New York: Columbia University Press, 1984), p. 226.
21. Rogers, *Report*, 1:60–1; see also testimony of Arnold Thompson, Design Engineer, Morton Thiokol, Inc., in ibid., 4:693–7; Robert S. Ryan, "The Role of Failure/Problems in Engineering: A Commentary on Failures Experienced – Lessons Learned," NASA Technical Paper 3213, George C. Marshall Flight Center, Huntsville, AL, March, 1992, p. 15.
22. Ryan, "The Role of Failure," p. 15.
23. Rogers, *Report*, pp. 122–3.
24. Ibid., p. 124.
25. John Q. Miller, memorandum to Glenn Eudy, January 9, 1978, in ibid., pp. 234–5.
26. John Q. Miller, memorandum to Glenn Eudy, January 19, 1979, in ibid., p. 236.
27. Ibid., p. 123.
28. Ibid., pp. 124–5; "Space Shuttle Verification/Certification Propulsion Committee," July 10, 1980; Lt. Gen. Thomas W. Morgan, Chairman, in ibid., 5:1649–56.
29. Ibid., 1:125.
30. Ibid., p. 133.
31. Ibid., pp. 126, 157–8.
32. Ibid., 5:1658–9.
33. Roger Boisjoly, "Interview: Whistle-Blower," *Life Magazine*, March 1988, pp. 17–19.
34. Rogers, *Report*, 1:128.
35. Ibid., p. 128.
36. Ibid., pp. 128, 132.
37. Ibid., pp. 132–4.
38. Ibid., pp. 135–6.
39. Ibid., pp. 136, 247.

40. Ibid., 5:1590.
41. Ibid., 1:136.
42. Ibid., p. 147.
43. Rogers, *Report*, 2:H-39, "Volumetric and Thermal Analysis and Test Results."
44. Ibid., 1:66, 137–8.
45. Ibid., p. 147.
46. Ibid., pp. 139, 249–50.
47. Brian Russell, letter to James Thomas, in ibid., 5:1568–9.
48. Ibid., pp. 1587–8.
49. Ibid., p. 1588.
50. Ibid., p. 1541.
51. Ibid., 1:139–40.
52. Ibid., p. 256.
53. Ibid., pp. 140–1; 252, 254–5.
54. Ibid., 5:1579.
55. Ibid., 1:142–3, 5:1576.
56. Ibid., 1:144.
57. Ibid., pp. 85–6.
58. Ibid., pp. 87, 106.
59. Ibid., 4:811–2.
60. Ibid., 1:90.
61. Ibid., p. 89.
62. Ibid., 4:604–5.
63. Ibid., 1:92–3.
64. Ibid., 4:624.
65. Ibid., pp. 626–7.
66. Ibid., p. 811.
67. Ibid., p. 605.
68. Ibid., p. 607.
69. Ibid., pp. 631–2.
70. Ibid., p. 655.
71. Ibid., 1:95.
72. Ibid., 4:704.
73. Ibid., p. 633.
74. Ibid., pp. 632–5.
75. Ibid., 1:145.
76. Ibid., p. 155.
77. Ibid., p. 148.
78. Ibid., pp. 19–21.
79. McConnell, *Challenger*, p. 181.
80. Debra Bulkeley, "The Making of a Hero," *Design News*, February 15, 1988, pp. 86–92.
81. Ibid.
82. Hans Mark, *The Space Station: A Personal Journey* (Durham, NC: Duke University Press, 1987), pp. 221–2.
83. Lawrence Mulloy, taped interview in Management Operations Office, MSFC, *Oral Interviews: Space Shuttle History Project*, Transcript Collection, December 1988, p. 195.

84. Ibid., pp. 198–9.
85. Ibid., pp. 199–200.
86. Lawrence J. Kamm, *Successful Engineering: A Guide to Achieving Your Career Goals* (New York: McGraw-Hill, 1989), p. 132.
87. Rogers, *Report*, 1:134.
88. Kamm, *Successful Engineering*, p. 132.
89. James Kingsbury, taped interview in Management Operations Office, MSFC, *Oral Interviews: Space Shuttle History Project*, Transcript Collection, December 1988, pp. 437–8.
90. Rogers, *Report*, 1:123–4, 237–8
91. Bulkeley, "The Making of a Hero."
92. Joseph J. Trento, *Prescription for Disaster* (New York: Crown, 1987).
93. Trudy E. Bell and Karl Esch, "The Space Shuttle: A Case of Subjective Engineering," *IEEE Spectrum*, 26, June 1989, pp. 42–5.

Appendix A:

Description of a Space Shuttle Mission

This description of the typical shuttle mission has taken from Ryan[1] and several official NASA documents.

In the launch configuration, the orbiter and two solid rocket boosters (SRB) are attached to the external tank in a vertical (nose-up) position on the launch pad. Each SRB is attached at its aft skirt to the mobile launcher platform by four bolts. Approximately 10 hours before launch, the liquid oxygen and hydrogen fuel is transferred to the external tank. Due to cryo-propellant temperatures of the liquid fuel, the vehicle shrinks. The struts linking the aft (rear) solid rocket booster to the external tank have been designed to account for this shrinkage. The weight of the propellant places additional loads on the vehicle-to-mobile launch platform interfaces and the shuttle element-to-element interfaces. This causes the solid rocket motors to bow laterally and the total vehicle to bend in the pitch plane. The cryo-shrinkage moves the struts 7 degrees, so that they are now perpendicular to the solid rocket booster and the external tank (which have been designed to account for this movement). A punch load toward the tank is created, which is counteracted by the radial shrinkage of the tank, storing energy in the structure that will be released at lift-off.

The SSMEs are ignited 6.6 seconds before lift-off and must reach 90 percent power level by T minus 3 seconds, which stores even more energy in the structure. At T minus zero the SRBs are ignited. Lift-off occurs almost immediately because of the extremely rapid thrust buildup of the boosters. The process is designed so that the vehicle base bending loads return to minimum by T minus zero.

In order to reduce these lift-off loads, a study was conducted of performance versus loads when SRB ignition is delayed until a minimum stored energy point is reached. The study results supported the delayed solid rocket motor ignition. The dynamics are such that the SSME thrust bends the vehicle and lifts it, pushing the orbiter and tank between the SRBs in a gear train mode and setting up an oscillation which produces minimum stored energy approximately 6 seconds after the SSME ignition, at which time the SRMs are ignited. Internal pressure in the SRBs above 900 psi stretch the vehicle in a longitudinal transient mode simultaneously with the release of stored potential energy. The holddown bolts are blown and the aft skirts released from the pad. Large vehicle dynamic motions result.

As the vehicle clears the tower, it is rolled up to 180 degrees (mission dependent) to place the orbiter down and produce a more optimal total thrust angle. At 20 seconds flight time, the vehicle velocity (performance) is assessed. If performance is low, the main engines are not throttled as deeply as planned to improve performance. If performance is nominal, preprogrammed throttling occurs. If performance is high, deeper throttling occurs, if possible to the 65 percent limit. Engines are throttled to keep the dynamic pressure within design limits as the vehicle traverses through the maximum dynamic pressure regime, which occurs during the time of high winds. Pitch, yaw, and elevon load relief are used to reduce aerodynamic loads, at the expense of performance. This also introduces high thermal loads after max Q (nominally approximately 60 seconds after lift-off) when the vehicle is moving back to its optimum path (large side-slip angles introduced).

According to Ryan, during the design phases, major trade-offs involved loads versus performance losses. The option was to beef up the structure to handle loads (performance loss) or deviate time and trajectory path and reduce loads (performance loss and loss of launch probability). The decision was generally to take the deviations and not to beef up the structure, although in some specific areas structure was beefed up.

Approximately 2 minutes into the ascent phase, when the two SRBs have consumed their propellant, they are jettisoned from the external tank. The boosters briefly continue to ascend, while small motors fire to carry them away from the space shuttle. The boosters then turn and descend, at a predetermined altitude. Parachutes are deployed to achieve a safe splashdown in the ocean, approximately 141 nautical miles (162 statute miles) from the launch site. The boosters are recovered and reused.

Concomitantly, the orbiter and external tank continue to ascend, using the thrust of the three space shuttle main engines. Approximately 8 minutes after launch and just short of orbital velocity, the main engines are shut down (main engine cutoff), and the external tank is jettisoned on command from the orbiter and breaks up on reentering the atmosphere. The resulting debris footprint is critical and must be controlled. The external tank continues on a ballistic trajectory and enters the atmosphere, where it disintegrates.

The orbiter then fires its orbiter maneuvering system (OMS) engines to achieve final desired orbit. These forward and aft reaction control system engines provide attitude (pitch, yaw, and roll) and the translation of the orbiter away from the external tank at separation and return to attitude hold prior to the OMS thrusting maneuver. Normally, two thrusting maneuvers using the two OMS engines at the aft end of the orbiter are used in a two-step thrusting sequence: to complete insertion into Earth orbit and to circularize the spacecraft's orbit. The OMS engines are also used on orbit for any major velocity changes.

The orbital altitude of a mission is dependent upon that mission. The nominal altitude can vary between 100 to 217 nautical miles (115 to 250 statute miles). The forward and aft reaction control system (RCS) thrusters (engines) provide attitude control of the orbiter as well as any minor translation maneuvers along a given axis on orbit.

After completing its mission, the orbiter reenters by firing the OMS engines, slowing the vehicle; then it reenters the atmosphere and lands. All of these are very fine-tuned, critical events. The reentry drives the thermal protection system design, which is very critical to survival. The on-orbit phase is unique for each mission. One additional complication occurs during the ascent phase – abort options early in the mission engine failures mean ditching. Next comes return to launch site, a very critical maneuver.

At the completion of orbital operations, the orbiter is oriented in a tail first attitude by the reaction control system. The two OMS engines are commanded to slow the orbiter for deorbit. The reaction control system turns the orbiter's nose forward for entry. The reaction control system controls the orbiter until atmospheric density is sufficient for the pitch-and-roll aerodynamic control surfaces to become effective.

Entry interface is considered to occur at 400,000 feet altitude approximately 4,400 nautical miles (5,063 statute miles) from the landing site and at approximately 25,000 feet per second velocity. At 400,000 feet altitude, the orbiter is maneuvered to zero degrees roll and yaw (wings level) and at a predetermined angle of attack for entry. The angle of attack is 40 degrees. The flight control system issues the commands to roll, pitch, and yaw reaction control system jets for rate damping.

Entry guidance must dissipate the tremendous amount of energy the orbiter possesses when it enters the Earth's atmosphere to assure that the orbiter does not either burn up (entry angle too steep) or skip out of the atmosphere (entry angle too shallow) and that the orbiter is properly positioned to reach the desired touchdown point. Energy is dissipated by the atmospheric drag on the orbiter's surface. Higher atmospheric drag levels enable faster energy dissipation with a steeper trajectory. Normally, the angle of attack and roll enables the atmospheric drag of any flight vehicle to be controlled. However, for the orbiter, angle of attack was rejected because it creates surface temperatures above the design specification. The angle of attack scheduled during entry is loaded into the orbiter computers as a function of relative velocity, leaving roll angle for energy control. Increasing the roll angle decreases the vertical component of lift, causing a higher sink rate and energy dissipation rate. Increasing the roll rate does raise the surface temperature of the orbiter, but not nearly as drastically as an equal angle of attack command.

If the orbiter is low on energy (current range-to-go much greater than nominal at current velocity), entry guidance will command lower than nominal drag levels. If the orbiter has too much energy (current range-to-go much less than nominal at the current velocity), entry guidance will command higher than nominal drag levels to dissipate the extra energy.

Roll angle is used to control cross range. Azimuth error is the angle between the plane containing the orbiter's position vector and the heading alignment cylinder tangency point and the plane containing the orbiter's position vector and velocity vector. When the azimuth error exceeds a computer-loaded number, the orbiter's roll angle is reversed. Thus, descent rate and down ranging are controlled by bank angle. The steeper the bank angle, the greater the descent rate and the greater the drag. Conversely, the minimum drag attitude occurs when the wings are level. Cross range is controlled by bank reversals.

The entry thermal control phase is designed to keep the backface temperatures within the design limits. A constant heating rate is established until below 19,000 feet per second. The equilibrium glide phase shifts the orbiter from the rapidly increasing drag levels of the temperature control phase to the constant drag level of the constant drag phase. The equilibrium glide flight is defined as flight in which the flight path angle, the angle between the local horizontal and the local velocity vector, remains constant. Equilibrium glide flight provides the maximum downrange capability. It lasts until the drag acceleration reaches 33 feet per second squared. The constant drag phase begins at that point. The angle of attack is initially 40 degrees, but it begins to ramp down in this phase to approximately 36 degrees by the end of this phase.

In the transition phase, the angle of attack continues to ramp down, reaching the approximately 14-degree angle of attack at the entry terminal area energy management (TAEM) interface, at approximately 83,000 feet altitude, 2,500 feet per second, Mach 2.5 and 52 nautical miles (59 statute miles) from the landing runway. Control is then transferred to TAEM guidance.

TAEM guidance steers the orbiter to the nearest of two heading alignment cylinders, whose radii are approximately 18,000 feet and which are located tangent to and on either side of the runway centerline on the approach end. In TAEM guidance, excess energy is dissipated with an S-turn; and the speed brake can be used to modify drag, lift-to-drag ratio, and flight path angle in high-energy conditions. This increases the ground track range as the orbiter turns away from the nearest heading alignment circle (HAC) until sufficient energy is dissipated to allow a normal approach and landing guidance phase capture, which begins at 10,000 feet altitude. The orbiter also can be flown near the velocity for maximum lift over drag or wings level for the range stretch case. The spacecraft slows to subsonic velocity at approximately 49,000 feet altitude, about 22 nautical miles (25.3 statute miles) from the landing site.

At TAEM acquisition, the orbiter is turned until it is aimed at a point tangent to the nearest HAC and continues until it reaches way point 1 (WP-1). At WP-1, the TAEM heading alignment phase begins. The HAC is followed until landing runway alignment, plus or minus 20 degrees, has been achieved. In the TAEM prefinal phase, the orbiter leaves the HAC, pitches down to acquire the steep glide slope, increases airspeed, banks to acquire the runway centerline, and continues until on the runway center-line, on the outer glide slope, and on airspeed. The approach and landing guidance phase begins with the completion of the TAEM prefinal phase and ends when the spacecraft comes to a complete stop on the runway.

The approach and landing trajectory capture phase begins at the TAEM interface and continues to guidance lock-on to the steep outer glide slope. The approach and landing phase begins at about 10,000 feet altitude at an equivalent airspeed of 290, plus or minus 12, knots 6.9 nautical miles (7.9 statute miles) from touchdown. Autoland guidance is initiated at this point to guide the orbiter to the minus 19- to 17-degree glide slope (which is over 7 times that of a commercial airliner's approach) aimed at a target 0.86 nautical mile (1 statute mile) in front of the runway. The spacecraft's speed brake is positioned to hold the proper velocity. The descent rate in the later portion of TAEM and approach and landing is greater than 10,000 feet per minute (a rate of descent approximately 20 times higher than a commercial airliner's standard 3-degree instrument approach angle).

At 1,750 feet above ground level, a preflare maneuver is started to position the spacecraft for a 1.5-degree glide slope in preparation for landing with the speed brake positioned as required. The flight crew deploys the landing gear at this point.

The final phase reduces the sink rate of the spacecraft to less than 9 feet per second. Touchdown occurs approximately 2,500 feet past the runway threshold at a speed of 184 to 196 knots (213 to 226 mph).

A complex launch constraint system is in place to ensure that the vehicle is not launched in unsafe conditions. Part of this involves measuring wind speed and calculating structural loads and performance margins. Approximately 2 hours prior to launch, a choice is made between several I-loads (trajectory shape to reduce loads, and increase performance margins). At 30 minutes, a decision is made relative to launch, whether to launch or not based on these data. In the last part of the launch sequence, all systems are monitored. Most redline cutoffs are automatic (launch stopped if system out of specs).

Note

1. R. S. Ryan, "The Role of Failure/Problems in Engineering: A Commentary on Failures Experienced – Lessons Learned," NASA Technical Paper 3213, George C. Marshall Space Flight Center, Huntsville AL, March 1992, pp. 9–10.

Appendix B:

A. O. Tischler's Personal Commentary

Your comments on my ethics are very laudatory (to me). I recognize that your understanding of what was happening within NASA is based on a relatively few pieces of written material, and that your purpose was (is) to give credence to predetermined hypotheses, but it's difficult to penetrate complex issues from what appears externally. Therefore, I'll provide a few remarks here to enhance your understanding.

First, I was somewhat of a fish from a different pond when I first came to NASA headquarters, as were most of those drawn from the NACA. In the early 1950s I had been assigned to work on the (even now) perplexing problem of combustion-driven oscillations in rocket chambers. Since space rocket equipment was at that time in short supply I was obliged to design, build, and operate my own equipment. In the course of time and concomitant experience I became reasonably expert in rocket equipment and in the various oxidizer–fuel combinations of interest. My associates at Lewis were among the first to use liquid hydrogen as a rocket fuel (successfully). That made me one of the few people within NASA able to take on managing NASA's initial rocket developments.

In Washington, where many managerial decisions have the first objective of making the boss look good, mine were dictated primarily by technical considerations. Rather than accept calculated final numbers, I was able to examine the basic assumptions of an analysis with a view to their effect on the outcome. I tried to foster this ability in the people I worked with as well. All engineering involves compromises but they should be based on the best facts that can be ascertained, and consensus should not determine the result.

You say that I resigned because I was disillusioned. Wrong word; it implies that I didn't understand my situation, and that certainly wasn't true. Frustrated is closer to it. Here's why.

As director of the Shuttle Technologies Office in the 1969–1972 period, I may have had the opportunity to advocate a fully reusable recoverable launch vehicle. *My* choice would have been a single-stage-to-orbit vehicle because that concept was (and is) the only one in which direction and (cost) control could be confined to a single management string. At the time, however, the performance of such a vehicle could not be ascertained from the scant real data available. Since then computer-aided structure design and performance calculations show the SSTO performance to be real and acceptable, albeit not on a par with a two–stage vehicle. Remember now that it's cost, not mass or bulk, that matters. Meanwhile the costs of the partially recoverable shuttle system selected escalated, in large part due to the complexity of working four major contractors through two NASA centers. That management arrangement, in my opinion, also contributed to NASA's biggest mission failure.

Realistically, I could not have overcome the self-induced anxiety of NASA and contractor personnel to get a successor to the Apollo program started, but my failure to insist on more complete evaluation of the SSTO concept may have set the progress of space exploration and exploitation back by a full generation. My further failure to persuade NASA to reduce program costs, or even to provide real support to do so, eroded the rest of my resolve.

It's all still there to be done. But it is now unlikely to be realized within my lifetime. Furthermore, I, more than anyone else in the organization, then or now, realize the consequences of these failures. That's frustration.

Now you understand a little more.

Appendix C:

Summary of Technical Issues Raised by the Covert Committee

Technical Issues

By March 1978, development work on the engine had been in progress officially for almost six years. However, in terms of accumulated test experience it was still "immature" according to the Ad Hoc Committee. Through February 1978, only 660 seconds of testing had been accumulated at 100 percent of rated power, while 109 percent of rated power would be required to launch the full payloads scheduled for later flights. NASA had planned to accumulate a total of 80,000 seconds throughout the SSME's entire development phase and Rocketdyne expected to have half of that, or 40,000 seconds, on a flight-configured engine before the launch of the first manned orbital flight. (Note that one shuttle flight constituted approximately 480 seconds of operating time.)[1] The initial flight, at the time of the hearing, was at least one year away. Considering only accumulated test time as a measure of readiness, Rocketdyne had accomplished less than 2 percent of its goal. It was noted that in large-scale development projects such as this, the rate at which experience is gained increases with time. During the beginning stages, in other words, accumulating test time is a slow process but, as some experience is gained, the process accelerates.[2] Although this possible explanation was offered, the Ad Hoc Committee finally concluded that it may be necessary to delay the first manned orbital flight since there were numerous changes yet to be implemented and tested.[3]

Turbopumps

The major technical problems that caused delays were those associated with the high-pressure turbopumps and other rotating machinery. The

336

highest pressure required in a properly operating shuttle engine, 7,650 pounds per square inch (psi), was found in the high-pressure oxidizer turbo-pump (HPOTP).[4] In order to satisfy the overall shuttle design require-ments, severe weight restrictions were applied to the engine as a whole. This, in turn, caused the shaft and housing of the HPOTP to be relatively flexible and have complicated load paths.[5] The reusability and performance specifications added another set of technical dilemmas, leading the Ad Hoc Committee to recommend that NASA develop a parallel backup shaft and housing design.[6] Still hoping to meet the launch date, Dr. Frosch, in his written reply, concluded that improving the oxidizer pump was not critical to the success of the first manned orbital flight although a parallel design was already under way. Indicative of what Dr. Covert labeled "Band-Aid fixes," NASA believed that the pump's deficiencies were relevant only when considering the reusability and extended life issue.[7] This position was taken even though the engine tests were marked by a series of dramatic high-pressure turbopump failures (see Chapter 5).

These longer-term issues were highlighted in the Ad Hoc Committee's second review. In fact, the committee recommended that the alternate HPOTP design continue to receive attention with the goal of installing the new design in operational engines in five or six years. NASA replied that "it would not cost too much more" to design an entire second generation engine instead of merely this pump since testing was a large portion of the total cost.[8] NASA then introduced the possibility that it may actually request additional funds from Congress to develop this second-generation engine.

The high-pressure fuel turbopump (HPFTP) blades were another sig-nificant concern of the Ad Hoc Committee in 1978. High cycle fatigue had apparently caused three major blade failures. The first two incidents were confirmed by fractographic analysis but the reason for the third could only be inferred. Highlighting the dangers of using the engine as the test bed, a fire destroyed the engine during the third failure. Considering that the amount of energy released during combustion is equivalent to five million horsepower, the committee was not surprised at these failures.[9]

However, the Ad Hoc Committee, maintaining the philosophy of com-ponent testing, was concerned that the properties of the material were not adequately understood and suggested that characterization be carried out with respect to the number of cycles to failure for high and low stresses. These tests were to be performed as a function of age, since grain struc-tures change after experiencing repeated high-temperature cycles. NASA reported that this type of program had begun.[10] The Ad Hoc Committee was apparently not completely satisfied with the progress of this program by the time of the second review in 1979. It felt that, before the first manned orbital flight, more needed to be understood about turbine blade cracks and cracks in the turbine platforms. A recommendation was made that stated that NASA and Rocketdyne should prepare a detailed technical case to include a method for determining crack growth rate in turbine platforms, inspection procedures, and criteria for replacing blades in the HPFTP.[11] The committee was concerned because NASA, opposing tradition, was intending to launch the first flight with cracks in the platforms. Although cracks in the leading edges of the blades were considered intolerable by NASA and Rocketdyne, platform cracks were viewed as acceptable until they propagated into the blade. The committee felt that NASA needed to better understand the crack growth rates through experimentation before an HPFTP could be permitted to fly with cracks of any type or severity.

Heat exchanger

The heat exchanger was also singled out in 1978 as being of technical concern to the Ad Hoc Committee, although no failures had been experienced with this component to date. A complicated welded assembly, the heat exchanger is located downstream of the oxidizer turbopump and thus is difficult to inspect. Failure of this component could be catastrophic.[12] Its complexity renders redundancy impractical and the committee recommended that the ongoing efforts toward relocating or reconfiguring it be continued.[13] NASA shared the committee's concern but also noted that its engineers knew of no better location for the component. Instead, their precautionary defense was to build higher structural margins into this component than any other in the engine.[14] However, in December after the first review, a leak in the coil tubing of the heat exchanger was found to be the source of a major fire. The welding of an adjacent bracket during rework had caused a weakness in the tubing which went undetected since no procedures were in place to inspect the component after rework. As a result, the Ad Hoc Committee listed this item as one of its continuing concerns in the 1979 review. Additionally, new tests had indicated that some of the heat exchanger's tubes had been rubbing against their supports; small holes could have been developing in the tubes. Because of the hot hydrogen-rich environment, a catastrophic explosion was a possibility.[15]

Preliminary flight certification

Another real concern that the Ad Hoc Committee focused on involved the preliminary flight certification partial tests. The committee recommended that the Main Propulsion Test Article tests of a flight-configured engine should consist of at least two uninterrupted full-duration runs at the 102 percent power level.[16] Also, at least ten successful starts and 3,000 seconds should be accumulated on a single flight-configured engine with at least 25 percent of the time at the 102 percent level. NASA reported that its testing schedule already included these recommendations and, in some instances, were even more stringent.

Postflight inspection

One testing issue over which there was considerable disagreement involved the complete teardown of engines. The Ad Hoc Committee believed that the engine should undergo a teardown and inspection after the first and sixth flights and, for the second through fifth, the usual boroscope and visual inspection would suffice if there was no indication of trouble. (A boroscope is a device that uses fiber optics technology to illuminate and visually inspect otherwise inaccessible areas.) Dr. Frosch recognized what he believed was a "difference of philosophy with regard to reliability."[17] NASA, he said, was in the process of determining whether reliability was improved by tearing down and rebuilding an engine or by replacing it with a new one.

NASA's current plans called for a teardown of at least one engine after the fourth and sixth flights and of all three engines after the fifth flight; only boroscope and visual inspections were planned for the second and third. Dr. Frosch discussed the evident trade-off between reliability for the second flight and the accumulation of age and wear data. Moreover, he argued that NASA needed to determine if normal inspections gave accurate and sufficiently extensive data. At this point, NASA's plans had not been confirmed, and its current strategy was still being evaluated.[18] By 1979, NASA had not yet agreed to accept the Ad Hoc Committee's recommendation for the complete teardown and inspection of an engine after the first flight. Some engine experts believed that a closely coupled operating engine should not be taken apart unless a boroscope inspection indicated problems. The committee, however, argued that if the preliminary flight certification program was conducted properly, there would be no cause for concern in removing one engine.[19] Several technical issues have been discussed and, for the most part, the Ad Hoc Committee was neither shocked nor even surprised that NASA and Rocketdyne were experiencing difficulties. The committee members' collective experiences contributed to their understanding from the outset that advances as complex as the space shuttle main engine take a considerable amount of time, money, and effort to develop a safe and reliable system. Failures, they recognized, were an unavoidable aspect of development.

Notes

1. Eugene E. Covert, *Technical Status of the Space Shuttle Main Engine: A Report of the Ad Hoc Committee for Review of the Space Shuttle Main Engine Development Program* (Washington, DC: Assembly of Engineering, National Research Council, National Academy of Sciences, March 1978), pp. 9, 19.
2. U.S. Congress, Senate, Subcommittee on Science, Technology and Space of the Committee on Commerce, Science and Transportation, *Report of the National Research Council's Ad Hoc Committee for Review of the Space Shuttle Main Engine Development Program*, 95th Congress, 2d session, March 31, 1978, Serial 95-78, 28–027 O, p. 9.
3. Covert, *Report*, p. 3.
4. Ibid., p. 7.
5. Subcommittee on Science, Technology and Space, March 1978, p. 5.
6. Ibid., p. 14.
7. Ibid., p. 73.
8. Subcommittee on Science, Technology and Space, February 1979, pp. 1097, 1130.
9. Subcommittee on Science, Technology and Space, March 1978, p. 5.
10. Ibid., p. 12.
11. Subcomittee on Science, Technology and Space, February 1979, p. 1096.
12. Subcommittee on Science, Technology and Space, March 1978, p. 5.
13. Ibid., p. 6.
14. Ibid., p. 61.
15. Subcommittee on Science, Technology and Space, February 1979, p. 1091.
16. Subcommittee on Science, Technology and Space, March 1978, p. 6.
17. Ibid., p. 60.
18. Ibid.
19. Subcommittee on Science, Technology and Space, February 1979, p. 1107.

Appendix D:

Post-Challenger *Redesign Activities of the Solid Rocket Boosters*

This description of the redesign, testing, and oversight activities related to the field and nozzle joints following the *Challenger* accident is taken from official NASA documents.

Redesign of the Field Joint

Through the last flight of the *Challenger*, application of actuating pressure to the upstream face of the O-ring was essential for proper joint-sealing performance. This was necessary because large sealing gaps were created by pressure-induced deflections, and compounded by significantly reduced O-ring sealing performance at low temperature. After *Challenger* the major change in the motor case is a new tang capture feature that provides a positive metal-to-metal interference fit around the circumference of the tang and clevis ends of the mating segments. The interference fit limits the deflection between the tang and clevis O-ring sealing surfaces caused by motor pressure and structural loads. The joints are designed so that the seals will not leak under twice the expected structural deflection and rate.

The new design, with the tang capture feature, the interference fit, and the use of custom shims between the outer surface of the tang and inner surface of the outer clevis leg, controls the O-ring sealing gap dimension. The sealing gap and the O-ring seals are designed so that a positive compression (squeeze) is always on the O-rings. The minimum and maximum squeeze requirements include the effects of temperature, O-ring resiliency and compression set, and pressure. The clevis O-ring groove dimension

341

has been increased so that the O–ring never fills more than 90 percent of the O–ring groove and pressure actuation is enhanced.

The new field joint design also includes a new O–ring in the capture feature and an additional leak check port to ensure that the primary O–ring is positioned in the proper sealing direction at ignition. This new or third O–ring also serves as a thermal barrier in case the sealed insulation is breached.

The field joint internal case insulation was modified to be sealed with a pressure–actuated flap called a J–seal, rather than with putty as in the space transportation system. Longer field–joint–case mating pins, with a reconfigured retainer band, were added to improve the shear strength of the pins and increase the metal parts' joint margin of safety. The joint safety margins, both thermal and structural, are being demonstrated over the full ranges of ambient temperature, storage compression, grease effect, assembly stresses, and other environments. External heaters with integral weather seals have been incorporated to maintain the joint and O–ring temperature at a minimum of seventy–five degrees Fahrenheit. The weather seal also prevents water intrusion into the joint.

The solid rocket motor case–to–nozzle joint, which experienced several instances of O–ring erosion in flight, has been redesigned to satisfy the same requirements imposed on the case field joint. Similar to the field joint, cast–to–nozzle joint modifications have been made in the metal parts, internal insulation, and O–rings. Radial bolts with Stato–O–Seals were added to minimize the joint sealing gap opening. The internal insulation was modified to be sealed adhesively, and a third O–ring was included. The third O–ring serves as a dam or wiper in front of the primary O–ring to prevent the polysulfide adhesive from being extruded into the primary O–groove. It also serves as a thermal barrier in case the polysulfide adhesive is breached. The polysulfide adhesive replaces the putty used in the 51–L joint. Also, an additional leak check port was added to reduce the amount of trapped air in the joint during the nozzle installation process and to aid in the leak check procedure.

The internal joints of the nozzle metal parts have been redesigned to incorporate redundant and verifiable O–rings at each joint. The nozzle steel fixed housing part has been redesigned to permit the incorporation of the 100 radial bolts that attach the fixed housing to the case's aft dome. Improved bonding techniques are being used for the nozzle nose inlet, cowl/boot, and exit cone assemblies. The distortion of the nose inlet assembly's metal-part-to-ablative-parts bond line has been eliminated by increasing the thickness of the aluminum nose inlet housing and improving the bonding process. The tape-wrap angle of the carbon cloth fabric in the areas of the nose inlet and the throat assembly parts were changed to improve the ablative insulation erosion tolerance. Some of these ply-angle changes were in progress prior to STS 51-L. The cowl and outer boot ring has additional structural support with increased thickness and contour changes to increase their margins of safety. Additionally, the outer boot ring ply configuration was altered.

Minor modifications were made in the case factory joints by increasing the insulation thickness and lay-up to increase the margin of safety on the internal insulation. Longer pins were also added, along with a reconfigured retainer band and new weather seal to improve factory joint performance and increase the margin of safety. Additionally, the O-ring and O-ring groove size was changed to be consistent with the field joint.

Design Analysis Summary

Improved, state-of-the-art analyses related to structural strength, loads, stress, dynamics, fracture mechanics, gas and thermal dynamics, and material characterization and behavior were performed to aid the field joint, nozzle-to-case joint, and other designs. Continuing these analyses will ensure that the design integrity and system compatibility adhere to design requirements and operational use. These analyses will be verified by tests, whose results will be correlated with pretest predictions.

Redesigned SRM certification is based on formally documented results of development motor tests, qualification motor test, and other tests and analyses. The certification tests are conducted under strict control of environments, including thermal and structural loads; assembly, inspection, and test procedures; and safety, reliability, maintainability, and quality assurance surveillance to verify that flight hardware meets the specified performance and design requirements.

Independent Oversight

As recommended in the Presidential Commission Report and at the request of the NASA administrator, the National Research Council established an Independent Oversight Panel chaired by Dr. H. Guyford Stever, who reports directly to the NASA administrator. Initially, the panel was given introductory briefings on the shuttle system requirements, implementation, and control, the original design and manufacturing of the SRM, Mission 51-L accident analyses, and preliminary plans for the redesign. The panel has met with major SRM manufacturers and vendors, and has visited some of their facilities. The panel frequently reviewed the RSRM design criteria, engineering analyses and design, and certification program planning. Panel members continuously review the design and testing for safe operation, selection, and specifications for material, and quality assurance and control. The panel has continued to review the design as it progresses through certification and to review the manufacturing and assembly of the first flight RSRM. Panel members have participated in major program milestones, project requirements review, and preliminary design review; they also will participate in future reviews.

In addition to the NRC, the redesign team has a design review group of twelve expert senior engineers from NASA and the aerospace industry. They have advised on major program decisions and serve as a "sounding board" for the program. Additionally, NASA requested the four other major SRM companies – Aerojet Strategic Propulsion Co., Atlantic Research Corp., Hercules Inc., and United Technologies Corp.'s Chemical Systems Division – to participate in the redesign efforts by critiquing the design approach and providing experience on alternative design approaches.

Appendix E:

Glossary

abort. To cut short or break off an action, operation, or procedure with a vehicle, especially because of equipment failure.

accelerometer. Instrument that measures the rate at which the velocity (speed) of an object is changing.

anisotropic. Material that has different properties in different directions.

axial loads. Force on an object in the direction of the length of the object.

baffle. An obstruction in a combustion chamber used to prevent combustion instability by maintaining uniform mixtures and equalizing pressures; in a fuel tank, used to prevent sloshing by damping propellant oscillation.

cast form. A material (liquid metal, for example) formed into a particular shape by pouring into a mold.

chafing. Wearing away or irritating by rubbing.

cryogenic. Requiring or involving the use of very low temperature.

damping. Suppression of oscillation.

delta wing configuration. An aircraft with swept-back wings that give it the appearance of an isosceles triangle.

expansion ratio. The ratio of the gas pressure in a rocket combustion chamber to the gas pressure at the nozzle outlet.

external tank. The only major component of the shuttle that is not recovered and reused. When the 790 tons of propellants that it carries are consumed, the external tank is jettisoned from the orbiter. It breaks up upon atmospheric entry and, about an hour after lift-off, falls into the Indian or Pacific Ocean. The external tank absorbs the thrust of lift-off because the orbiter and the two solid rocket boosters are attached to it.

FMEA (failure modes and effects analysis). A "bottom-up" approach to risk analysis, which starts at the lowest levels of each subsystem, and determines how each device/part can fail and what the effects and consequences of such a failure on the component and all other interfacing, interacting components would be. The consequences of each identified failure mode are then classified according to its severity.

gimbal. A device with two mutually perpendicular and intersecting axes of rotation, thus giving free angular movement in two directions, on which an engine or other object may be mounted; also, to move a reaction engine about on a gimbal so as to obtain pitching and yawing correction moments.

hot test. A propulsion system test conducted by actually firing the propellants; also called hot firing.

hydrogen embrittlement. Hydrogen gas infuses into the metal alloy, causing alloy to become rigid. Consequently, the alloy will fracture rather easily.

impeller. A device that imparts motion to a fluid; rotating blades of a centrifugal pump.

Inconel (INCO). Trade name for nickel-based heat-resistant alloy.

Mach number. The ratio of the speed of a body with respect to the surrounding fluid to the speed of sound in the medium.

man-rating. Process of assuring that a launch vehicle is safe for manned space flight.

mockup. A full-sized replica, sometimes incorporating actual functioning equipment.

Narloy-Z. A copper alloy used in the SSME main combustion chamber.

orbiter. The "airplane" of the shuttle system and approximately the size of an average airline transport. The major structural components are the pressurized crew compartment (forward fuselage), the payload bay (mid fuselage), the aft fuselage, and the vertical tail. The orbiter's aluminum skin must be protected from the heat generated by friction with the Earth's atmosphere during reentry. This thermal protection system consists of about 30,000 individual ceramic tiles, which are installed on the orbiter.

pneumatic valves. Valves run by or using compressed air.

pogo. Term coined to describe longitudinal dynamic oscillations generated by the interaction of vehicle structural dynamics with propellant and the engine combustion process.

shuttle main engines. Three high-performance rocket engines attached to the aft section of the orbiter; they are in use only for about the first eight and a half minutes of flight. The propellants, liquid hydrogen and oxygen, are carried in the external tank.

solid rocket boosters. Each consists of three components: the nose cone, the solid rocket motor, and the nozzle assembly. A solid rocket motor is composed of four casting segments into which the propellant is poured.

The casting segments are joined with field joints during the final assembly process at Kennedy. The boosters' fuel is exhausted about two minutes after lift-off; the solid rockets are separated from the external tank, drop into the ocean, and are recovered by two specially designed recovery vessels.

spall. Chip or flake off the surface of a metal.

specific impulse. A performance parameter of a rocket propellant, expressed in seconds, equal to the thrust divided by the weight flow rate.

standard operating procedures. Fixed or standardized set of instructions for carrying out a particular task.

superalloy. A complex temperature-resistant material of a homogeneous mixture or solid solution, usually of two or more metals, the atoms of one replacing or occupying interstitial positions between the atoms of the other.

thermal fatigue. Weakness in a material, such as metal, resulting from prolonged stress caused by heat.

thrust. The pushing or pulling force developed by an aircraft engine or a rocket engine.

turbine. A machine consisting principally of one or more turbine wheels (a multivaned rotor rotated by the impulse from or the reaction to a fluid passing across the vanes) and a stator (the stationary casing and blades surrounding an axial-flow compressor or a turbine wheel).

wrought form. A material (metal for example) shaped by hammering with tools.

XLR-129. An experimental reusable liquid oxygen – liquid hydrogen rocket engine developed by Pratt & Whitney for the Air Force development program.

Appendix F

NASA Personnel and Field Centers

Personnel[1]

Abrahamson, James

Associate administrator for Space Flight, NASA, 1984; Air Force general

Beggs, James

Administrator, 1981–5

Coates, Keith

MSFC engineer, 1983–4

Cook, Richard C.

Budget analyst, 1985–6

Crippen, Robert L.

Astronaut; deputy director, Flight Crew Operations, Johnson Space Center; 1985–6; astronaut since 1969

Disher, John H.

Director of advanced programs, Office of Space Transportation Systems, 1960s

Donlan, Charles

Deputy associate administrator for MSFC, 1971

Driver, Orville

Shuttle Projects Office, 1969–86; Main Propulsion Test deputy manager, National Space Technology Laboratory, 1977–86

Eudy, Glenn

Chief engineer, Solid Rocket Division, MSFC, 1977–9

Fletcher, James

Administrator, 1971–7 and 1986–9

Freis, Sylvia

Director, NASA History Office

Frosch, Robert	Administrator, 1977–81
Graham, William	Acting administrator, 1985–6
Hardy, George	Deputy director of science and engineering, MSFC, 1982–6; project manager, Solid Rocket Booster, 1974–82; joined NASA in 1960
Kingsbury, James	Science and Engineering Division, MSFC, 1960–86; Propulsion and Vehicle Engineering Lab, materials division chief, 1966–9; Astronautics Lab deputy director, 1969–73; associate director for engineering, 1973–5; director for engineering, 1975–86
Lovelace, Alan	Deputy administrator, 1977–81
Lovingood, Judson A.	Deputy manager, Marshall Shuttle Projects Office, 1983–6; manager, shuttle main engine project, MSFC, 1982–3; joined NASA in 1962
Low, George	Deputy administrator, 1970
Lunney, Glenn R.	Manager, National Space Transportation Program Office, 1983
Lucas, William R.	Program development director, MSFC, 1969–71; MSFC deputy director, technical, 1971–4; MSFC director, 1974–86
Malkin, M. S.	Director, space shuttle program, 1978
Mann, Michael B.	Chief, Space Transportation System Resources Analysis Branch, Office of Comptroller
Mark, Hans	Deputy administrator, 1981–4
Marshall, Robert	Program Development/Preliminary Design Office, 1969–74; Chief Engineer's Office, MSFC, 1974–9; program development director, 1980–6; Shuttle Projects Office manager, 1986–8

McCool, Alex — Director, MSFC Structures and Propulsion Laboratory, 1977–9

Miller, John Q. — Chief, Solid Rocket Motor, MSFC, 1977–9

Moore, Jesse W. — Associate administrator for space flight, 1985–6.

Mueller, George — Associate administrator for MSFC, 1970–1

Mulloy, Lawrence — Assistant to director, science and engineering, MSFC, 1986; SRB project manager, 1982–6; inertial upper stage project chief engineer, 1980–2; External Tank project chief engineer, 1974–80; Analytical Mechanics Division, system/products chief engineer, 1974; deputy chief, MSFC, 1970–73; joined NASA in 1960

Myers, Dale — Associate administrator for MSFC, 1971–4

Odom, James B. — Shuttle Projects Office, MSFC, 1972–82; External Tank project manager, 1972–83; Space Telescope Office manager, 1983–86; science and engineering director, 1986–8

Paine, Thomas — Administrator, 1970 (resigned September 1970)

Petrone, Rocco — Director, Apollo Program, 1969; associate administrator, 1975, resigned April 1975

Ray, Leon — MSFC engineer, 1977–84

Reinartz, Stan — Manager, MSFC Projects Office, 1985–6; joined NASA in 1960

Rodney, George — Director, Office of Associate Administrator for Safety, Reliability, Maintainability and Quality Assurance, 1986–87

Ryan, Robert S.	Chief, Structures and Dynamics Division, MSFC, 1974-94; joined NASA in 1956
Shapley, Willis	Associate deputy administrator, 1975
Silveira, Milton	NASA Chief Engineer, 1986
Sneed, William	Program planning director, 1969–81; assistant director for policy and review, 1982–8
Spears, Luke	Program Development, Advanced Systems Analysis, deputy director, 1969–73; Payload Studies Office, deputy director, 1973–7; Advanced Systems, deputy director, 1977–86; Advance Projects Office, STS Manager, 1986
Stewart, Frank	FI/HI Engine Projects Manager, 1969–70; Special Missions Office, manager, 1970–72; MSFC deputy manager – management, 1972–3; Shuttle Projects Office, MSFC, 1973–80; SSME project deputy manager, 1973–4; Major Test Management Office, manager, 1974–80
Tischler, A. O.	Chief, Liquid Rocket Engines, 1958–61; assistant director for propulsion, Office of Manned Space Flight, 1961–3; director, Chemical Propulsion Division, Office of Advanced Research and Technology, 1964–9; director Shuttle Technologies Office (OART), 1969–73
Thompson, James R.	Manager, SSME Project, MSFC
Thomson, Jerry	SSME chief engineer, MSFC
Ullian, Louis J.	Range safety officer, Kennedy Space Center, 1985–6.
von Braun, Wernher	Deputy associate administrator, retired July 1972

Willoughby, Will   Head of Reliability and Safety, 1960s–
                                   70s

Winterhalter, David   Acting director, Shuttle Propulsion
                                   Group NASA Headquarters, 1986

Weeks, L. Michael   Deputy associate administrator for
                                   space flight – technical, 1985–6

White, George   Director, quality and reliability, 1970

Yardley, John   Associate administrator for MSFC,
                                   1974

Field Centers

Johnson Space Center   Houston, Texas (management of the
                                   orbiter)

Marshall Space Flight Center   Huntsville, Alabama (orbiter's main
                                   engines, external tank, solid rocket
                                   boosters)

Kennedy Space Center   Merritt Island, Florida (assembly of
                                   shuttle components, conducting
                                   launches)

Note

1. For a number of cases, dates of employment refer to period of time noted in book.

Bibliography

Abbreviations

AIAA	American Institute of Aeronautics and Astronautics
AAAS	American Association for the Advancement of Science
SAE	Society of Automotive Engineers
ASME	American Society of Mechanical Engineers
ASEE	American Society for Engineering Education
IEEE	Institute of Electrical and Electronics Engineers

Letters and Memoranda,
NASA History Office, Washington, D.C.

Grey, Edward Z., Assistant Administrator for Industry Affairs and Technology Utilization, NASA. Letter to Adelbart O. Tischler, July 12, 1974.

Low, George M. Letter to Adelbart O. Tischler, July 25, 1975.
Memorandum to Adelbart O. Tischler, November 13, 1972.
Memorandum to H, Director, Space Cost Evaluation [Tischler], February 20, 1973.

NASA memorandum from EA, Director, Engineering and Development to ES, Chief, Structures and Mechanics Division, "SSME Hot Gas Manifold Materials," June 2, 1975.

NASA memorandum from LA2, Manager for Systems Integration, Space Shuttle Program to SA51, Manager, Main Engine Project, MSFC, "SSME Critical Items List," September 9, 1974.

Tischler, Adelbart O. Letter to Robert C. Seamans, Secretary of Air Force, January 2, 1969.
Letter to Robert Seamans, June 2, 1969.

Letter to John Newbauer, Editor, *AIAA*, June 26, 1969.

Memorandum for the Record, November 14, 1969.

Memorandum to George Low, September 14, 1972.

Thompson, James R., Jr., Manager, SSME Project. Memorandum MSFC NASA to Paul D. Castenholz, Vice-President and SSME Program Manager, Rocketdyne Division of Rockwell International Corp., "SSME Proposed Cost Reduction Items," June 20, 1974.

Wilmotte, Raymond M. "The Structure of Risk Technologies in the Decision Making Process," Memorandum to the Aerospace Safety Advisory Panel, July 1971.

Young, John W. NASA memorandum circulated by Chief Astronaut John Young, "One Part of the 51-L Accident – Space Shuttle Flight Safety," March 4, 1986.

Reports

Baum, Robert J. *Ethics and Engineering Curricula*. Hastings-on-Hudson, NY: Institute of Society, Ethics and the Life Sciences, 1980.

Clouser, K. Danner. *Teaching Bioethics: Strategies, Problems, and Resources*. Hastings-on-Hudson, NY: Hastings Center, 1980.

Committee on Shuttle Criticality Review and Hazard Analysis Audit of the Aeronautics and Space Engineering Board. *Post-Challenger Evaluation of Space Shuttle Risk Assessment and Management*. Washington, DC: National Academy Press, January 1988.

Control Dynamics Company. *Analysis of SSME HPOTP Rotordynamics Subsynchronous Whirl*. Final report, Huntsville, AL, January 24, 1983.

Covert, Eugene E. *Technical Status of the Space Shuttle Main Engine: A Report of the Ad Hoc Committee for Review of the Space Shuttle Main Engine Development Program*. Washington, DC: Assembly of Engineering, National Research Council, National Academy of Sciences, March 1978.

Kustas, F. M., and M. S. Misra. *Ion Implantation of 440C RCF Test Specimens*. Martin Marietta Aerospace, Denver, CO, Final Report for Contract NAS 8-35055, July 1983.

Ladenson, Robert F., James Choromokos, Ernest d'Anjou, Martin Pimsler, and Howard Rosen. *A Selected Annotated Bibliography of Professional Ethics and Social Responsibility in Engineering*. Chicago: Center for the Study of Ethics in the Professions, Illinois Institute of Technology, 1980.

Matejczyk, D. E. "Retardation Analytical Model to Extend Service Life," monthly technical progress narrative. Rockwell International, Canoga Park, CA, October 8, 1985.

Nagy, Peter, Ken Seitz, and Kim Bowen. "Evaluation of Tailored Single Crystal Airfoils, Final Report." Williams International, Walled Lake, MI, March 1986.

Pritchard, Michael S. "Teaching Engineering Ethics: A Case Study Approach." Paper presented at the Illinois Institute of Technology, Chicago, June 12–13, 1990.

Robinson, J. K., G. A. Teal, and C. T. Welch. "Space Shuttle Main Engine, Powerhead Structural Modeling, Stress and Fatigue Life Analysis." Lockheed Missiles & Space Company, Huntsville Research and Engineering Center, vol. 4, December 1983.

Taniguchi, M. H. "Failure Control Techniques for the SSME, NAS836305, Phase I Final Report." Rockwell International, Canoga Park, CA, RI/RD86-165 (revised), undated.

Vroman, G. A. "Validation Testing of Shallow-Notched Round-Bar Screening Test Specimens." Rockwell International, Canoga Park, CA, R-9743, June 6, 1975.

Weatherwax, R. K., and C. W. Colglazier. *Review of Shuttle/Centaur Failure Probability Estimates for Space Nuclear Mission Applications.* Teledyne Energy Systems, Timonium, MD, AFWL-TR-83-61, December 1983.

Weil, Vivian. *Beyond Whistleblowing: Defining Engineers' Responsibilities.* Chicago: Center for the Study of Ethics in the Professions, Illinois Institute of Technology, 1984.

"Engineering Ethics in Engineering Education: Report of a Conference," June 12–13, 1990. Illinois Institute of Technology, Chicago, May 1992, p. i.

ed. *Report of the Workshops on Ethical Issues in Engineering.* Proceedings of the Second National Conference on Ethics. Chicago: Center for the Study of Ethics in the Professions, Illinois Institute of Technology, 1979.

Whitbeck, C. "The Engineer's Responsibility for Safety: Integrating Ethics Teaching into Courses in Engineering Design." Presented at the ASME Winter Annual Meeting, Boston, 1987.

NASA Reports and Technical Documents

Aerospace Safety Advisory Panel. *Annual Report.* Washington, DC: NASA, March 1992.

Aerospace Safety Advisory Panel. *Annual Report*. Washington, DC: NASA, March 1993.

Aerospace Safety Advisory Panel. *Annual Report*. Washington, DC: NASA, March 1994.

Allaway, Howard. *The Space Shuttle at Work*. Washington, DC: NASA Division of Public Affairs, 1979.

Bhat, B. N. "Fracture Analysis of HPOTP Bearing Balls." MSFC, NASA TM-82428, May 1981.

Bhat, B. N., and F. J. Dolan, "Past Performance Analysis of HPOTP Bearings." NASA Technical Memorandum, March 1982.

Boyce, Lola. "Probabilistic Analysis for Fatigue Strength Degradation of Materials." NASA Grant, January 1989.

Chen, S. S., J. A. Jendrzejczyk, and M. W. Wambsganss, "Flow-Induced Vibration of SSME Injector Liquid-Oxygen Posts." Argone National Laboratory, NASA Scientific and Technical Facility N85-27951.

Ezell, Linda Neuman. *NASA Historical Data Book*. Vol. III. Washington, DC: Scientific and Technical Information Division, NASA, 1988.

Fries, Sylvia. "NASA: Safety Organization and Procedures in Manned Space Flight Programs." NASA History Office, March 24, 1986. Unpublished manuscript.

Glover, R. C., B. A. Kelly, and A. E. Tischer. *Studies and Analyses of the Space Shuttle Main Engine: SSME Failure Data Review, Diagnostic Survey and SSME Diagnostic Evaluation*. BCD-SSME-TR-86-1, Contract No. NASA-3737. Columbus, OH: Battelle, December 15, 1986.

Hansen, James R. *Engineer in Charge*. Washington, DC: NASA Scientific and Technological Information Office, 1987.

Kaufman, Albert, and Jane M. Manderscheid. "Cyclic Structural Analyses of SSME Turbine Blades." Lewis Research Center, Cleveland, N8527963, no date.

"Simplified Cyclic Structural Analyses of SSME Turbine Blades." In *Advanced Earth-to-Orbit Propulsion Technology 1986*, edited by R. J. Richmond and S. T. Wu, vol. 2, pp. 107–24. NASA Conference Publication 2437. Huntsville , AL, 1986.

Kielb, Robert E., and Jerry H. Griffin. "SSME Blade Damper Technology." NASA Technical Report N87-22798. Cleveland, no date.

Kukowski, James. "Second Test Firing of Challenger Engines Set for January 25." NASA News Release No. 83-4, January 20, 1983.

Levine, Arnold S. *Managing NASA in the Apollo Era*. NASA History Series. Washington, DC: NASA, 1982.

Logsdon, John M. "Opportunities for Policy Historians: The Evolution of the US Civilian Space Program." In *A Spacefaring People: Perspectives on Early Spaceflight*, edited by Alex Roland, pp. 81–107. NASA History Series. Washington, DC: NASA, 1985.

Marshall Space Flight Center. *Chronology: MSFC Space Shuttle Program: Development, Assembly and Testing, Major Events*. Huntsville, AL, 1988.

McCarty, John P., and Byron K. Wood. "Space Shuttle Main Engine: Interactive Design Challenges." In *Space Shuttle Technical Conference Papers*, edited by Norman Chaffee, pp. 600–18. Washington, DC: NASA Scientific and Technical Information Branch, 1985.

News Article M&R, May 18, 1964. NASA History Office.

Richmond, R. J., and S. T. Wu, eds. *Advanced Earth-to-Orbit Propulsion Technology 1986*. Vol. 2. NASA Conference Publication 2437. Huntsville, AL, 1986.

Roland, Alex. *A Guide to Research in NASA History*. Washington, DC: NASA History Office, NASA Headquarters, 1983.

A Spacefaring People: Perspectives on Early Spaceflight. NASA History Series. Washington, DC: NASA, 1985.

Ryan, Robert S. "Practices in Adequate Structural Design." NASA Technical Paper 2893, George C. Marshall Flight Center, Huntsville, AL, 1989.

Ryan, Robert S. "The Role of Failure/Problems in Engineering: A Commentary on Failures Experienced – Lessons Learned." NASA Technical Paper 3213. George C. Marshall Flight Center, Huntsville, AL, March 1992.

Sanchini, Dominick J. *The Space Shuttle Main Engine*. A74-42369. New York: American Institute of Aeronautics and Astronautics, 1974.

Space and Shuttle News Reference. NASA. Washington, DC: U.S. Government Printing Office, 1981-341-570/3256.

Space Shuttle Data for Planetary Mission RTG Safety Analysis. Johnson Space Center, Houston, February 15, 1985.

"Space Shuttle Engine Negotiations." NASA News Release No. 71-131. July 13, 1971. NASA History Office, Washington, DC.

"Space Shuttle Engine RFP." NASA News Release No. 70-26. February 18, 1970. NASA History Office, Washington, DC.

"Space Shuttle Main Engine – Interactive Design Challenges." NASA
    Contract Publication 2342, pt. 2, 1985.
Swenson, Lloyd S. *This New Ocean: A History of Project Mercury.* Wash-
    ington, DC: NASA, 1966.
Wheeler, John T. "Statistical Analysis of 59 Inspected SSME HPFTP Tur-
    bine Blades (uncracked and cracked)." NASA technical memoran-
    dum, January 1987.
Williams, Walter C. *Report of the SSME Assessment Team.* Washington, DC:
    NASA, January 1993.

NASA Taped Interviews

Management Operations Office, Marshall Space Flight Center. *Oral Inter-
    views: Space Shuttle History Project.* Compiled by Jessie E. Whalen
    and Sarah L. McKiney, Transcript Collection, December 1988.
    Huntsville, AL (Contract No. NAS8-35900):
    Brown, Richard L., pp. 233–60.
    Driver, Orville, pp. 1–18.
    Kingsbury, James, pp. 434–49.
    Marshall, Robert, pp. 108–54.
    Mulloy, Lawrence, pp. 180–203.
    Odom, James B., pp. 75–94.
    Sneed, Williams, pp. 155–79.
    Stewart, Frank, pp. 56–74.

Other Government Documents

Reactor Safety Study. *An Assessment of Accident Risks in U.S. Commercial
    Nuclear Power Plants.* WASH-1400 (NUREG-75/014). Nuclear Reg-
    ulatory Commission, October 1975.
Rogers, William P. *Report of the Presidential Commission on the Space Shuttle
    Challenger Accident.* Vols. 1–5. Washington, DC: Government Print-
    ing Office, June 6, 1986.
Stafford, Thomas P., Lt. General USAF (Ret.), Statement to the House
    Science, Space and Technology Committee; Space Subcommittee on
    the Future of the U.S. Space Launch Capability, February 17, 1993.
U.S. Congress. House. Committee on Science and Astronautics. *Space
    Shuttle-Skylab Manned Space Flight in the 1970's: Status Report for*

the Subcommittee on NASA Oversight. 92d Congress, 2d session, Serial N, January 1972.

Committee on Science and Astronautics. *1973 NASA Authorization.* 92d Congress, 2d session, H.R. 12824, No. 15, Part 1, February 8, March 12, 1972.

Committee on Science and Astronautics. *Space Shuttle, Space Tug, Apollo-Soyuz, Test Project-1974: Status Report.* 93rd Congress, 2d session, Serial K, February 1974.

Committee on Science and Astronautics. *1975 NASA Authorization Hearings before the Subcommittee on Space Science and Applications,* Vol. I, Part 1, 93d Congress, 2d session, 1974.

Committee on Science, Space, and Technology. *NASA Multiyear Authorization Act of 1992.* Report No. 102-500, April 22, 1992; February 9, 1993.

Committee on Science and Technology. *Space Shuttle 1975: Status Report.* 94th Congress, 1st session, Serial B, February 1975.

Subcommittee on Manned Space Flight of the Committee on Science and Astronautics. *1975 NASA Authorization Hearings.* 93rd Congress, 2d session, 1974.

Subcommittee on Space Science and Applications of the Committee on Science and Technology. 94th Congress, 1st session, H.R. 2931 [No. 2], Vol. II, Part 3, February 19–20, 1975. Serial 49-296 O, p. 1905.

Subcommittee on Space Science and Applications of the Committee on Science and Technology. *1976 NASA Authorization.* 94th Congress, 1st session, H.R. 2931, No. 2, Vol. II, Part 3, February 19–20, 1975.

Subcommittee on Space Science and Applications of the Committee on Science and Technology. *Space Shuttle 1976: Status Report.* 94th Congress, 1st session, October 1975. Serial N.

Subcommittee on Space Science and Applications of the Committee on Science and Technology. *1977 NASA Authorization.* 94th Congress, 1st and 2d sessions, December 1975. Serial 666.

Subcommittee on Space Science and Applications of the Committee on Science and Technology. 94th Congress, 2nd session, H.R. 11573, No. 63, Vol. I, Part 3, January 28, 29, February 3–5, 13, 14, 17, 19, 1976.

Subcommittee on Space Science and Applications of the Committee on Science and Technology. *1978 NASA Authorization.* 95th Congress, 1st session, H.R. 2221, No. 10, Vol. I, Part 2, February 2–7, 9, 1977.

Subcommittee on Space Science and Applications of the Committee on Science and Technology. *1978 NASA Authorization.* 95th Congress,

1st session, H.R. 2221, No. 11, Vol. I, Part 3, February 16, 17, 23, March 4–7, 1977.

Subcommittee on Space Science and Applications of the Committee on Science and Technology. *Space Shuttle 1978: Status Report.* 95th Congress, 2nd session, January 1978. Serial U.

Subcommittee on Space Science and Applications of the Committee on Science and Technology. *Space Shuttle 1979: Status Report.* 95th Congress, 2nd session, December 1978. Serial 22.

Subcommittee on Space Science and Applications of the Committee on Science and Technology. *Oversight: Space Shuttle Cost, Performance, and Schedule Review.* 96th Congress, 1st session, June 28, 1979. Serial 31, 50365 O.

Subcommittee on Space Science and Applications of the Committee on Science and Technology. *Oversight: Space Shuttle Cost, Performance, and Schedule Review.* 96th Congress, 1st session, August 1979, serial U, 49-320 O.

Subcommittee on Space Science and Applications of the Committee on Science and Technology. *1980 NASA Authorization (Program Review).* 95th Congress, 2d session, No. 100, Vol. I, Part 1, September 25—7, 1978.

Subcommittee on Space Science and Applications of the Committee on Science and Technology. *1980 NASA Authorization.* 96th Congress, 1st session, H.R. 1756, No. 21, Vol. I, Part 4, February 15, 21, 28, March 9, 12, 1979.

Subcommittee on Space Science and Applications of the Committee on Science and Technology. 96th Congress, 1st session, H.R. 1786, No. 11, Vol. I, Part 2, February 6, 1979.

Subcommittee on Space Science and Applications of the Committee on Science and Technology. *Space Shuttle 1980: Status Report.* 96th Congress, 2d session, January 1980. Serial AA.

Subcommittee on Space Science and Applications of the Committee on Science and Technology. 96th Congress, 2d session, H.R. 6413, No. 111, Vol. IV, February 5–7, 11, 14–16, 18, 1980.

Subcommittee on Space Science and Applications of the Committee on Science and Technology. *1982 NASA Authorization (Program Review).* 96th Congress, 2d session, No. 164, Vol. II, September 16–18, 1980.

Subcommittee on Space Science and Applications of the Committee on Science and Technology. *1982 NASA Authorization.* 97th Congress, 1st session, No. 7, Vol. IV, January 28, February 20, 23, 27, March 2, 4, 5, 10–11, 1981.

Subcommittee on Space Science and Applications of the Committee on Science and Technology. *1984 NASA Authorization*. 97th Congress, 1st session, No. 17, Vol. II, February 3, 8, 10, 22–6, 28, March 1–3, 1983.

U.S. Congress. Senate. Committee on Aeronautical and Space Sciences. *NASA Authorization for Fiscal Year 1975*. 93d Congress, 2d session, 1974.

Committee on Appropriations. *Department of Housing and Urban Development, Space, Science, Veterans, and Certain Other Independent Agencies Appropriations for Fiscal Year 1973*. 92d Congress, 2d session, H.R. 15043, 1972.

Committee on Appropriations. *Department of Housing and Urban Development, Space, Science, Veterans, and Certain Other Independent Agencies Appropriations for Fiscal Year 1975*. 93d Congress, 2d session, 1974.

Subcommittee on Science, Technology and Space of the Committee on Commerce, Science, and Transportation. *Report of the National Research Council's Ad Hoc Committee for Review of the Space Shuttle Main Engine Development Program*. 95th Congress, 2nd session, March 31, 1978. Serial 95-78, 28-027 O.

Subcommittee on Science, Technology and Space of the Committee on Commerce, Science, and Transportation. *NASA Authorization for Fiscal Year 1980, S. 357*. 96th Congress, 1st session, February 21, 22, 28, 1979, Part 2. Serial 96-1, 43-135 O.

Papers

Baum, Robert J. *Ethics and Engineering Curricula*. Hastings-on-Hudson, NY: Institute of Society, Ethics and the Life Sciences, 1980.

Bella, David A. "Ethics, Values and Organizations: Beyond Functional Behavior." In *American Association for the Advancement of Science Annual Meeting, San Francisco, CA, January, 1989*. Washington, DC: AAAS, 1989.

Bilardo, Izquierdo, and Smith. "Development of the Helium Signature Test for Orbiter Main Propulsion System Revalidation between Flights." New York: AIAA, 1987.

Cannon I., A. Norman, and M. Olsaky. "Application of SSME Launch Processing Lessons Learned to Second Generation Reusable Rocket Engines Including Condition Monitoring." In *AIAA/ASME/SAE/*

ASEE 24th Joint Propulsion Conference, Boston, July 11–13, 1988. New York: AIAA, 1988.

Cikanek, Harry A., III (MSFC). "Characteristics of SSME Failures." Paper presented at the AIAA/SAE/ASME/ASEE 23d Joint Propulsion Conference, San Diego, June 29–July 2, 1987.

Colbo, H. I. "Development of the Space Shuttle Main Engine." Paper presented at the AIAA/SAE/ASME 15th Joint Propulsion Conference, New York, June 18–20, 1979.

Gibson, H. M. "Rocket Engine Design and Development Methodology." In *AIAA/SAE/ASME/ASEE, 21st Joint Propulsion Conference, July*. New York and Monterey, CA: AIAA, 1985.

Koen, Billy V. "Toward a Definition of the Engineering Method." In *Engineering Education*, December 1984: 150–5.

Whitbeck, C. "The Engineer's Responsibility for Safety: Integrating Ethics Teaching into Courses in Engineering Design." Presented at the ASME Winter Annual Meeting, Boston, 1987.

White, G. C., Jr. "Apollo Quality through Predictive Testing." *Testing for Prediction of Material Performance in Structures and Components*. Philadelphia: STM, 1972.

Wilmotte, Raymond M. "An Introduction to the Basic Problem of Risk Management." Paper presented to the Aerospace Safety Advisory Panel, July 6, 1971.

Newspaper and Magazine Articles

"Aerojet/NR/UAC Bid for Shuttle Main Engine." *Space Business Daily* 49(20); March 27, 1970: 129.

Alexander, Tom. "The Unexpected Payoff of Project Apollo." *Fortune*, July 1969.

Biddle, Wayne. "Crippling of 2d Space Shuttle Tied to Design Flaw." *New York Times*, March 2, 1983, pp. A1, A24.

Broad, William J. "High Risk of New Shuttle Disaster Leads NASA to Consider Options." *New York Times*, April 9, 1989.

"Booster Rockets in Shuttle Fleet Cause Concern." *New York Times*, December 4, 1993, pp. 1, 6.

"Some Feared Mirror Flaws Even before Hubble Orbit." *New York Times*, December 7, 1993, pp. B7, B11.

Broome, Taft H., Jr. "The Slippery Ethics of Engineering." *Washington Post*, December 28, 1986.

Cook, Richard C. "Why I Blew the Whistle on NASA's O-ring Woes." *Washington Post*, March 16, 1986.

"Costly Slowdown for Space Shuttle." *U.S. News & World Report*, February 7, 1983.

Davis, Bob. "Bush Fires NASA Administrator Truly after Dispute over Management, Policy." *Wall Street Journal*, February 13, 1992, p. B10.

Diamond, Stuart. "NASA Cut or Delayed Safety Spending." *New York Times*, April 23, 1986.

"NASA Wasted Billions, Federal Audits Disclose." *New York Times*, April 23, 1986.

"Engine Repairs Delay Launch of Challenger." *Washington Post*, March 2, 1983, p. A5.

Fairlie, Henry. "Fear of Living." *The New Republic* 14, 1989: 15–17, 19.

"GAO Asked to Investigate Space Shuttle Engine Contract." *Space Business Daily* 57(16): July 26, 1971: 106.

Jones, Alex S. "Journalists Say NASA's Reticence Forced Them to Gather Data Elsewhere." *New York Times*, February 9, 1986.

"Limited Shuttle Engine Effort Set Pending GAO Review." *Space Business Daily* 58(1); September 9, 1971: 2.

Lyons, "Propriety of Space Shuttle Contract Is Questioned." *New York Times*, August 22, 1971, sect. 5, p. 24.

Masley, Peter. "Space Shuttle Contract: A Techno-Political Snag." *Washington Post*, January 16, 1972, pp. A1, A15.

"NR Challenges P & W Engine Experience Claim." *Space Daily* 59, November 3, 1971, pp. 16–17.

"NR's Rocketdyne Picked for Shuttle Main Engine." *Space Business Daily* 57(8); July 14, 1971: 52.

O'Toole, Thomas. "Space Shuttle Friction Is Worsened." *Washington Post*, March 16, 1983, p. A21.

"P & W Files Formal Protest on Shuttle Engine Award." *Space Business Daily* 57(34), August 19, 1971: 230.

Sanger, David E. "How See-No-Evil Doomed Challenger." *New York Times*, June 29, 1986.

"Shuttle Engine Deliveries to Begin in 1975." *Space Business Daily* 57(9), July 15, 1971; 60.

"Space Shuttle Struggle: NASA Award to Rocketdyne Is Challenged." *Washington Post*, August 8, 1971, p. F1.

Warden, Philip. "Ask Report on Space Shuttle Pact." *Chicago Tribune*, August 27, 1971.

Books

Bayless, Michael. *Professional Ethics.* 2d ed. Bellmont, CA: Wadsworth, 1988.

Beauchamp, Thomas L., and James F. Childress. *Principles of Biomedical Ethics.* 4th ed. New York: Oxford University Press, 1994.

Beauchamp, Thomas L., and James F. Childress, *Principles of Biomedical Ethics.* 3d ed. New York: Oxford University Press, 1989.

Beauchamp, Thomas L., and Laurence B. McCollough. *Medical Ethics: The Moral Responsibilities of Physicians.* Englewood Cliffs, NJ: Prentice-Hall, 1984.

Bosk, Charles. *Forgive and Remember: Managing Medical Failure.* Chicago: University of Chicago Press. 1978.

Brock, D., and A. Buchanan. *Deciding for Others: The Ethics of Surrogate Decision Making.* Cambridge: Cambridge University Press, 1989.

Brody, Baruch. *Life and Death Decisionmaking.* Oxford: Oxford University Press, 1988.
  Moral Theory and Moral Judgements in Medical Ethics. Dordrecht: Kluwer Academic Publishers, 1988.

Burke, John G. "Technology and Government." In *Technology and Social Change in America*, edited by E. T. Layton Jr. New York: Harper & Row, 1973.

Burke, John P. *Bureaucratic Responsibility.* Baltimore: Johns Hopkins University Press, 1986.

Calabresi, Guido, and Philip Bobbitt. *Tragic Choices: The Conflicts Society Confronts in the Allocation of Tragically Scarce Resources.* New York: W. W. Norton, 1978.

Chaisson, Eric J. *The Hubble Wars: Astrophysics Meets Astropolitics in the Two-Billion-Dollar Struggle over the Hubble Space Telescope.* New York: Harper Collins, 1994.

Churchill, Larry R. *Rationing Healthcare in America: Perceptions and Principles of Justice.* South Bend, IN: University of Notre Dame Press, 1987.

Collins, Michael. *Liftoff.* New York: Grove Press, 1988.

Cullen, Francis T., William J. Maakestad, and Gray Cavender. *Corporate Crime Under Attack: The Ford Pinto Case and Beyond.* Cincinnati, Ohio: Anderson. 1987.

Drane, James F. *Clinical Bioethics: Theory and Practice in Medical Ethical Decision Making.* Kansas City, MO: Sheed & Ward, 1994.

Ehrenrich, Barbara, and John Ehrenrich. *American Health Empire: Power, Profits and Politics.* New York: Vintage Books, 1978.

Feynman, Richard P. *What Do You Care What Other People Think? Further Adventures of a Curious Character,* as told to Ralph Leighton. New York: W. W. Norton, 1988.

Flores, Albert, ed. *Designing for Safety: Engineering Ethics in Organizational Contexts.* Troy, NY: Rensselaer Polytechnic Institute, 1982.

Flores, Albert, and Robert J. Baum, eds. *Ethical Problems in Engineering.* 2d ed., Troy, NY: Center for the Study of the Human Dimensions of Science and Technology, Rensselaer Polytechnic Institute, 1980.

Florman, Samuel C. "Moral Blueprints: On Regulating the Ethics of Engineers. Harper's." In *Blaming Technology,* pp. 162–80. New York: St. Martin's Press, 1981.

French, Peter A. *Responsibility Matters.* Lawrence: University Press of Kansas, 1992.

Fulford, K. W. M. *Moral Theory and Medical Practice.* Cambridge: Cambridge University Press, 1989.

Gerth, H. H., and C. Wright Mills, eds. *Max Weber: Essays in Sociology.* New York: Oxford University Press, 1946.

Gillroy, John M., and Maurice Wade, eds. *The Moral Dimensions of Public Policy Choice: Beyond the Market Paradigm.* Pittsburgh, PA: University of Pittsburgh Press, 1992.

Harris, Charles E., Jr., Michael Pritchard, and Michael J. Rabins. *Engineering Ethics: Concepts and Cases.* New York: Wadsworth Publishing, 1995.

Hoffmaster, Barry, Benjamin Freedman, and Gwenn Fraser, eds. *Clinical Ethics: Theory and Practice.* Clifton, NJ: Humana Press, 1988.

Joint Commission for the Accreditation of Health Care Organizations. *Accreditation Manual for Hospitals.* 1992; Oakbrook Terrace, IL, 1995.

Albert R. Jonsen, and Stephen Toulmin. *The Abuse of Casuistry: A History of Moral Reasoning.* Berkeley: University of California Press, 1988.

Kamm, Lawrence J. *Successful Engineering: A Guide to Achieving Your Career Goals.* New York: McGraw-Hill, 1989.

King, N. M., L. R. Churchill, and A. Cross. *The Physician as Captain of the Ship: A Critical Reappraisal.* Boston: Reidel Publishers, 1986.

Kjellstrand, C. M., and J. B. Dossetor, eds. *Ethical Problems in Dialysis and Transplantation.* London: Kluwer Academic Publishers, 1992.

Kuhn, Thomas S. *The Essential Tension: Selected Studies in Scientific Tradition and Change.* Chicago: University of Chicago Press, 1977.

Lawrence, Paul R., and Jay W. Lorsch. *Organization and Environment*. Cambridge, MA: Harvard University Press, 1967.

Layton, Edwin T., Jr. *The Revolt of the Engineers: Social Responsibility and the American Engineering Profession*. Baltimore: Johns Hopkins University Press, 1986.

Lewis, Richard S. *The Voyages of Columbia: The First True Spaceship*. New York: Columbia University Press, 1984.

Mark, Hans. "The Challenger and Chernobyl." In *Traditional Moral Values in the Age of Technology*, edited by W. Lawson Taitte. Andrew R. Cecil Lectures on Moral Values in a Free Society. Austin: University of Texas Press. 1987.

   The Space Station: A Personal Journey. Durham, NC: Duke University Press, 1987.

Markowitz, Gerald E., and David Rosner. "Doctors in Crisis: Medical Education and Medical Reform During the Progressive Era, 1985–1915." In *Health Care in America: Essays in Social History*, edited by Susan Reverby and David Rosner, pp. 183-5. Philadelphia: Temple University Press, 1979.

Martin, Mike W., and Roland Schinzinger. *Ethics in Engineering*. 2d ed. New York: McGraw-Hill, 1989.

McConnell, Malcolm. *Challenger: A Major Malfunction*. Garden City, NY: Doubleday, 1987.

McCurdy, Howard E. *The Space Station Decision: Incremental Politics and Technological Choice*. Baltimore: Johns Hopkins University Press, 1990.

   Inside NASA: High Technology and Organizational Change in the American Space Program. Baltimore: Johns Hopkins University Press, 1993.

McDougall, Walter A. *The Heavens and the Earth: A Political History of the Space Age*. New York: Basic Books, 1985.

Meisel, Alan. *The Right to Die*. New York: Wiley, 1989; suppl. 1994.

Mohler, Peter Ph., and Cornelia Zuell. *TEXTPACK, Version V*. Mannheim: ZUMA: August, 1982.

Munson, Ronald, ed. *Intervention and Reflection: Basic Issues in Medical Ethics*. 4th ed. Belmont, CA: Wadsworth, 1992.

Murray, Charles, and Catherine B. Cox. *Apollo: The Race to the Moon*. New York: Simon and Schuster, 1989.

Nagel, Thomas. *Mortal Choices*. Cambridge: Cambridge University Press, 1983.

Noble, David. *America by Design: Science, Technology and the Rise of Corporate Capitalism*. NY: Knopf, 1977.

Perrow, Charles. *Normal Accidents: Living with High-Risk Technologies.* New York: Basic Books, 1984.

Complex Organizations: A Critical Essay. 3d ed. New York: McGraw-Hill, 1986.

Petroski, Henry. *To Engineer Is Human: The Role of Failure in Successful Design.* New York: St. Martin's Press, 1985.

Powell, Walter W., and Paul J. DiMaggio, eds. *The New Institutionalism in Organizational Analysis.* Chicago: University of Chicago Press, 1991.

Ross, J. W., John W. Glaser, D. Rasinski-Gregory, Joan M. Gibson, and Corine Bayley. *Health Care Ethics Committees: The Next Generation.* Chicago: American Hospital Publishing, 1993.

Schaub, James H., and Karl Pavlovic. *Engineering Professionalism and Ethics.* New York: Wiley, 1983.

Rothman, David J. Strangers at the Bedside: A History of How Law and Bioethics Transformed Medical Decision Making. New York: Basic Books, 1991.

Sidel, Phil. *KWX Users Manual.* Pittsburgh, PA: Social Science Computer Research Institute, University of Pittsburgh, 1987.

Simon, Herbert A. *Administrative Behavior.* 3d ed. New York: Free Press, 1976.

Sinclair, Bruce, with the assistance of James P. Hull. *A Centennial History of the American Society of Mechanical Engineers, 1886–1980.* Toronto: University of Toronto Press, 1980.

Smith, Robert W. *The Space Telescope: A Study of NASA, Science, Technology and Politics.* Cambridge: Cambridge University Press, 1989.

Thompson, Dennis F. *Political Ethics and Public Office.* Cambridge, MA: Harvard University Press, 1987.

Thompson, James, *Organizations in Action.* New York: McGraw-Hill, 1967.

Trento, Joseph J. *Prescription for Disaster.* New York: Crown, 1987.

Unger, Stephen H. *Controlling Technology: Ethics and the Responsible Engineer.* New York: Holt, Rhinehart and Winston, 1982.

Veatch, Robert M. *A Theory of Medical Ethics.* New York: Basic Books, 1981.

Weber, Max. *Theory of Social and Economic Organization.* Translated by A. M. Henderson and Talcott Parson. New York: Oxford University Press, 1947.

Weil, Vivian, ed. *Moral Issues in Engineering, Selected Readings.* Chicago: Illinois Institute of Technology, 1989.

Weinberg, Alvin M. "Impact of Large Scale Science." In *Reflections on Big Science*, pp. 161–4. Cambridge, MA: MIT Press, 1967.

Wood, M. Sandra, and Suzanne Shultz. *Three Mile Island: A Selectively Annotated Bibliography.* New York: Greenwood Press, 1988.

Journal Articles

American Health Consultants. "Managed Care Brings a Demand for Institutional Ethics Policies." *Medical Ethics Advisor* 10(10), 1994: 125–8.

Anderton, David. "J-2, M-1 Engine Design Details Reported." *Aviation Week and Space Technology* 78, May 6, 1963: 56–9.

Arnold, Greg, and Clyde Jones. "Welding the Space Shuttle Engine." *Robotics Today,* October 1986: 13–16.

Arras, John D. "Getting Down to Cases: The Revival of Casuistry in Bioethics." *Journal of Medicine and Philosophy* 16(1), 1991: 29–51.

Beauchamp, T. "The Principles Approach." *Hastings Center Report,* special suppl. November–December 1993: 59.

  "Principlism and Its Alleged Competitiors." *Kennedy Institute of Ethics Journal* 5 (3), September 1995: 181–98.

Beecher, Henry K. "Ethics and Clinical Research." *New England Journal of Medicine* 274(24), 1966: 1354–68.

Bell, Trudy E. "Managing Murphy's Law: Engineering a Minimum-risk System." *IEEE Spectrum* 26, June 1989: 23–5.

Bell, Trudy E., and Karl Esch, "The Space Shuttle: A Case of Subjective Engineering." *IEEE Spectrum* 26, June 1989: 42–5.

Broome, Taft H., Jr. "Engineering the Philosophy of Science." *Metaphilosophy* January 16(1), 1985: 47–57.

  "Engineering Responsibility for Hazardous Technologies." *Journal of Professional Responsibility in Engineering* 113(2), 1987: 139–49.

  "Can Engineers Hold Public Interest Paramount." *Research in Philosophy and Technology* 9 (1989): 3–11.

Bulkeley, Debra. "The Making of a Hero." *Design News,* February 15, 1988: 86–92.

Cambel, Ali Bulent, and Sandra A. Schuch. "Proposed: An Academic Program in Engineering Ethics." *Engineering Education,* April 1989: 413–4.

Caplan, Arthur L. "Ethical Engineers Need Not Apply: The State of Applied Ethics Today." *Science Technology and Human Values* 33, 1980: 32–74.

Cassel, Christine K. "Deciding to Forego Life-Sustaining Treatment: Implications for Policy." *Cardoso Law Review* 6(2), 1984: 268–87.

Churchill, Larry R. "Tacit Components of Medical Ethics: Making Decisions in the Clinic." *Journal of Medical Ethics* 3, 1977: 129—32.

"Reviving a Distinctive Medical Ethic." *Hastings Center Report* 9(3), 1989: 28–34.

Churchill, Larry R., and R. L. Pinkus, "The Use of Anencephalic Organ Donors Historical and Ethical Dimensions." *Milbank Quarterly* 68(2), 1990: 147–69.

Clouser, K. Danner. "Medical Ethics: Some Uses, Abuses, and Limitations." *New England Journal of Medicine* 293(8), 1975: 384–7.

"Bioethics and Philosophy." *Hastings Center Report* 23(6), special suppl. November–December 1993: 510–11.

Clouser, K. Danner, and B. Gert. "A Critique of Principlism." *Journal of Medicine and Philosophy* 15, April 1990: 219–36.

"Common Morality as an Alternative to Principlism." *Kennedy Institute of Ethics Journal* 5(3), September 1995: 219–37.

Covault, Craig. "New Challenger Engine Cracks Found." *Aviation Week and Space Technology*, March 7, 1983: 23–5.

"Shuttle Engines Spark Broad Review." *Aviation Week and Space Technology*, February 28, 1983: 18–20.

Cruse, Thomas A. "Designing for Uncertainty." *Aerospace America* 26, November 1988: 36–9.

Dankhoff, Walter, Paul Herr, and Melvin C. McIlwain. "Space Shuttle Main Engine (SSME) – The Maturing Process." *Astronautics and Aronautics*, January 1983: 26–32, 49.

Davis, Michael. "Thinking Like an Engineer: The Place of a Code of Ethics in the Practice of a Profession." *Philosophy and Public Affairs*. 20(2), 1991: 150–67.

Fries, Sylvia. "2001 to 1994: Political Environment and the Design of NASA's Space Station System." *Technology and Culture* 29, July 1988: 568–93.

Goldberg, Steven. "The Space Shuttle Tragedy and the Ethics of Engineering." *Jurimetrics Journal* 27(2), 1987: 155–9.

Gioia, Dennis A. "Pinto Fires and Personal Ethics: A Script Analysis of Missed Opportunities." *Journal of Business Ethics* 11(5–6), May 1992: 379–89.

Hart, David. "Conference Report: Science, Technology, and Individual Responsibility." 4th Biennial Student Pugwash International Conference. *Science, Technology, and Human Values* 11(2), 1986: 63–7.

Hieronymus, William S. "$1-Billion Shuttle Engine Program Seen." *Aviation Week and Space Technology* 94, June 21, 1971: 60–3.

Hollander, Rachelle. "Conference Report: Engineering Ethics." *Science, Technology, and Human Values* 8(1), 1983: 25–9.

Johnson, Debra. "The Social/Professional Responsibility of Engineers." *Annals of the New York Academy of Sciences* 577, 1989: 106–14.

Johnson, Katherine. "Shuttle Engine Speeded by GAO Decision." *Aviation Week and Space Technology*, April 10, 1972: 14–5.

Johnson, Thomas H. "The Natural History of the Space Shuttle." *Technology and Society* 10, 1988: 417–24.

Jonsen, A. R. "The Birth of Bioethics." *Hastings Center Report*, special suppl., November–December 1993: 51–4.

    "Casuistry: An Alternative or Complement to Principles?" *Kennedy Institute of Ethics Journal* 5(3), September 1995: 237–52.

Kolcum, Edward H. "NASA Managers Debate Need for Flight Readiness Firing Test." *Aviation Week and Space Technology* 126, February 23, 1987: 23–24.

    "Pratt Reviving Propulsion Plants to Attract Future Engine Contracts." *Aviation Week and Space Technology*, June 6, 1988: 26–7.

    "Shuttle Engine Pump Issue Develops after Firing Test." *Aviation Week and Space Technology* 129, August 22, 1988: 20.

Kovach, K. A., and B. Render. "NASA Managers and *Challenger:* Profile and Possible Explanation." *IEEE Engineering Management Review* 16, 1988: 2–6.

Kranzel, Harald. "Shuttle Main Engine Story." *Spaceflight* 30, 1988: 378–80.

Ladd, John. "Collective and Individual Moral Responsibility in Engineering: Some Questions." *IEEE Technology and Society* 1(2), 1982: 3–10.

Lerner, Eric J. "An Alternative to "Launch on Hunch."" *Aerospace America* 25, May 1987: 40–4.

Levy, Edwin, and David Copp. "Risk and Responsibility: Ethical Issues in Decision-making." *IEEE Technology and Society* 1(4), 1982: 3–8.

Lewis, Richard S. "Whatever Happened to the Space Shuttle?" *New Scientist* 87, July 31, 1980: 356–9.

Lidz, C. W., and R. M. Arnold. "Rethinking Autonomy in Long Term Care." *University of Miami Law Review* 47, 1993: 603–33.

Logsdon, John M. "The Space Shuttle Program: A Policy Failure?" *Science* 232, 1986: 1099–1105.

Lombardo, Joseph A., and John P. McCarty. "Chemical Propulsion: The Old and the New Challenges." *AIAA Student Journal* 11, December 1973: 22–5.

Luegenbiehl, Heinz C. "Codes of Ethics and the Moral Education of Engineers." *Business and Professional Ethics Journal* 2(4), 1983: 41–61.

Lynn, Frances M. "The Interplay of Science and Values in Assessing Regulating Environmental Risks." *Science Technology and Human Values* 11(2), 1986: 40–50.

McCarty, John P., and Joseph A. Lombardo. "Chemical Propulsion: The Old and the New Challenges." *AIAA Student Journal* 11, December 1973: 24–5.

Meisel, Alan, and Loren Roth. "Toward an Informed Discussion of Informed Consent: A Review and Critique of the Empirical Studies." *Arizona Law Review* 25, 1983: 272.

Meiskins, F. Peter. "The 'Revolt of the Engineers' Reconsidered." *Technology and Culture* 29(2), 1988: 219–46.

Meiskins, Peter F., and James M. Watson. "Professional Autonomy and Organizational Constraint: The Case of Engineers." *Sociological Quarterly* 30(4), 1989: 561–85.

Mitcham, Carl. "Industrial and Engineering Ethics: Introductory Notes and Annotated Bibliography." *IEEE Technology and Society* 3(4), 1984: 8–12.

Morreim, Haavi, F. "Cost Containment: Issues of Moral Conflict and Justice for Physicians." *Theoretical Medicine* 6, 1985: 258–79.

Mueller, George E. "The New Future for Manned Spacecraft Development." *Astronautics and Aeronautics* 7(3), 1969.

"NASA Studies Gas Problem on Challenger." *Aviation Week and Space Technology*, January 3, 1983: 21.

Pellegrino, Edmund. "The Metamorphosis of Medical Ethics: A 30-Year Retrospective." *Journal of the American Medical Association* 269(9), 1993: 1159–62.

Pritchard, Michael. "Beyond Disaster Ethics." *Centennial Review* 34(2), (1990): 295–318.

Roland, Alex. "The Shuttle: Triumph or Turkey?" *Discover*, November 1985: 29–49.

"Priorities in Space for the USA." *Space Policy*, May 1987: 104–11.

Shrader-Frechette, K. S. "The Conceptual Risks of Risk Assessment." *IEEE Technology and Society* 5(2), 1986: 4–11.

"Shuttle Engine Still on Schedule." *Aviation Week and Space Technology* 95, September 13, 1971: 21.

"Shuttle Engineer Protest Detailed." *Aviation Week and Space Techonology*, August 23, 1971: 23.

"Shuttle 6 Rescheduled for Mid-March." *Aviation Week and Space Technology*, February 7, 1983: 28–9.

Stiles, Dan. "Engine Production Soars with Blend of CAD/CAM & Robotics." *Production Engineering* 34, September, 1987: 5459.

Thomson, Jerry. "Shuttle: The Approach to Propulsion Technology." *Astronautics and Aeronautics* 9(2), February 1971: 64–7.

Tischler, Adelbart O. "A Commentary on Low-Cost Space Transportation." *Astronautics and Aeronautics* 7, August 1969: 50–64.

"Developing the Technological Base for the Space Shuttle." *Aerospace Management* 5(1), 1970: 59.

"Low Cost Space." *Astronautics and Aeronautics*, May 1973: 23.

"Defining a Giant Step in Space Transportation: Space Shuttle." *Astronautics and Aeronautics* 9, February 1971: 23–5.

"Which Way to Shuttle Upper Stages?" *Astronautics and Aeronautics*, July–August 1975: 28.

Toner, Mike. "It's Pay Off or Perish for the Shuttle." *Science Digest*, May 1985: 64–7, 87–8.

Weil, Vivian. "Moral Issues in Engineering: An Engineering School Instructional Approach." *Professional Engineer*, October 1977: 35–77.

Wilson, Richard. "Commentary: Risks and Their Acceptability." *Science Technology and Human Values* 11(2), 1986: 11–22.

Winsor, D. A. "Communication Failures Contributing to the Challenger Accident: An Example for Technical Communicators." *IEEE Transactions on Professional Communication* 31, 1988: 101–7.

Yaffee, Michael L. "P&W Shuttle Engine Based on XLR129." *Aviation Week and Space Technology* 94, June 14, 1971: 51–7.

"Reusable Rocket Motor Unveiled." *Aviation Week and Space Technology* 93, August 31, 1970: 38–44.

Index